D1640374

Klaus Haefner (Ed.)

Evolution of Information Processing Systems

An Interdisciplinary Approach for a New Understanding of Nature and Society

With 46 Figures

Springer-Verlag
Berlin Heidelberg NewYork
London Paris Tokyo
Hong Kong Barcelona Budapest

Professor Dr. Klaus Haefner (Editor)

Fachbereich Mathematik und Informatik
Universität Bremen
2800 Bremen 33
Federal Republic of Germany

ISBN 3-540-55023-2 Springer-Verlag Berlin Heidelberg New York
ISBN 0-387-55023-2 Springer-Verlag New York Heidelberg Berlin

Library of Congress Cataloging-in-Publication Data
Evolution of information processing systems: an interdisciplinary approach for a new understanding of nature /
Klaus Haefner (ed.). p. cm. Includes bibliographical references., ISBN 3-540-55023-2 (Berlin). -- ISBN
0-387-55023-2 (New York) 1. Electronic data processing. I. Haefner, Klaus, 1936- . QA76.E945 1992
003'.54--dc20   92-5332

© Springer-Verlag Berlin Heidelberg 1992
Printed in Germany

The use of general descriptive names, registered names, trademarks, etc. in this publication does not imply, even
in the absence of a specific statement, that such names are exempt from the relevant protective laws and regulations
and therefore free for general use.

Product liability: The publishers cannot guarantee the accuracy of any information about dosage and application
contained in this book. In every individual case the user must check such information by consulting the relevant
literature.

Typesetting: Thomson Press (India) Ltd., India; Printing: Mercedes-Druck, Berlin;
Binding: Lüderitz & Bauer GmbH, Berlin
61/3020-5 4 3 2 1 0 – Printed on acid-free paper.

Contents

Overview VII
Participants in the Project Evolution
of Information Processing Systems IX

I Basic Concept
    I.1 K. Haefner: Evolution of Information
        Processing – Basic Concept   1

II Contributions to the Concept of Information
    II.1 G. Kampis: Information: Course and Recourse 49
    II.2 E. Laszlo: Aspects of Information 63
    II.3 A. Semenov: Mathematical Aspects
        of PANINFORMATICA 72
    II.4 G. Kampis: Process, Information Theory and the Creation
        of Systems 83
    II.5 E. Oeser: Mega-Evolution of Information Processing
        Systems 103

III Information Processing Systems at the Physical Level
    III.1 M. Requardt: From "Matter-Energy" to "Irreducible
        Information Processing" – Arguments for a Paradigm Shift
        in Fundamental Physics 115
    III.2 H.-W. Klement: Inorganic Matter as One of Four Levels
        of Natural Information Processing 136
    III.3 H. Haken: The Concept of Information Seen from the Point
        of View of Physics and Synergetics 153
    III.4 G. Nicolis: Dynamical Systems, Instability of Motion
        and Information Processing 169
    III.5 J. Kurths, U. Feudel and W. Jansen: Pragmatic Information
        in Nonlinear Dynamo Theory for Solar Activity 185

IV Information Processing in Biological Systems
    IV.1 S.W. Fox: Thermal Proteins in the First Life
        and the "Mind-Body" Problem 203
    IV.2 J. Kien: Remembering and Planning: A Neuronal Network
        Model for the Selection of Behaviour and its Development
        for Use in Human Language 229
    IV.3 V. Csányi: Nature and Origin of Biological
        and Social Information 257

**V The Evolution of Information Processing Systems
at the Social Level**
V.1 E. Laszlo: Unitary Trends in Sociocultural Evolution 281
V.2 M. Pantzar and V. Csányi: The Replicative Model
of the Evolution of Business Organization 288

**VI The Evolution of Information Processing Systems
at the Sociotechnical Level**
VI.1 K. Haefner: Information Processing
at the Sociotechnical Level 307
VI.2 E. Oeser: From Neural Information Processing
to Knowledge Technology 320

Epilogue 343

Bibliography 347

Overview

Assuming that the invention of information technology and its widespread use are neither singular nor accidental events, we have to consider technical information processing in its historical and evolutionary contexts. It makes sense to understand technical "information processing" as a very recent approach to coping efficiently with information processing, the storage, the exchange and the usage of information having been carried out in earlier phases of evolution by physical, genetic, neural and social systems.

This point of view leads to the following questions: Is there a way of understanding information processing in nature and in society in a way that we get more insights into its *basic principles*? Are there *general mechanisms for the evolution of information and its processing* in nature and society? Does it make sense to propose a general theory of *information processing* which holds true for all levels of evolution? Can we establish a scientifically fruitful *set of hypotheses about information processing*, allowing a transdisciplinary understanding of the universe and its evolution?

These and similar questions are dealt with in this volume. It is an attempt to find answers by looking at these problems from the viewpoint of various disciplines. In Chap. I, Basic Concept, a paradigmatic approach is proposed which tries to give the foundations for a basic understanding of information processing in nature and society. This proposal is deliberately dogmatic and strict. It is going to set the scene for a necessarily intensive scientific discourse.

The contributions by scientists of various disciplines given in this book show how difficult it is to work with the basic hypotheses put forward in the Basic Concept. All contributions are considered a challenge to sciences to understand information, information processing and its evolution in depth.

In the Epilogue it is made quite clear that the questions posed above cannot easily be answered at present, since we still do not have a "general information theory" which allows us to deal in a common format with the broad variety of phenomena of information processing in nature and soceity. Maybe we will be able to outline such a theory in the future. This might help to answer many open problems tackled within this book.

The project "Evolution of Informaton Processing Systems" has been generously supported by the Bertelsmann Foundation, Gütersloh. Thus, it has been possible to bring together such a interdisciplinary international group of scientists.

Participants in the project have been pre-selected because of their previous interest in this area. All members of the project met at a workshop in Bremen in October 1990 and exchanged their ideas on the basis of first drafts of their contributions. The final papers are presented in this book.

The Basic Concept stems from ideas which originated in the 1970s and which I have published in an early version as a series of ten papers in the German journal "Computer Magazin" in 1986. The basic hypotheses are the output from many fruitful discussions within the project.

I have to thank Mrs. Ute Riemann-Kurz for all her work in getting people together, organizing the information exchange, setting up the workshop, and organizing the literature.

Klaus Haefner   Bremen, February 1991

Participants in the Project Evolution of Information Processing Systems

Professor Dr. Vilmos Csányi
Department of Behaviour Genetics, ELTE University of Budapest,
H-2131 Göd Javorka, S.-u. 14, Hungary

Professor Dr. M. Euler
Department of Electrical Engineering, FH Hannover, Ricklinger Stadweg 120,
D-3000 Hannover, Germany

Professor Dr. Sidney W. Fox
Department of Plant Biology, Southern Illinois University, Carbondale,
Illinois 621901-6509, USA

Professor Dr. Klaus Haefner
Department of Mathematics and Informatics, University of Bremen,
D-2800 Bremen 33, Germany

Professor Dr. Hermann Haken
Institute for Theoretical Physics and Synergetics, University of Stuttgart,
Pfaffenwaldring 57/4, D-7000 Stuttgart 80, Germany

Dr. George Kampis
Department of Behaviour Genetics, ELTE University of Budapest,
H-2131 Göd Javorka, S.-u. 14, Hungary

Priv.-Doz. Dr. Jenny Kien
Institute for Zoology, University of Regensburg, Universitätsstr. 31,
D-8400 Regensburg, Germany

Dr. Hans-Werner Klement
Füssmannstr. 16, D-4600 Dortmund 30, Germany

Dr. Jürgen Kurths
Institute of Astrophysics, Former Academy of Sciences of the GDR,
O-1501 Tremsdorf, Germany

Professor Dr. Ervin Laszlo
Vienna International Academy, Vienna, Austria

Professor Dr. Gregoire Nicolis
Service de Chimie Physique, Code Postale no. 231. Campus Plaine U.L.B.,
Boulevard du Triomphe, B-1050 Bruxelles, Belgium

Professor Dr. John S. Nicolis
Department of Electrical Engineering, University of Patras, Patras, Greece

Professor Dr. Erhard Oeser
Institute f. Wissenschaftstheorie u. Wissenschaftsforschung, Abteilung Termino-
logie und Wissenstechnik, University of Vienna, Sensengasse 8, A-1090 Vienna,
Austria

Dr. Mika Pantzar
Labour Institute for Economic Research, University of Helsinki,
Hämeentie 8A, 00530 Helsinki, Finland

Priv.-Doz. Dr. Manfred Requardt
Institute of Theoretical Physics, University of Göttingen, Bunsenstr. 9,
D-3400 Göttingen, Germany

Professor Dr. Alexei Semenov
Institute for New Technologies, 11 Kirovogradskaya, Moscow 113587, USSR

I Basic Concept

Evolution of Information Processing – Basic Concept

Klaus Haefner

1	From Subatomic Interaction Towards the Computerized Society	2
1.1	Overview	2
1.2	Working Hypotheses	4
2	Fundamental Principles of Information Processing and Its Evolution	5
2.1	Information	5
2.2	Information Processing	7
2.3	Information Processing Systems	8
2.4	Information Processing As a New Paradigm	8
2.5	Goals in Understanding the Evolution of Information Processing	9
2.6	Making the Definitions Explicit	10
2.7	Problems with the Meta-Theory Approach	11
3	Information Processing in Nature Described by Physics	12
3.1	Introduction	12
3.2	Gravitation	13
3.3	Atomic Organization	13
3.4	Electromagnetic Interactions	14
3.5	Essential Properties	15
4	Genetic Information Processing	16
4.1	The Principle	16
4.2	The Complex Machinery of the GenIPS	17
5	Neural Information Processing: Dealing with Complex Environments Directly	21
5.1	General Concept	21
5.2	Details of the NeurIPS	21
5.3	The Human Information Processing System	25
6	Social Information Processing	29
6.1	Cooperative Neural Information Processing	29
6.2	Essential Characteristics	32
7	Technical Information Processing	33
7.1	Introduction	33
7.2	Origin of TechIPSs	33
7.3	Is the TechIPS Really a New Level in Mega-Evolution?	38
8	Computers and Society – Sociotechnical Information Processing	39
8.1	The Present Level in Mega-Evolution of IPSs	39
8.2	Basic Features	40
9	The Heuristic Value of the Information Processing Paradigm	44

1 From Subatomic Interactions Towards the Computerized Society

1.1 Overview

At the moment we are in the middle of dramatic developments in information processing: whereas up to the 1950s *all* information processing was done by natural systems only, there has been, since the invention and the coming into widespread use of computers, a continuously changing balance between human thinking and technical information processing. This is emphasized by the political notion that we are on the way towards an "information society", a type of society in which merely commodities in the form of information and services for the social sphere are produced by people, but hardly any material goods. Classical goods and services are produced by computers in fully automated firms.

No matter how strongly one feels for or against increasing computerization, no party – employees, unions, politicians or scientists – is in possession of an historically founded perspective regarding the currently expanding use of the computer. Systems built by *Homo sapiens* for carrying out information processing technically have suddenly moved into the centre of public interest although, as we shall show, storage, transport and processing of information have always been of great significance in nature, ever since the beginning of the universe and the earth.

The purpose of this paper is to provide the basic concept which will attempt to describe and analyse central principles used in nature for storage, transport and processing of information at the different levels of material, biological and social organization, from the elementary particle, via the atom, the single cell and the human being up to the fifth generation computer. It is proposed that the evolution of methods for handling information occurs in accordance with eight basic hypotheses, which are presented in Sect. 1.2 of this chapter.

An understanding of the evolution of information processing will permit forecasts of future developments and trends to be made, since evolution will not stop in 1991. In particular, this may result in consequences for the mastering of the challenge of modern information technology in society, a challenge which is often interpreted today as if there were no continuous linkages through time at all. This, however, is an incorrect assessment if the methods of information processing and the context of evolution are considered. Information processing has been developing continually over billions of years and will, most probably, be further refined in the future.

After introducing a general framework for information processing, this paper and the other contributions in this book will, step by step, deal with the most important methods of handling information at the various levels of evolution. At the first level, transport and processing of information in physical structures will be described. In this area it is necessary to re-examine classical physical concepts such as the physical field, and to discuss to what extent they can be interpreted theoretically in an information theory context. In attempting this, it becomes quite clear that we know very little about the mechanisms for dealing with information within physical systems. The concepts of information and information processing have, up to now, not been used as a central paradigm in physics.

At the molecular level, it becomes clear that with a "larger" physical body, new structures of information processing are established. Here, we encounter for the first time the phenomenon of transport of "internal" information across larger distances, namely along molecular chains. Information storage and processing is achieved, e.g. for the specification of secondary, tertiary and quaternary structures.

At the second level of the evolution of information processing, we encounter the principles of molecular genetics. In this context we will discuss how complex structures can be specified by means of the DNA code. Material transport of internal information at this level in the form of molecular messages (e.g. messenger-RNA) was the first to appear in evolution.

At this point an important principle in the development of information processing becomes obvious: whenever a new method is "discovered" in evolution it is based on "the level which came before it". When carrying out identical replication the cell uses all previously developed methods for handling information, e.g. the "atomic communication" or the storage of information in chain molecules.

At the third level of the evolution of information processing, cells are going to use efficient mechanisms for the handling of "external" information. New types of storage evolve for information that is acquired or learned. Processing of messages within complex environments (e.g. chemical stimuli, light signals) takes place. This level is called neural information processing. In metazoans we encounter the internal organization of "knowledge" for the processing, recognition and memorization of structured representations of the environment, as well as complex and indirect interaction between acquired and genetic information and its processing in the organism.

Cognitive psychology has given insight into human information processing and personal communication at the "highest level" of neural processing of information. Thus, models can be developed for central functions of the human brain.

At the fourth level of evolution, we encounter social systems with common external information and common procedures. *Homo sociales* has refined the techniques of external storage by means of painting pictures and, later, using letters and scripts. In addition to genetic and neural information processing, "externally organized knowledge" evolved which finally, by means of printing technology, became what is called "global information."

At the fifth level within the evolution of information processing, technical (i.e. extrasomatic) processing of information originates by means of mastering models and abstractions of the world in man-made and man-programmed physical structures.

Since both use the computer extensively, technical and societal information processing are closely linked. I shall discuss these as the sixth emerging level of information processing, the sociotechnical level.

1.2 Working Hypotheses

The following working hypotheses are postulated, which summarize my present understanding of information processing in nature.

Hypothesis 1 All natural systems (matter/energy) are Information Processing Systems (IPSs). Each IPS can receive, store, process and transmit information. Information processing is an essential internal feature of all systems. The universe as a whole may be viewed as a gigantic IPS.

Hypothesis 2 Information processing systems are open systems with energy and information flows. Information processing is crucial for the self-organization of all natural matter/energy structures.

Hypothesis 3 Information is a system variable which exists in relation to a particular IPS only. We have to distinguish between (a) *internal* information, concerning only the particular system itself, and (b) *external* information (measured in a given observational system), which is pragmatic information sent out by other IPSs. Internal information is an actual and essential component of every natural structure. External information is only a potential for a given natural system, which might or might not be used.

Hypothesis 4 In evolution nature has given rise to *various completely distinct levels* of IPSs; this is called *mega-evolution*. In the mega-evolution of information processing systems six distinct levels of IPSs can at present be identified in nature: *physical IPSs* (physIPS), *genetic* IPSs (genIPS), *neural IPSs* (neurIPS), *social IPSs* (socIPS), *technical IPSs* (techIPS) and *sociotechnical IPSs* (sotecIPS).

Hypothesis 5 At every distinct level of an information processing system, evolution takes place. This is designated *level-evolution* and it results in differentiation of a given type of IPS into *further* subsystems having specialized functions for receiving, processing, storing and transmitting information. In level-evolution there is a permanent increase in the amount of a typical "class" of information used *internally*. Thus, there is a spectrum of manifestations of the particular basic structures of a given IPS level.

Hypothesis 6 *Mega-evolution* gives rise to the emergence of new levels of IPS with new properties and functions, through variations or aggregation and cooperation of lower-level IPSs. Mega-evolution is the result of a sophisticated integration of "old levels" of IPSs "cooperating" at a new level of IPS. New properties, e.g. the ability to process new kinds of information, emerge at a higher level of information processing. In accordance with the emergence principle these new properties and functions are *not* reducible to lower-level IPSs.

Hypothesis 7 In nature, the emergence of a new IPS level has always happened with a "leap" within a relatively short period of time. (At least, no widely spread intermediate structures are known.)

Hypothesis 8 Evolution of IPSs is the story of a steady increase in the capacity of internal memory and "processing power". Whenever in level-evolution a given IPS reaches its limits in coping with a complex "informational environment" new structures and functions come into existence, giving rise to a new level in mega-evolution.

These hypotheses point to an understanding of nature as a complex arrangement of interacting information processing systems, matter and energy. They form the beginning of a new paradigm, namely the consideration that information processing is a very important and essential feature of *all* natural processes, and hence, they formulate the basic principles of a new meta-theory of nature. This has, however, to be developed *much* further to become really scientifically fruitful. It is hoped that the application of this new paradigm may also supply, in the end, a new understanding of evolution.

My basic hypotheses may help us, in addition, to look at the *future* of evolution and to formulate some trends concerning what the next steps of "refinement" of the IPSs will look like. It is not unlikely that in sociotechnical structures there will be repetitions of what we have found earlier in level-evolution: differentiation, specialization and, following this, new integration. It thus appears very likely that human beings will be able to create fundamentally new sociotechnical structures within the framework of an integration of neural and technical procedures for processing information (see also Chap. VI. 1).

2 Fundamental Principles of Information Processing and Its Evolution

This basic concept is meant to provide a first assessment of the evolution of information processing that has occurred within the past few billion years. In making this assessment, it is first of all necessary to create a *conceptual framework* for information, information processing and information processing systems. This is the aim of this section, and it will form the basis of considerations in the following sections. (In some contributions in this book, however, my definitions are not always used consistently.)

2.1 Information

Information is to be understood as a message which has a significance for a given "system to be informed". This means that the message allows the system to initiate "meaningful activities" which could *not* be performed without it. In this way information changes the system from a state of relative uncertainty to more certainty, it supports self-organization.

A "system to be informed" is usually an *open* matter/energy system, allowing the influx of matter/energy and messages. Systems can be subsystems of a "higher-level system". In this context an atom, for example, is an information processing system which may be a subsystem of a cellular structure; a tree is an IPS as well, however, composed of many cellular systems.

We define messages as concatenated signals which are transferred immaterially (e.g. as electromagnetic waves) *or* materially (e.g. as molecules, printed paper) and which can be organized according to particular codes. There is no evidence that messages (and information) can be passed on without using matter/energy. Thus information and energy seem to be closely linked.

Whenever we define a *separated system* we have to distinguish between *external* and *internal* information: *external information* is defined as a set of messages available to a system from outside, i.e. from the environment. From this the system can "pick up" information and process it. (This interpretation of external information corresponds with the widespread use of the term "information" particularly in relation to "information transmission".) In studying the evolution of information processing we are going to emphasize, however, *internal information*. This term is used for messages "understood" inside a system interpreted on the basis of its "substructures". Thus, internal information is constitutive for the self-organization of any system (e.g. DNA in a cell).

For external as well as for internal information we have to take into consideration the *syntactic, semantic* and *pragmatic* aspects of information. The *syntactic* aspect involves the encoding and decoding of a message which allow systems to "recognise" a message or not. The *semantic aspect* encompasses the meaning of an encoded message for a particular system. Within a system, the *pragmatic* aspect of information causes possible changes in its "behaviour"; this may result in it sending off new messages.

A book, for example, is external information to a human being. Reading the book means

(i) decoding the patterns of the letters and processing the syntax of the text,
(ii) deriving a meaningful sense from the book, which may influence the mind and
(iii) possible stimuli within the human information processing system which may lead to an activity, for example talking to another person or solving a problem.

At present, due to the syntactic, semantic and pragmatic aspects of information, it is very difficult to measure external and internal information. "Information per se" doesn't make *sense at all!* Information can only be measured within a *specific* system, namely as "information at a given time in a given status". However, this is practically *impossible* because all systems change in the course of time.

Thus, at present, information is a highly system-dependent term which is far from being measured either qualitatively or quantitatively. In particular, all the approaches to measuring information in terms of "certainty" or "entropy" do not produce any statements that are meaningful with respect to the internal "value" of messages or information *within* a given system.

This makes the situation very complicated. The "traditional" concept of information must be replaced by completely new approaches to the definition of information which, however, are not at hand at present. Today we cannot prove information to be an "explicit variable". Rather, we have to consider information as being a *hidden variable*. What can be observed in all experimental and empirical circumstances is solely the "interaction" of matter/energy with matter/energy observation systems. The term information can only be used in an heuristic sense.

This insight is in *sharp contrast* to our general "understanding" of information. Everybody is talking about information; we even use wordings like "information jungle" or "information society". All this is, under present conditions, a discourse on a very high level of interpretation. It has a scientific *touch*, but it is far away from any concise information theory and its application to *real* systems.

Still, we think that the notions of information, information processing, information processing systems and their evolution are worth studying as a challenging approach. If we could derive an understanding of crucial principles, this might help to give rise to a "Grand Unified Theory of Information Processing" which must be developed, but which is not at hand yet.

2.2 Information Processing

In the following, *information processing* will be defined as the totality of activities performed in (a) "decoding" the syntax of messages, (b) "understanding" semantic information and (c) "acting" as a consequence of dealing with the pragmatic aspect of information. Some "basic", "pre-existing" internal information always needs to be available within a given matter/energy system for it to be able to perform all these tasks.

Although every system must have some status of "initial" information, internal information can be changed continuously by information processing. This means that the actual status of information in a system is rarely stable, nor reproducible. All systems are dynamic.

All systems practise self-organization by means of information processing over and over again using internal information as well as the steady influx of external information. Thus, in *my way of interpretation*, the whole universe as well as all of its "subsystems" can be understood as information processing systems. Every matter/energy structure is controlled by internal and external information and its "proper" internal processing.

Information processing requires *declarative* and *procedural* information as internal information. Whereas declarative information is understood as the essential component for "understanding" the information "flowing" within a system or coming in form "outside", procedural information is crucial for processing the information "flowing" within the system to produce "pragmatic output". However, it is quite difficult to separate these two types of information. If e.g. an animal hears the sound of a mate then it "recognises" this sound (declarative information), but it simultaneously has to process the information in order to take proper action, e.g. moving closer or running away (procedural information).

2.3 Information Processing Systems

Information processing in my understanding takes place in information processing systems. These matter/energy structures enable the following to occur:

1. receiving and decoding "relevant" messages,
2. storing internal information,
3. producing "new" information from "old" information (at all levels),
4. altering the structure of a system, and
5. sending information outside the system in a given code.

From my point of view information processing systems are always material structures. And, vice versa, all material structures are information processing systems! However, using the experimental and empirical methods available today, we can usually only observe the matter/energy structure. The important "informational patterns" are (still) hidden because of the *lack* of proper observation systems.

The most explicit form of information processing in nature may be what goes on in a physical field (e.g. the electromagnetic field). A particle in this field "knows" the acting force (external information) at a given point at a given time, and therefore "behaves properly" (pragmatic information). However, orthodox physics, for example, is neither able to measure information, nor does it consider information processing within its standard theories!

2.4 Information Processing As a New Paradigm

By introducing the fundamental new notions of information, information processing and information processing systems as essential structures of all natural systems I postulate a *new paradigm*. I argue that information processing plays an essential role at *all* levels of evolution. I postulate that it is possible to find, at *every level* of evolution, structures within a system which form the four basic components of every information processing system: receiver, storage (with internal declarative as well as procedural information), processor and sender.

Complexity of information processing and information processing systems increases from one mega-level of evolution to the next. However, at every level the structures of the "lower level" of information processing are used in a certain way.

2.5 Goals in Understanding the Evolution of Information Processing

The term evolution was originally introduced to aid understanding of the origin of the vast variety of living things on earth. More recently, the expression has been used for processes happening in a time sequence (e.g. "evolution of the universe"). At present there are several aspects in evolution research which are of high scientific interest:

- Understanding in which *sequence* structures have evolved (e.g. the evolution of the horse).
- Understanding *why* a heterogeneous (complex) set of entities has evolved from a simpler one (e.g. the evolution of man within the hominids).
- Understanding *general mechanisms* responsible for (biological) evolution (e.g. the notion of mutation and natural selection).
- Understanding the *"unfolding" of the universe* in terms of structures far from thermodynamic equilibrium (e.g. the concept of "creation" against that of "unfolding").

In this paper we are mainly concerned with two questions:

A *Which distinct levels of information processing systems can be found in nature?*
B *How did they come into existence?*

(In the long run answers to these questions might also help to explain the "unfolding of the universe").

To answer (A) we have to study nature and obtain answers to the following subquestions at the physical, genetic, neural, social, technical and sociotechnical levels respectively.

(A1) What class of messages is received and used as *internal* information?
(A2) What *new information* can be produced by an information processing system?
(A3) How does the system *function internally* (e.g. interaction between storage and processor; storage of external information ("learning"))?
(A4) What is the *contribution of information processing to self-organization?*
(A5) Which *material structures* form an IPS at a given level?
(A6) What *performance* is the information processing system capable of (e.g. procedural and declarative information, speed, capacity)?
(A7) How does the system *integrate the "lower-level IPSs"?*

To answer question (B) we have to study closely the emergence of new, distinct levels of information processing systems in nature ("mega-evolution"). Furthermore, it is necessary to consider the evolution of information processing systems taking place at a given level of a distinct class of information processing systems (*level-evolution*).

2.6 Making the Definitions Explicit

In order to make the reader more familiar with my terminology and analytical approach which will be applied to the various stages of information processing systems in the following sections, I am going to present a short analysis of *technical information processing*, since it is not too difficult to answer questions (A1) to (A7) in this field.

Considering (A1), signals which are transported and used in technical information processing systems are electromagnetic currents forming digital impulses. This means that technical information processing systems (techIPSs) recognise internally only "0–1 conditions". However, only those 0–1 messages that have been organized according to a "known" code (e.g. ASCII) can be used *as information*. In techIPSs, 0–1 messages can be "understood" according to the internal information (data, programs) already available. This results in a "meaning" (semantics) within the framework of a program, resulting in output (pragmatics). All techIPSs require proper analog–digital/digital–analog conversion for communications with the information environment.

Considering (A2), present-day techIPSs can produce many "forms" of external information comparable to those which they receive and process, e.g. numbers, text, graphics, series of tones. Thus, the internal organization of information processing in electronic circuitry permits a wide variety of communication activities to be dealt with. Computers, however, are considered to be cognitive slaves working under the control of and for human beings only. Thus, "new information" is a special class of outputs which make sense to human beings only. It is open to debate whether syntactic information and its processing within the computer results in semantic information ("conscious computer"). However, the output of a computer can be considered an activity. In other words, computers do final processing of pragmatic information.

Considering (A3), syntactic information (data) to be processed as well as the rules related to the processing of information (programs) are stored in techIPSs in a physical structure. Programs are usually sequentially processed by means of a single processor (or a few) (von Neumann machine). Thus, for a particular sequence of basic operations a techIPS requires a finite amount of time which increases according to the amount of information to be processed as well as the complexity of the processing. The maximal internal performance of the techIPS is limited by the internal speed of information transfer into storage and between storage and the processor (which is finally determined by the velocity of electromagnetic waves).

Considering (A4), in computer science, we are *not at all* concerned with information for "self-organization" of the computer hardware itself. Thus, we know very little – aside from physical principles – about the nature of the stability of computer hardware. At the level of micro-miniaturization we are at the borderline of quantum mechanical interpretations (which, however, are not considered "rules for self-organization" by orthodox physics!). Operating systems are organizing the flow of data and programs. They can be considered self-organizing structures.

Considering (A5), techIPSs today function internally by essentially exploiting physical semiconductor effects and magnetic phenomena in "basic elements" whose volumes are in the region of 10^{-3} to 10^{-5} mm^3. The main basic substances used today are silicon, and metal and various plastic compounds. TechIPSs are being increasingly manufactured by means of complicated automated processes (e.g. chip production).

It is interesting to note that in the design of techIPSs humans have made (up to now) practically no explicit use of biological concepts of information processing, even though their own outstanding cognitive performance results from a neurIPS. This, however, will change with the "biocomputer" presently under research, which will use biological structures and might even be produced by means of gene technology within biological systems. Using artificial neural networks will allow for "copying" the neurIPS.

Considering (A6), the highest performance of techIPSs today corresponds to the processing of some billion (10^9) "simple" operations per second. (Special chips permit, e.g. 100,000 words of text to be "read" per second by recognising predefined word patterns.) Many billions of characters can be stored in a device by means of magnetic procedures. The use of optical mass storage techniques (e.g. CD-ROM) permits storage densities of some 10^6 characters per mm^3.

Generally speaking, "simple algorithms" as well as "complex heuristics" can be stored and processed in the computer. Therefore permanent evaluations of how "advanced computers" compare to human intelligence can be made. They show that for a long time to come computers will be "stupid" in terms of *general behaviour in an open informational environment*. However, they are quite "clever" at processing "narrow" cognitive tasks.

Considering (A7), techIPSs are different from all other IPSs because they have been consciously created by humans. In computer engineering humans make heavy use of physical effects, e.g. at the level of semiconductor circuits and elementary magnetic and optical processes. Thus, the computer is a physical machine. It works with physical information processing while, however, simultaneously using programs which are constructed by human information processing, the most advanced form of neurIPS.

2.7 Problems with the Meta-Theory Approach

Although it is possible to give some detailed specification of information and its processing at an *abstract* level, there are serious problems in measuring information, its storage, its transport and its processing in detail in *real* systems.

With the observation systems at hand today we can only study matter and energy interactions. As a result, we cannot expect *direct* measurements of information processing structures. Thus, my entire framework is aimed at a meta-theory (only). This has to be broken down in the future to an operational level. At present, however, there is little insight into how this could be done.

My approach of an evolution of information processing is a "high-level" meta-theory which has to be considered in *parallel* with "orthodox" theories. It by no means "outdates" other classical disciplines. But, in the long run, it might prove to be an approach which presents a more general point of view for all natural structures.

The information processing paradigm and its consequences is the first attempt at answering serious "why questions". Its heuristic advantage is based on two aspects: (i) it is an approach which in the long run might unify theories across traditional disciplines and (ii) it is a way of looking at evolution in a manner which allows *extrapolation to the future* of the "sociotechnical society" as being the highest level information processing on earth (see also Sect. 9).

3 Information Processing in Nature Described by Physics

3.1 Introduction

"Physical nature" is the area most difficult to deal with in the mega-evolution of information processing systems. In the following, the *quite controversial* attempt will be made to explain basic laws of physics in terms of the information processing paradigm.

The reason such an approach, up to now, has been neither fully formulated nor sufficiently examined for its consequences to become clear seems to result mainly from the fact that in physics the most important laws and theories were formulated at a time when concepts such as information and information processing – in the sense used here – were hardly used or even unknown. Furthermore, due to the historical separation of sciences into the "physical/chemical", "biological/medical" and "social/human" fields, no elaborate attempts have been made to arrive at a unified approach. The laws of physics were and are formulated by physicists; biological phenomena are explained by biologists; anthropologists, psychologists, physicians, philosophers are responsible for the arts and social sciences; information technology is dealt with by computer scientists, etc.

No attempt will be made here to discuss the *very early phases of the evolution of the physical cosmos* where, most likely, there was only *one basic force*, which subsequently split into the four basic forces known today. The analysis of "very early" information processing in matter is, however, of great significance and will have to be carried out in the future.

3.2 Gravitation

I shall begin with the most important physical phenomenon, namely the force of gravity (which is prerequisite for the origin and existence of the planet earth). The phenomenon was described by Newton in his famous gravitational law in the year 1666: $F = cm_1 m_2/r^2$. This law states that two masses m_1 and m_2 which are at a distance r from each other are attracted by a force F acting in the direction of a straight line connecting the centres of gravity of the masses m_1 and m_2. Nothing (on earth) can shield this gravitational force; it is always there and it is everywhere, i.e. it is universally present. In Newton's approach the "action" is considered to be "immediate", without any time dependence.

As a consequence of the ever-present effects of gravitation (and some other important findings), Albert Einstein described gravitation as a property of the space-time continuum. In the general theory of relativity he combined gravitation and inertia as the property of bodies moving on space-time trajectories.

As an alternative, the information paradigm offers the following *hypothetical* interpretation: fundamentally, all physical "basic structures" which are affected by gravitation are physical Information Processing Systems (physIPS) exchanging gravitational information. They evaluate it continuously (syntactic information) and transform it into (pragmatic) information for action (e.g. forces). Hence, for example, each neutron continuously determines its "gravitational status" by means of vector addition of the gravitational information relating to all other masses in the universe, and then "behaves properly".

All elementary particles transmit and receive "gravitational information" at the speed of light. They process information internally in their material "fine structure". However, atomic and nuclear physics is at present not capable of providing an explanation of the details of this "internal computer", which must be quite different from the "traditional" quantum theoretical interpretations.

In physIPSs storage of received information does *not* take place. There is just an "extremely quick" procedure which continually determines the vector sum of forces and provides a result which ensures the "correct behaviour" of the particle in the universe. (Gravitation may be one of the "outputs" of the "computing space" argued for by Requardt in Chap. III.1.)

As far as the evolution of gravitation is concerned, we only know that it must have come into existence at the very beginning of the cosmos. However, whether gravitation has remained *quantitatively* constant and independent of time or not has not been shown physically.

3.3 Atomic Organization

The physical phenomenon that "everything is informed about everything else" and that "everything refers to everything else", which can be observed in the case of gravitation, exists – in a modified form – in other physical principles as well. The Pauli principle states that *within* one atom two particles having "the same status" (i.e. the same quantum number combination) are never allowed to exist simultaneously. Thus, from the point of view of an information processing interpretation, information is continually exchanged within the components of an atom. If an electron breaks the rule in an atom it is "kicked out". (For details see Chap. III.2.)

Comparable rules also apply within the nucleus of an atom, even though it is more complicated. The weak and the strong forces obviously apply only within the radius of the nucleus, i.e. they are "local procedures" of the physIPS.

What can be speculated about these "particle information processing systems"? Firstly, it is obvious that messages about the structure of all particles are continually exchanged within the atom. Secondly, every particle "stores" its "characteristic value" (e.g. quantum numbers) and continually transmits these. Thirdly, we have no detailed knowledge about how the actual processing of the "status information" takes place in electrons or nuclear particles. Fourthly, it is undisputed that the exchange of information between nuclear particles in their specific interaction (weak and strong forces) is limited to the dimensions of the nucleus of the atom. Fifthly from the point of view of the extremely small distances, it is impossible to determine the rate of transfer of information. If messages in an atom were transferred at the speed of light ($3 \times 10^8 \, ms^{-1}$), then the highest transfer times would be about $(3 \times 10^{-10} \, m)/(3 \times 10^8 \, ms^{-1}) = 10^{-18} \, s$ (the best atomic clocks, however, can only measure precisely up to approximately $10^{-12} \, s$). For the nucleus the transfer times would be about $10^{-23} \, s$. Sixthly, self-organization of physical structures does not need incoming energy. Information exchange and processing "works" without "external energy".

In the atom, on the one hand the physIPS has to determine the current quantum-mechanical status by means of taking into account the relevant information concerning the other particles. On the other hand, "decisions" must be made regarding whether the status is "allowed" or not.

The cooperation of a large number of physIPSs finally allows even large macroscopic material structures to be "constructed" and "maintained".

Elementary particles (e.g. electrons, protons) "know how to behave", but in doing so they have a certain "liberty" within a defined framework described by Heisenberg's principle of uncertainty. This shows the "inadequacy" of physical observation systems, which are not made for understanding physical information processing. The "uncertainty" might just be an expression of the (finite) speed of the physIPS: if one measures the position of a particle its impulse has to be computed by the physIPS; therefore it is not "available" at that time.

3.4 Electromagnetic Interactions

A third important aspect of physical information processing to be investigated is that of electromagnetic interactions. These are described in terms of physics by Maxwell's equations. One of the central statements there is that (moving) electrical or magnetic particles (or their electromagnetic waves) cause certain effects. Thus the principle is observable that "everything that is electrically charged or magnetized is informed about everything that is electrically charged or magnetized" and "adjusts itself according to other electrical or magnetic particles". A charge is a typical form of "declarative" information. The law of magnetic induction, for example, is then the result of pragmatic information produced by the proper procedural information.

3.5 Essential Properties

Our basic questions (A1) to (A7) from Sect. 2.5 can be answered for physIPSs as follows:

(A1) Messages are transported at the velocity of light between all physical "particles". Inside the "particles", however, we do not know the representation of information nor its processing.

(A2) Internally, new information is generated by "adding up" the incoming messages and producing pragmatic information which controls the action taken by the "particle". The "adding-up procedures" can be quite simple (vectorial).

(A3) We do not understand the internal apparatus of the physIPS. The experimental approaches available cannot help us to get to an appropriate understanding (imagine a person trying to analyse and understand a computer by shooting at it using various types of ammunition and then studying the outcome!). However, we know that physIPSs *cannot store* and reuse external information at a later time.

(A4) Information processing – to my way of looking at physical phenomena – plays an essential and crucial role for the existence of all physical structures. I consider them as being self-organizing systems. Physical structures are dynamic and unstable (e.g. dissociation). Therefore, they have to establish the structure by steady self-organization via information processing. However, no "external energy" seems to be required to do physical information processing.

(A5) We do not know how "elementary components" form the physIPS. There may be, however, a continuous "computing space", forming "clusters" (see Chap. III.1).

(A6) There is declarative information within all elementary particles which specifies their *specific* "behaviour" (their "type"). Processing capacity must be extremely high, particularly for gravitational information. There is no storage of external information inside physIPSs for "longer" periods; after processing external information is "forgotten" immediately.

(A7) Physical information processing is *the* basic information processing in nature. Thus, it does not need integration of another – lower – level. (Although there might be a sub-elementary particle space; see Chap. III.1.) However, as the notion for a "Grand Unified Theory" argues, there might be just *one* type of force, a "basic procedure in the physIPS", which is altered according to the "circumstances" (e.g. temperature).

The level-evolution of physIPSs must have taken place in the "very early" phase of the origin of the cosmos. Otherwise our present material world would hardly have come into being without dramatic upheavals.

Summarizing my very first attempt to interpret physical systems from the point of view of the paradigm of information processing, I have to point out once more that there are *very essential unanswered questions*. This is because, on the one hand, the traditional particle and field approach is used in physics only. On the other hand, there is the obvious deficit of the information paradigm that no quantitative mathematical formulations can be given at present.

However, and this is of enormous significance for further considerations about the evolution of information processing, we have to accept that at the physical level "(almost) everything is informed about everything else". There is a "global knowledge" (gravitation), a "local knowledge" (nuclear forces) and "specific knowledge" (electromagnetic forces).

Thus, messages are continuously exchanged which leads to a reduction in the uncertainty of all participating physical "partners". In this universe it looks as if everything has to "organize" itself appropriately according to the received information and has to "participate" in the organization of dynamic physical structures.

4 Genetic Information Processing

4.1 The Principle

Simple molecules were formed by means of physical information processing some billion years ago followed later on by crystals and chain molecules. The latter offered a new option for information processing: they allowed the storage of a new set of information by means of a sequential lining-up of different components. This fundamental property was, and still is, the basis of the genetic Information Processing System (genIPS) which will be analysed in the following.

For reasons of chemical stability there are two particularly interesting candidates for this kind of information storage: chain molecules of carbon and silicon compounds. Today it is impossible to determine if, and to what extent, evolution "tried" the latter. Hydrocarbons were obviously more "successful". Nucleic acids organized molecular information storage which determined the performance of the genIPS and that of all observed forms of life on earth. The fundamental principle of genetic information processing is based on "reading" and "processing" information stored in the double-helix-shaped chain molecule deoxyribonucleic acid (DNA) which seems to be about three million years old.

GenIPS is basically quite simple even though in its level-evolution it has become "refined" by means of "specialization" forming thousands of different "versions" of the fundamental concept ranging from the virus via the plant up to the mammalian cell. I shall explain the principle of genIPSs using bacteria as a well-researched example of a simple biological system. The genIPS interprets as messages a coded, sequential series of deoxyribonucleic acid bases of only four different types: adenosine (A), cytosine (C), guanine (G), and thymine (T). Groups of three bases, called a triplet or codon (e.g. the series CGC or AGC), are used to code an amino acid. Some hundreds or thousands of triplets in sequential order specify a protein consisting of many amino acids. (In turn, different proteins form, by means of various complex mechanisms, cells or viruses.) The way the DNA is "read" is regulated by sophisticated mechanisms.

By means of a biochemically very elaborate procedure the DNA is first transcribed into messenger ribonucleic acid (m-RNA), which has the complementary components adenosine (A), cytosine (C), guanine (G), and uracil (U) forming triplets. These sequences are translated into a particular protein molecule using special components, the ribosomes, as well as t-RNA (an abbreviation of transfer-RNA) which "carries along" the necessary amino acids.

Since $4^3 = 64$ different triplets can be formed from the four different DNA bases, this enables a precise coding of 64 different amino acids. However, in biological systems there are only 20 important amino acids. The code which defines the relationship between the triplets and the amino acids is therefore degenerate. Often the same amino acid is specified by means of different, even though similar, m-RNA triplets, e.g. the amino acid leucine is coded by CUU, CUC, CUA and CUG.

The genIPS has to "know", of course, where the information for a given protein starts in the DNA. A starter triplet serves this purpose. There are also stop triplets which mark the end of the information necessary for a certain protein. The sequence of triplets "responsible" for one protein is called a gene.

4.2 The Complex Machinery of the GenIPS

It is a fact that it is possible to derive all necessary information for a living cell from a chain molecule. Is this, however, really an information processing system as we have defined it above, or are we simply observing a "replication machine" at work which merely creates biological systems from the DNA code?

Three decades of intensive genetic and macromolecular biological research allow these questions to be answered rather precisely. An elaborate genIPS is at work in the cell which by no means produces only copies. On the contrary, complicated control and regulation procedures are used. New results in recent research even show that in some cases various genes contribute to a single m-RNA ("RNA-editing").

These processes, the details of which are very complicated and differ between organisms, can be described in principle by using as an example the infection of a bacterial cell by means of a phage, and its replication in the cell. Since phages do not possess ribosomes of their own they are not able to multiply without a suitable host cell. They do have, however, the necessary information in their nucleic acid base sequence to stimulate the genIPS of a cell to create new phages instead of carrying on with the normal bacterial cell "household". The mechanism by which a bacterium is infected by a phage is the following. After the phage – by means of a specific adsorption – has attached itself to the cell wall of its host, it transfers its DNA through a hole in the cell wall. This external information "creeps into" the bacterial DNA sequence. In this way, it becomes internal information. It is finally transcribed from the genIPS of the host cell into messenger nucleic acid. This is the internal information stopping the bacterial metabolism and producing phages. When all of the phages are "ready" a special gene of the phage DNA is read which leads to the formation of a specific protein, resulting in lyses of the bacterial cell wall. The host cell is destroyed and the phages are set free.

The central significance of the genIPS for living systems is its ability for identical replication. At the macromolecular level a phenomenon can be observed which the chip manufacturers up to now have not even dared to dream of: a genIPS not only functions reliably, it is also capable of replicating itself (almost) without errors and in quantities which seem to be *inconceivable* to us in information processing technology. A bacterium and its corresponding genIPS can, for example, replicate itself in approximately 20 minutes. Hence, by using a suitable bioreactor, after only 12 hours $2^{3 \times 12} = 2^{36} =$ approx. 10^{10} (almost) identical functional replicates can be obtained. This has to be compared to the fact that today only several billions chips can be produced worldwide yearly!

The genIPS is intensely active when replicating cells. It produces, e.g. suitable proteins which can be used for the synthesis of new DNA. It organizes by means of various kinds of regulation the construction of new cell structures. Components of the cell which are "in the way" are broken up. A minimum of some dozen genes participate in this process, which is switched on and off according to certain complicated procedures.

The advantage of the genIPS over physical information processing are considerable:

1. A broad variety of messages (triplet sequences) can be stored and processed.
2. Newly formed internal information represents the organization of complicated structures.
3. A genIPS is formed from relatively simple macromolecular basic structures which, in turn, are constructed through the use of physical information processing.
4. A universal code and a uniform "storage medium" are used for the broad variety of life.
5. The density of storage in the genIPS is unbelievably high (almost all necessary information for the ontogenesis of a human being is stored in DNA molecules having a total length of approximately 2 m and a diameter of only 2×10^{-9} m, i.e. corresponding to the volume of a cube whose edges are only $1.3/1000$ mm long.

6. The speed of information processing is high. The complete processing of all information of approximately 1000 genes in a bacterial cell takes place in about 20 minutes. This speed, however, is a lot lower than in physical information processing.
7. The basic procedures of transcription and translation are simple. There are, however, very complicated "controlling structures" that make sure that exactly the "proper information" is processed at the "right time".
8. The genIPS functions only locally in volumes of approximately $(1/100 \times 1/100 \times 1/100)$ mm^3.

In answering questions (A1) to (A7) for Sect. 2.5, genIPS can be specified as follows:

(A1) External information cannot be used by the genIPS. It relies on internal information (DNA/RNA) only. (The infection of a cell by means of the DNA/RNA of a virus/phage is a special case.)
(A2) GenIPSs produce all the internal information necessary to build a cell or even a human being. In the first step m-RNA is the central internal information. This, however, is embedded in other structures, which also can be considered as internal information. They represent the necessary declarative and procedural information.
(A3) Internally, genetic information processing can be described as a highly sophisticated biochemical factory. This has been, up to now, merely understood in principle and for simple components. Storage of internal information and its processing are linked by several mechanisms.

At this point the information processing paradigm shows a dilemma: it is not possible to pinpoint *in detail* which information is really going where and what all of the cellular components are doing in terms of information processing. However, we know that a genIPS is organized very efficiently. (Taking out a component usually results in a complete breakdown of the entire structure.)
(A4) Genetic information processing is absolutely crucial for stability and reproduction of cells, plants, animals and human beings. Stopping genIPSs results in the breakdown of living structures. The genIPS is the central "facility" for self-organization. In addition it establishes a "containment" in which neural information processing is possible and thus provides the basis for very flexible reactions to variations of the "informational environment".
(A5) The basic material structures of internal information are ribonucleic acids. However, many more or less specialized enzymes have to be "available" to run the genIPS. The biochemical structure of the cell and its genIPS are inseparable.
(A6) The quantitative aspects of the performance of the genIPS have been illustrated before. Here it can be stated clearly that the genIPS is able to build an extremely sophisticated organization from a single cell (usually a zygote) within hours, days, weeks or many months. Considering the number

of atoms and molecules organized by the genIPS in a living cell it is obvious that the genIPS is much more powerful than the physIPS in establishing "material order".

Declarative information is very obviously stored primarily in the DNA. Procedural information is supplied by the DNA as well. Functionally it is, however, spread "all over" the cell, it is part of the cellular structure. Living beings *cannot* come into existence by means of a naked DNA; some "basic containment" must always be available. This shows the sophisticated relationship between declarative and procedural information in gen-IPSs.

(A7) GenIPSs use heavily biochemical mechanisms which can be traced back to chemical, and finally to physical, processes. However, a reductionist approach fails, and genIPSs are not determined by physIPSs directly. Rather, a genIPS is a very specific selection of "initial conditions" from a very broad variety of physical possibilities. A given set of conditions which has formed a very complex physIPS is the basis for the genIPS. These "initial conditions", however, are *not* determined by physical laws. They emerge as a step from the level of physIPS to genIPS. A very complex physIPS forms the basis for the genIPS but the latter is not reducible to physical properties. With respect to the "emergence principle", a genIPSs is a new level with new characteristics.

The level-evolution of the genIPS can be traced back over at least two billion years of evolution in which its performance has improved noticeably. At the same time, however, the basic principles have remained essentially stable. Thus, at present "old", "lower" forms of organisms do not have a fundamentally different genIPS compared to "modern", "higher" organisms. (Admittedly, there seems to be an even "older" genetic code which provides for the coding of the amino acids by the t-RNA, which is being investigated at present.)

An important, and limiting, property of the genIPS, as opposed to the physIPS, is that it can use only the (genetic) information available within the system. This information is kept extremely stable. It is only changed randomly (in genetic terms, through mutations). A genIPS therefore, does not have the ability to react *specifically* to variations of the environment. Only its "physical subcomponents" can process "simple" physical environmental messages by means of physical information processing, but they are not transmitted to the level of the genIPS, as far as we know.

External energy is necessary for "running" the genIPS. However, self-organization of the cell can also be stabilized without an energy supply (e.g. in a frozen status).

At present the mechanisms for the leap from the physIPS to the genIPS are still unclear. Is the genIPS a fundamentally new system? Or is the performance of a "composite" physIPS sufficient to produce this mega-level of IPS? Why is there – obviously – only one basic type of genIPS? (For some answers see Chap. IV.1.)

5 Neural Information Processing: Dealing with Complex Environments Directly

5.1 General Concept

As we have seen, the genIPS works internally only: it uses only the information stored in DNA to organize and replicate cells and biological systems. In doing so it is characterised by strict procedures and is unable to process external information which is in the environment of the biological system. Hence, the genIPS cannot adapt itself directly to changing environments. It does not possess any "sensors" for external information and has no memory system able to store "external" information for later usage. (Geneticists classify this as the inability to inherit acquired properties.)

I shall call all biological IPSs which are able, through some type of (special) receiver, to pick up a given spectrum of external information **neural** Information Processing Systems (neurIPSs). (Neural because the basic mechanisms rely on neural structures embedded in biological systems.) These systems can store external information. They use very complex and variable procedures to "handle" the external information properly, they can "learn".

The first forms of neurIPSs must have occurred at about the same time as the genIPS in evolution, or perhaps a little bit later. We know that even very "simple" and "old" organisms are able to perceive and store external information temporarily ("short-term memory"). Even "simple" neurIPSs are able to carry out procedures in order to deal with particular messages properly. In doing so, various forms of external information are selectively transformed into internal information through the use of genetically prespecified declarative information, or by information that has already been "learned" from the environment. "Organized" internal information can be stored in "long-term memory".

Typical characteristics of primitive neurIPSs are, for example, the ability of individual cells to move towards a source of light ("phototaxis") or to use chemical gradients ("chemotaxis"). Certain worms, for example, can, by means of electric shocks, be "induced" to choose a particular path through a simple maze even though they have never been confronted with electric shocks in their natural surroundings before. Procedural information of the neurIPS can also be changed in worms: for example, a different kind of electrical stimulus can induce the same worm to choose a new path. Due to its neurIPS a bee, for example, is able to find its way back to its hive across even a long distance.

5.2 Details of the NeurIPS

Using results from intensive investigation of many types of neurIPSs the fundamental structures of the systems can be summarized as follows.

All neurIPSs continually use special sensors which selectively receive physical or chemical "stimuli" from the environment as external information and decode them into internal information. They are stored in a specific form in the neurIPS: one can generally distinguish between "short-term" and "long-term" memory. The stored internal information is only available for seconds or minutes in the former, whilst in the latter it is available much longer.

In the neurIPS more or less complicated procedures are available which fulfil four tasks: (i) the direct processing of external information, (ii) the use of external information in combination with previously stored internal (declarative) information for the purpose of forming new information, (iii) "reorganization" of internal information, e.g. the generation of new internal information from previously stored (nongenetic) information (pigeons, for example, are able to recognise a person they have never seen before if they have seen a series of slides of different views of this person; thus their neurIPS is working from an abstract representation of the image of the person), (iv) initiation of direct activities of the biological system (pragmatic information), e.g. direct steering of a motor system.

The *basic structure* of all neurIPSs is, of course, determined by the genIPS. In all organisms "pre-fixed" information and procedures are internally "pre-organized" by the genIPS. Thus, there is a connection between neurIPS and genIPS, the former relying directly on "pre-programmed information". NeurIPSs are not limited in size to small volumes, like physIPSs in the atomic nucleus or genIPSs. On the contrary, in the case of higher animals they reach larger macroscopic dimensions (e.g. in the form of a long nerve cell).

"Sensor techniques" in the various neurIPSs are extremely many and varied. This has important consequences: different types of neurIPS are specialized to a particular "informational environment". The human auditive IPS, for example, cannot decode frequencies higher than approximately 20,000 Hz, whereas the dolphin's neurIPS works up to some 100,000 Hz. (This important phenomenon of selectivity has already been seen from the early physIPSs: electrical messages, for example, can only be received by magnetically or electrically charged particles. They either have only "receivers" for the appropriate external information or can only process some "types" of external information.)

Selectivity is heavily exploited in various neurIPS in order to achieve a specific communication system (e.g. the chemical communication system of the butterfly). "Highly developed" neurIPSs possess several sensors and processing systems which are able to receive, decode and process different "classes of external information" in parallel and cooperatively (e.g. the sense of seeing and hearing of mammals). In general, messages received by the neurIPS from sensors are initially "preprocessed" and "compressed" (e.g. in a mammal's eye) and then decoded through the use of internally stored information from the long-term memory (e.g. "pattern recognition").

Because of space limitations in this article I shall not be able to discuss in detail the various basic biochemical/electrophysiological structures of the information storage of neurIPSs. Different mechanisms can be used, e.g. coding of chain molecules or physiological/electrical "charging" of neural networks.

However, it should be noted that an important functional property lies in the fact that short-term memories are always volatile, long-term memories are stable. External information can be transformed by neurIPSs into a stable state (it can be learned). Under certain circumstances this internal information is available for a lifetime, e.g. defensive reactions or an animal's recognition of its mother.

However, one should not confuse the stable memory of the neurIPS with stable genetic storage (DNA). In neurIPSs, sensor systems are *required* to provide internal ("learned") information which is then stored in stable memories. DNA, however, contains *all* information absolutely necessary for the "basic operations" from the very beginning of a living organism. Furthermore, it must be noted that the complete set of genetic information is not used by all cells of an organism. On the contrary, only certain areas of the DNA are transcribed and translated into protein structures in a particular phase. In contrast to this, the stable memory in the neurIPS can (in general) be used as a complete unit. Performance of the neurIPS varies quite a lot in respect of the storage capacity and the processing of information internally.

Our basic questions from Sect. 2.5 can be answered for neurIPSs as follows:

(A1) Various types of external information can be used by different neurIPSs with the help of special sensors. However, a given neurIPS is quite specific in the spectrum of external information it is able to handle.

(A2) In every neurIPS declarative as well as procedural information can be formed by "learning". Internal information acquired in this way helps the system to organize itself in its environment. Life without neurIPS seems impossible.

(A3) The internal functions of a living system depend heavily on its neurIPS. Neurons (or other cell components) are the basis of all necessary operations. "Response time" in a neurIPS is of the order of fractions of a second up to many minutes. Memory plays an essential role. In "higher" neurIPSs there is a large volume of complex procedures (see e.g. details in Sect. 5.3). The neurIPS is able to store new external information, thus changing declarative and procedural information ("learning").

(A4) In all living organisms neural information processing is an essential feature of self-organization. Complex procedures and memory for external information allow for coping with the environment "actively". Beside the intensive use of external information neurIPSs take over the task of internal communications between "substructures" of a living system. (Here, we are getting into problems with the term "system" since substructures in a living organism can also be considered as "independent" systems.) In all cases proper input and processing of external information is a prerequisite for survival of living beings. Thus, neurIPSs "stabilize" plants, animals and human beings within their environments.

(A5) The basic material structures of all neurIPSs are nerve cells (or their precursor), using biochemical as well as electrical "signals" to transport and process external as well as internal information.

(A6) The performance of the various neurIPSs varies broadly, as has been discussed before. Most important is the insight that neurIPSs are able to "condense" external information in a way that is very effective for the adaptation and survival of a living being. NeurIPSs bring about the capacity of memory for external information; this is a big jump in mega-evolution compared to physIPSs and genIPSs. NeurIPSs are usually capable of setting up communications within a living system and between living systems, at least at the level of the same species (which, by the way, is true for physIPSs as well, but not for genIPSs).

(A7) NeurIPSs make heavy use of biochemical processes and thus, finally, of mechanisms of the physIPS. In *their* basic structure, neurIPSs are pre-determined by the DNA of the genIPS specifying the living system. However, after ontogenesis, the neurIPS is relatively "independent" of the genIPS. However, it cannot "feed back" the content of its memories to the DNA of the genIPS.

As we mentioned above, every "grown-up" neurIPS differs from its companion within a species. It is individual and quite specific because it has its own "experiences" which are aggregated in its memories. These form a crucial basis for all "activities", even though many "basic procedures" are predetermined by the genIPS (e.g. "instincts").

Here, a critical question arises which must be asked at this point: Who is actually "running" a living system, the neurIPS or the genIPS? From my point of view there is only one answer: Both systems do the business. They are able to "cooperate" in some way which is hidden in the complexity of the "biochemical factory" of a given organism. Both genetic and learned information are responsible for proper self-organization and adaptation to the environment.

Today we are very well informed about the level-evolution of structural and functional components of neurIPSs. The information is drawn from palaeonto-logical findings and anatomical comparisons: in many cases there are simple basic systems which have been "refined" over the course of millions of years. This, however, has nothing at all to do with teleological evolution. On the contrary, "better" neurIPSs only succeed in coming into existence when variations (mutations) have occurred in the DNA which "pre-specifies" the neurIPS. If mutations prove to be advantageous from a population genetic point of view, they spread in the population. (This mechanism alone might, however, not be sufficient to explain the origin of, e.g., a mammalian eye.)

Whereas only one "fundamental conception" for all terrestrial life has prevailed in genIPSs, the neurIPSs are extremely different in their construction and function. For example, a spider investigates its environment essentially through a sense of touch and via seismographic information; a bat investigates its environment through shortwave signals and the time required for echoes. In other words, in level-evolution extremely selective forms of "adaptation" of each neurIPSs to its (biological) environment can be observed.

5.3 The Human Information Processing System

In the following some of our insights into the best developed neurIPS, the human information processing system, are summarized. The data have been collected and interpreted mainly by cognitive psychologists. Scientists have been using the information processing approach in this domain for some 30 years; therefore, cognitive psychology is a good example of how to understand neurIPSs using, in general terms, my approach and terminology.

Up to now no adequate explanation has been provided why the cerebral cortex of the human brain developed so quickly and in so doing "overtook" all other neurIPSs in a relatively short period of some million years (this corresponds to only about one hundred thousand generations). It is probable that mutations took place in the early stage of the evolution of *Homo sapiens* which "destroyed" or totally "removed" genetically inherited instincts, reflexes and behavioural patterns specified before by the "old" genIPS. Thus, the human IPS was "suddenly" completely depleted of all standard procedures and therefore heavily dependent on learning processes. In such a situation mutations that influenced the learning capabilities positively (e.g. an increase of the size of the cortex of the human brain) had an enormous advantage. "Primitive man", capable of learning, was more successful than his "pre-programmed" ancestors because he was able to adapt himself much faster to new environments. He and she then populated the world whilst the "pre-programmed" primates remained "in the trees of the jungle".

In the following I shall focus mainly on two questions: (1) "What is the actual performance of the human information processing system?" and (2) "How is the system realised functionally?".

There is no simple and concise answer to (1). The most important point is that no *general* performance of *the* human information processing system exists. On the contrary, in reality an extremely broad spectrum of performances can be observed. We find the efforts of a mentally handicapped person trying to establish at least some kind of basic orientation in his or her informational environment (i.e. to be able to speak, make movements, etc.). And there is the splendid mental performance of a W. A. Mozart or an A. Einstein, the extreme physical performance of Steffi Graf, but also the political fanaticism of Napoleon, Hitler or Saddam Hussein.

All of our present-day efforts related to education (at home or in formal education) are aimed at ensuring a "minimum" performance standard. We are, however, quite well aware that this approach is successful only to a limited extent. This is reflected, for example, by the phenomenon that the industrial nations still have an approximately 30 percent dropout rate in secondary schools, and that even intensive educational efforts are not able to decrease this percentage significantly.

In spite of the "emptiness" of the human information processing system at birth, every individual IPS can apparently be developed only up to a genetically predetermined "maximal performance". Numerous studies indicate that the actual performance of the human IPS results from a complicated mixture of genetically determined "basic structures" and milieu-dependent learning, balancing off at about 60:40.

Investigations in cognitive psychology accomplished over the past decades give us some understanding of (a) the central functions of the human information processing system. Medical and physiological knowledge of the past one hundred years has provided us with (b) a good overview of the anatomy and morphology of the human information processing system. Neurological and molecular biological work (however, mainly on very simple neurIPSs) has led to (c) fundamental knowledge about the basis of neural, physiological and molecular processing of information. However, "high-level" functional operations are hardly explained by these findings.

What is completely missing is a *consistent* model for the human information processing system which is able to explain the typical performances at all three levels. For example, no one today can say where the image of a particular person is stored in the human brain in such a way that a moving silhouette observed by the eye can immediately be recognised as the person in question.

Taking into account this regrettable but understandable situation (also due to the relatively small amounts of money given to this research) I shall limit my arguments in the following to the *functional* level of "simple processes" studied by cognitive psychology. They can be summarized as follows (even though newer models are continually being developed and older ones rejected because the empirical material is manifold and contradictory).

(A) The human information processing system evaluates external information which is coming in from different sensory organs in the form of already "preprocessed" and "compressed" internal information. Essentially, in doing so, anticipatory pattern recognition is used which enables the representation, storage and processing of the high input of very complicated "primary signals" into conceptually "simple" forms to be carried out. The human IPS refers to previously stored "patterns" when it is necessary to "recognise" a new pattern. Learning of new patterns is, however, relatively slow.

(B) The human IPS has functional distinguishable areas of storage which, however, work together in an interdependent and varied form. These are semantic memory, episodic memory, procedural memory and "working memory" with "short-term memory".

Semantic memory is used to store knowledge individually in concepts of linguistic, acoustic or visual representations. The concepts are—according to the acquired state of knowledge—connected in a "meaningful" way in a complicated semantic network. (Such forms of networks can directly be mapped using reaction time experiments to test the "distances" of concepts.) New "entries" into the semantic network come from working storage at a low rate of transfer (of the order of magnitude of one concept or one new relationship every ten seconds approximately). Hence, for example, it is difficult to learn new words quickly.

The *episodic memory* is used for sequential storage of (all) "episodes" experienced during a lifetime. This by no means happens in the form of a "video tape", but once again in the form of conceptual structures. Entries into the episodic memory are made continually using semantic memory. (This has, for example, the effect that we very often know *when* we learned something even though we are no longer in a position to govern the proper procedure or information. This can easily be tested by first thinking about when one was taught to do square root calculations, and then trying to use the procedures necessary to determine the square root of, let's say, seven.)

Access to the episodic memory, as to all other kinds of memory, "fades" in a manner which is still largely unexplained. People are unable to recall all sequential details. In addition, episodic storage uses the concept of semantic networks internally. This is the reason, for example, for the numerous difficulties a judge has with witnesses when they attempt to recall one event using quite different conceptual structures.

Procedural memory has been least understood up to now. Essentially, it seems to be organized in the form of a set of production rules. These are "if... then..." relationships which are kept, according to the current situation of the IPS, in an "active list" and thus permit an essentially heuristic solution of cognitive problems. For example, with people waiting at a traffic light, the production "if red then stop" and, simultaneously, the production "if green then go" are active. However, there are other productions such as "if friend is seen then say hello" which can be activated rapidly. Productions activated by the if-clause "trigger" other productions. (Procedures for controlling the human motor system are more algorithmically structured.)

The *working memory* of the human information processing system has, in contrast to the almost unlimited storage capacity of the other three areas of memory, only a very limited capacity. Numerous investigations have shown that about ten chunks (conceptually uniform structures, e.g. digits, words, sentences, meanings) can be kept simultaneously in the working storage. If an additional chunk is added every second to the old ones, then one of the previously recorded ones "drops out" of the working storage and is no longer available for processing.

Most probably *all* information (internal as well as incoming external) processed by the human IPS has to "force" itself through this working storage. This is particularly true for auditory and visual bits of information received at the same time. They do not "run" parallel through the working storage but rather in a "time-slice" procedure and therefore are often interconnected, even misinterpreted.

The limited capacity of the working storage plays a central role in the number of productions usable for solving problems: working storage has to "hold" the proper memory items used, as well as the "addresses" of the necessary procedures with the sum of both being about 10. If one attempts, for example, to count "in the head" how often various letters of the alphabet occur in a long sentence, this turns out to be quite difficult a task because of the lack of storage for keeping the sentence, the production necessary to fulfil the task, and the list of letter frequencies.

The human information processing system is in a position to deal only with a small number of concepts simultaneously. If more information arrives than can be handled, then either the problem cannot be solved by the average human information processing system, or pencil and paper are necessary as an extremely important "extension of the working storage". "Addresses" of information are then placed in this "external storage" which thus enables the problem solving space of the human information processing system to be extended. (This is the reason for the rapid increase in problem solving capacity after people had been taught to use pencil and paper.)

(C) Human problem solving can be functionally understood as a sequential processing of (many) productions which use the semantic network for their activation. Productions are activated through coincidence in the if-clause by "incoming" information or internal information which has just become available in the working storage. All "addresses" of relevant productions and concepts have to be concurrently stored in the working storage. (All components involved in the solution of a problem are understood in this context as the problem solving space.)

(D) Preprocessing subsystems with anticipation are used continually to produce new information for three different tasks of the human IPS: (i) information for controlling the motor system (including, e.g. speech), (ii) "temporary" information which is only used in the working storage and (iii) information which leads to new structures in the semantic, the episodic or the procedural memory. The latter process, in turn, is composed of "intentional learning" and "spontaneous processing".

(E) An important finding of modern psychology is that the human IPS by no means reacts only to external information. On the contrary, it is continuously quite active when completely cut off from the external "informational environment". In the human IPS internal as well as external information is used continually to produce new information. There are no stable structures – a reason for the creativity as well as the unreliability of human beings.

(F) The "naked" – merely biologically conditioned – human IPS can only deal with a small fraction of all the information in the informational environment. For example, we cannot "see" infrared radiation, nor can we "hear" radio waves or "smell" most molecules, etc. However, by means of the human IPS *Homo sapiens sapiens* has finally succeeded in developing observational techniques which allow insight into more and more new areas of the human environment. Today, for example, we are able to investigate the total spectrum of electromagnetic waves by means of appropriate receiving equipment.

(G) The human brain is divided into two cerebral hemispheres which communicate intensively with one another but at the same time carry out "different" tasks.

In one hemisphere (the left one for about 85% of the population) emphasis is placed on the processing of cognitive-rational-verbal processes. The other one is used for spatial, affective, communicative and "vague" procedures. (This is known from detailed investigation into the functions of the human information processing system of patients who have undergone brain surgery, as well as from tricky investigations with healthy test persons.)

(H) Within the level-evolution of the neurIPS from *Homo sapiens to Homo sapiens sapiens* language has risen to a very high level. Language is a strong and powerful mechanism for organizing and communicating information. However, even people born deaf who have never learned a spoken language are able to live a "normal" life. The human information processing system seems to be the only one that can work with language, so it might be a special step in the level-evolution. (Whether dolphins have a language is still a matter of debate.)

(I) It has to be clearly emphasized that, in any case, a given human IPS works primarily for one individual only. It controls itself and one human body. The option for conscious suicide indicates the extreme strength of this control. On the other hand, it is also in a position to be "externally" controlled and can even be completely "taken over" via hypnosis by another person.

At this point I have to mention the controversy between "materialistic" and "mental" assessment of human performances. The functional (information oriented) description of the human IPS presupposes the existence of material (genetically determined) fundamental structures. However, since we still have no complete understanding of the morphological, physiological, biological and molecular structures which determine all functions in detail, the idea of an "immaterial mind" remains in principle a possibility. Admittedly, we do not have any kind of hard proof for it. The meta-theory of information processing proposed here argues for information *and* matter/energy as parallel entities within the human information processing system.

6 Social Information Processing

6.1 Cooperative Neural Information Processing

NeurIPSs process a broad variety of external information. For some of the inputs, there are proper procedures for the neurIPSs to interpret them specifically, for others there are only "vague" interpretations and limited processing. Within this broad spectrum of neural information processing activities there is always a special class of procedures which is (more or less) specific for *all* members of a given species only. There is also a specific set of information exchanged for well-defined purposes within a group or at least two members of a given species. (One might call this "the language of the species", although this wording is in almost all cases, aside from the human language, not in accordance with its usage by linguists.)

Whenever members of a given species use common procedures and their common "language", they show – to a larger or smaller extent – a specific set of actions. They "behave" properly (e.g. mating behaviour). Individuals involved in such a group of "communicative cooperations" are called a social group. We shall call the cooperative functioning of the neurIPSs of the members of such a social group social information processing. As a Social Information Processing System (socIPS) I define the *set of neurIPSs of the members of the social group working together*. Thus, the socIPS is a *composite structure* of several neurIPSs which, however, are not linked physically. They cooperate intensively in terms of common procedures and common sets of external information sent out and received.

At a first glance the socIPS level seems to be quite different from all other distinct levels in the mega-evolution of IPSs because it is composed of "independent" entities. However, this is basically true for physIPSs or neurIPSs as well. In those cases there are also many "substructures" which form the total system through "cooperation". But the distances in an atom, in the cell or in a single organism are much smaller than in socIPSs. However, gravitational information processing works even in the entire universe.

There is one feature in socIPSs which is quite unique within the mega-evolution of IPSs: the socIPS can "fall apart" (if there is no action among the members of the same species) and it can reassemble. SocIPSs can be formed in various ways according to the number and level of cooperation of participating neurIPSs and their specific performances. Thus, socIPSs are always – more or less – dynamic.

To make my (abstract) view of socIPSs better understandable, three examples are discussed first.

(A) NeurIPSs of male and female frogs react quite specifically to visual and auditory information received from the partner, generating the necessary pragmatic information as a base for the mating behaviour. After successful mating, however, the pair separates and the neurIPSs are not concerned with cooperative tasks any longer – until the next period of sexual activity.

In this case both frogs have certain procedures in common. There are common sets of information exchanged between male and female frog and there is a time-dependent formation of the socIPS. All frogs of a given species have almost the same socIPS. Within its performance there is sufficient flexibility to adapt to the environment in which mating takes place. However, mating of individuals of different species is impossible, since there is no common set of procedures and "understandable" internal information; the information available is "considered as external only".

(B) Termites live in large states. There is a spectrum of common procedures which allow for differentiation of tasks performed by the individuals. There is intensive communication between individuals which is the basis for the adaptability necessary to stabilize the termite state. There is, e.g. an active defence behaviour against enemies of any sort.

Here, the socIPS is quite stable; it works for years, even decades, in a given place. Basic procedures are well defined in the neurIPSs of all termites; however, there is sufficient flexibility to cope even with serious "accidents".

(C) The human family is a highly complex socIPS. There is, to a larger or smaller extent, a common base of many procedures ("habits") which control the daily life. Language is used heavily within the family for exchange of internal social information; there are many quite specific syntactic and semantic arrangements, which include, e.g. facial expressions and gestures.

In this human socIPS many cooperative activities can be performed, even if the members are far apart, as a consequence of the use of telecommunications. Procedures of this socIPS are learned intensively; there may be only some genetically determined "basic behaviour". The socIPS of the family can be extended to other areas of social behaviour (e.g. friends).

Thus, socIPSs are not necessarily "closed systems"; they are open for integration into "larger" structures of the same level of the mega-evolution of IPSs.

There are three important facts within the *level-evolution* of socIPSs which must be mentioned in this brief introduction:

(1) Social information processing (at least in more complex systems) uses forms of *external storage* of information (in addition to neural memory). In a simple form these are scent markers in an open territory; in early human societies primitive paintings or even simple scripts were used for this purpose.

(2) With increasing differentiation of the socIPSs' evolution the number and the complexity of common procedures increases. Whereas the mating behaviour of two crabs is "relatively simple", the internal organization of a termite state is very rich in terms of social procedures.

(3) In socIPSs adaptation to quite different situations is possible. Learning takes place as a *social* phenomenon with an impact on most members of the societal structure. In this way, procedures that have been adapted by a socIPS can be transferred to following generations. Thus, in parallel to genetic storage, there is a common stable "social memory" relying on somehow "synchronized" neurIPSs in several individuals. In human societies education (in its broadest sense) is a highly elaborate way of preserving and transferring declarative as well as procedural social information from one generation to the next. This is usually called "culture".

Finally, it is important to recognise that there is (usually) some interaction between different "types" of socIPSs; however, "understanding" one another is usually quite difficult (e.g. human societies and groups of dolphins both have sophisticated but very different neurIPSs and socIPSs).

6.2 Essential Characteristics

This chapter cannot go into the extremely broad variety of social information processing. In responding to questions (A1) to (A7) from Sect. 2.5 I shall present just a very brief summary of essential features.

(A1) Various types of external information can be used, e.g. chemical information, electromagnetic waves or sound as "primary" forms of internal information in the socIPS which might form some type of "language". Special sensory structures receive specific messages and process then as internal information. As at other levels of mega-evolution of IPSs, a given IPS uses only a *small* fraction of all external messages, and thus it is highly selective.

(A2) A broad variety of internal and external information can be generated by socIPSs. Again, a given socIPS is quite specific in the way pragmatic information is generated and distributed. There is the option of collective learning which gives rise to common information (declarative as well as procedural). Internal information is usually stored and processed in (many) neurIPSs. In "late" socIPSs storage can also be "organized" outside biological structures (e.g. the script).

(A3) SocIPSs function internally via an exchange of specific information within a social group and its use within common tasks. The processing is distributed between (many) individuals of a "society". SocIPSs can store and reuse external information ("social learning"). Processing speed is quite low compared to other IPS levels. This has (usually) to do with low information transfer rates within the socIPS. In addition, individual neurIPSs are not very fast in their activities. SocIPSs can "fall apart" for a given time and can be reorganized at a later time (e.g. the breeding behaviour of many birds). Thus, there are "virtual" socIPSs, which can be activated by specific mechanisms (e.g. activity of the sun).

(A4) There are no social systems without information processing. Thus, social information processing is the essential basis for self-organization. Social and behavioural sciences have tried to explain some of the characteristic mechanisms and procedures controlling self-organization at that level of mega-evolution.

(A5) The material structures are always living beings (only in some cases physical structures are used as external memory). Neural (sub)systems are the most important components. However, in many cases genIPSs play an important role as the basic source of procedural information (e.g. "instincts") needed to develop and stabilize social systems.

(A6) A given socIPS usually tries to give a social structure in a given environment a (long-term) advantage over other social systems. Thus, performance is directed towards the common processing of information coming from outside a given social system as well as from "partners" inside the social structure. Since most of the tasks are quite complex, many procedures are used in a given socIPS. They are usually distributed procedures, relying on an appropriate balance between all neurIPSs of the members of a social system.

(A7) Integration of "lower-level" IPSs within socIPSs is complex: neurIPSs directly serve as a basic structure of socIPSs. However, genIPSs play an important role by setting up "basic" structures as well as declarative and procedural information for all socIPSs (e.g. "instincts"). With increasing complexity of the socIPS in level-evolution, they rely more and more on "learned" information.

In level-evolution of socIPSs there is a long period of differentiation and specialization as well as an increase in size and performance. What presumably started billions of years ago through very simple "mating habits" (e.g. conjugation in bacteria) has today developed into complex units of human societies with a broad spectrum of information and procedures.

7 Technical Information Processing

7.1 Introduction

The most important aspects of Technical Information Processing Systems (techIPSs) have been discussed in Sect. 2.6. In the following, I can thus refer to this section in respect of the seven questions about information processing systems asked in Sect. 2.5. Thus, only some special aspects of techIPSs are outlined. In particular, I am interested in the interrelation of human information processing and techIPSs as a basis for their integration into sociotechnical systems. First, I am going to discuss the origin of techIPSs, since this is the *only* situation in mega-evolution of IPS where we have all the data at hand.

7.2 Origin of TechIPSs

Looking back at the history of technology one notices that people have traditionally followed three different lines of development to varying degrees of intensity: (i) the increasing extension and substitution of human muscle power, (ii) improvement and substitution of the human sense organs, and recently (iii) the improvement of human information processing by means of technical systems.

The first category consists of inventions such as the lever or the wheel in prehistoric times, the cart and the steam engine later on, followed by electromagnetic systems and, in particular, by machine tools as well as, in most recent times, energy technology. The second group of inventions started much later with, for example, pipe systems for the transfer of messages, signal rockets, telescopes and photography, and it culminated in the telephone and the media.

"Traditional" technology, however, merely gave rise to procedures and equipment which always had to be controlled *directly*: the plough had to be guided by the farmer, star constellations looked at with the telescope had to be evaluated by the astrologist, and later on by astronomers, the lathe had to be operated by the turner, the telegram had to be read by the receiver, etc. It was only in the last century that some technical systems could be controlled in such a way that continual human "participation" in all details of the processes was no longer required. Initially, it became possible through mechanical or electro-mechanical systems to provide a technical representation of primitive, and later on, even more complicated control algorithms. Examples are the (mechanically controlled) moving miniature temple of Heron of Alexander two thousand years ago, or the one-hundred-year-old combustion engine (valve regulation). Admittedly, we are talking about very simple and often cyclical "programs".

Indeed, the performance of the human brain was for a long time quite sufficient for solving everyday problems which were local and relatively simple (such as food, clothing, housing). This situation first changed a few thousand years before our recorded time when, on the one hand, the world population grew significantly because mechanical techniques (particularly in agriculture and the military) led to a drastic increase in efficiency, and, on the other hand, nations began to migrate, and seafaring across the oceans took place. It was at that time that the first "information techniques" originated; in particular writing was used as a procedure for storing information.

The printed word has to be considered as the most ingenious step on the path towards technical information processing. It enables a significant extension of the "working memory" and the "long-term memory" of the human brain. Writing allowed the manipulation of thoughts by means of texts in an "external storage device", as well as the ability to deal with numbers effectively. Something that had been written down did not get lost, as opposed to what had been learned by human beings. Information could be copied (written down many times and later on printed) and handed over from one generation to the next without it being necessary for the person handing it on to understand the context or even learn it. The use of the "slate pencil and papyrus" led to an extension of the very limited human working storage and thus to a significant increase in the capacity of humans to solve problems.

Technical procedures were also developed in the field of telecommunications — at the latest in antiquity. Aeschylos described signal chains for sending a message to Argus about the victory over Troy. The Greek historian Diodoros described a chain of acoustic signal stations arranged over a distance of approximately three hundred kilometres that supposedly enabled the mobilization of ten thousand Persian archers in one day. However, such technically interesting approaches were unable to replace the much more popular courier services (Roman "cursus publicus") since signal stations were simply too unreliable.

The requirement for improving information processing started to increase significantly in antiquity, since in empires extensive commercial commitments and numerous military problems had to be controlled. In particular, the processing of quantitative information became important for the determination of tribute payments, the calculation of troop strength or the calculation of a ship's course using observation of the sun or stars. Complicated number systems which had originated in earlier cultures were, therefore, further developed in order to enable rapid and reliable calculations to be made and they became the precursor of algorithmic techniques. The most significant "breakthrough" took place with the acceptance of the Arabic number system which is still in use today. The abacus was invented which was a further development of the calculation device used by the Persians. Arithmetic aids represented – from the point of view of human information processing – meaningful, structured forms of an extra-somatic number storage which supported the work of the human working memory.

Even though primitive basic principles of information storage (writing), "information processing" (e.g. the clock) and telecommunications (e.g. signal chains) had been known about since antiquity, powerful procedures and equipment for dealing with information were developed quite slowly throughout history. Deficits in human information processing performance evidently did not cause very many grave problems up to the end of the Age of Enlightenment; up to this time the life of the broad masses of the population was still simple, social integration loose, there was little division of labour and human labour was cheap.

The development of "modern" technical information processing systems started with the Age of Enlightenment, with the fading of relatively unreflective and uncritical information reception, with the expansion of the local and global division of labour, with the awareness of the significance of quantitative methods in the natural sciences and with the unfolding of a numerically oriented mathematical science.

There are four lines in mega-evolution of techIPSs which played an important role: (1) the development of calculators for the four basic rules of arithmetics, (2) the automation of production processes (e.g. control of the weaver's loom), (3) the use of printing techniques for the extensive spreading of information and (4) the development of procedures for using external storage (particularly pencil and paper) as a basis for increasing human performance in solving problems. All of this was supported by an expanding educational system which was extended to "lower" social classes of the population and thereby made popular the *systematic* handling of information. It is here that we find the roots of the development of techIPSs.

The requirements of efficient information processing increased dramatically during the Second World War. Demands for high precision in the calculation of ballistic data, the quality with which weapons should be manufactured, logistical work and the demands that resulted from continually increasing communication throughout the world surpassed the potential performance of mechanical and electromechanical machines. The "obsolete" techniques required fundamental modernization.

Actually Leibniz had already conceived of and built the first model of a genuine four-species calculating machine in 1673. Babbage designed the "analytical engine" in 1833 which corresponded, in principle, to present-day calculators. At the end of the last century Hollerith succeeded in extensively automating the national census by designing and using a machine with punched cards for counting. More than one hundred years ago Alexander Bell invented the telephone which then replaced the more complicated telegraph. Hertz published the principle of wireless telecommunication and thus created the prerequisites for the mass distribution media radio and television as a precursor for modern interactive networks.

However, the real breakthrough in the construction of programmable calculation machines first came about with Konrad Zuse's calculators Z1 to Z3 in the 1930s and 40s. After the end of the Second World War these were followed by modern developments of real electronic computers in the USA and England, e.g. by Eckert, Mauchley and Goldstein who in 1946 introduced the ENIAC. Von Neumann's consequent descriptions of the architecture of a digital computer with uniform treatment of data and procedures can be considered as the very beginning of modern techIPSs. A completely new technology was then developed in an extremely short period of time. (This happened, however, without any particular kind of reflection about the performance of human information processing.)

Various early lines of development were combined in the 1950s in the form of the commercial computer. The foundations for commercial systems resulted from the increasing appreciation of reliable hardware which finally, through the invention of the transistor, created the necessary conditions for the simple and economical processing of elementary procedures in the form of Boolean algebra in a binary form.

Replacement of "primitive" and very complicated programming using machine language and symbolic addressing by the languages of the third generation (e.g. FORTRAN) enabled first engineers and natural scientists, and later on office workers, to solve problems by using a computer directly. At that time in the USA "Computer Sciences" was established, followed later on by "Informatique", in Central Europe.

Broad availability of efficient integrated components for data processing technology resulting from mass production of subassemblies used in telecommunication finally led to an explosion of the market. IBM's study in 1944 proved to be completely wrong: here one could read that the global requirements for computers would be no more than just seven! In spite of the high price of the "old" hardware the mainframe computer represented the advent of a new instrument which was intensively used to satisfy the early rationalization requirements of the 1950s and 1960s.

The disadvantages at that time were, however, high unreliability, high costs and the "locking up" of computers in computer centres. The latter soon proved to be the driving force behind the creation of information processing directly available on the desk. This resulted in time-sharing facilities which, via data-lines, enabled "everyone" to have a processing facility at their disposal by the late 1960s and in the 1970s.

Computerization of firms first took place mainly in the financial and wages departments; banks and insurance firms recognised the potential and were able to realise business on a scale that had previously been thought impossible; space exploration became possible due to the feasibility of calculating orbits precisely, etc.

Exploitation of the new technology by the military went hand in hand with the boom in technical information processing in the civil sector. It was very strongly influenced by extensive government-subsidized programs and by initiatives directed towards "high-level" basic research supported by the "Defense Advanced Research Project Agency" (DARPA) of the US Department of Defense. It has thus been possible for many years to work on both military and civil problems by using the military budget. The "military-industrial complex" was and still is an important driving force behind developments in information technology as it has always been necessary to construct increasingly smaller, faster and more reliable as well as more closely integrated systems—in, for example the World Wide Military Command and Control Systems (WWMCCS) or the cruise missiles.

The broad expansion of microelectronics which resulted from the transistor and integrated circuits in the USA was accompanied rather unexpectedly by a new partner in the international business, namely, the Japanese. Within one decade the Japanese achieved a dominant position in the electronic mass-consumer markets for watches, pocket calculators and consumer electronics. They created and still do create a success out of the concept of easy-to-operate "intelligent" consumer goods. Whenever human intelligence seemed to be required to operate a product, the Japanese integrated and still introduce a microprocessor into products as an interface between the user and the primary process, and they invest a lot of effort in details to reduce a complicated function to something "naive-simple" (e.g. Japanese cameras).

The successful history of hardware – fascinating as it is – represents, however, only a *small* part of the revolution in the technical handling of information. A precise understanding of the structure of data and procedures was and still is essential for the transition from intelligent human information processing to pure technical processing. The concept of "structured programming" introduced in the early 1970s resulted in a wave of new data concepts, new programming languages and programming environments.

The times have finally passed when the user of a techIPS has to be concerned about a detailed representation of the problem in the computer. In many fields today "user-oriented languages" allow a description of the "what" of a particular problem solution without having to define "how" the solution should be accomplished in the computer. The success of this leap forward is shown, for example, by the sale of millions of spreadsheet programs which are used throughout the world today for solving quantitative problems – without any of the users having to reflect on how the spreadsheet calculations actually take place internally in the PC.

Hence, the super-special market for a few mainframe computers in the 1940s has today grown into a world-wide market of about $500 billion a year. The yearly growth of sales in this branch, ignoring fluctuations, is some 10 per cent. Furthermore, the information technology industry is in a position to invest about 10 per cent of its income in research and development. Taking into account investments allocated from the budgets of the various defence ministries in the western world for information technology, we arrive at a sum of about $55 billion a year presently invested in research and development world-wide.

The "real" challenge of techIPSs for human information processing began in the USA in the late 1950s with discussions about "artificial intelligence" (AI), when some computer scientists, psychologists, linguists and outsiders compared technical information processing with human intelligence.

Today it must be noted that with respect to performances in *special* sectors of human information processing, information technology has overtaken human cognitive performance by several orders of magnitude from the point of view of speed and cost. Who is capable of carrying out billions of calculations per second in his/her mind without making mistakes? Who is capable of operating a lathe twenty-four hours a day at the same high level of precision as a control computer? Who is capable of evaluating all measurements from a nuclear power station in milliseconds in order to ensure that the reactor does not become critical? Who is capable of translating specialized Russian literature into English day and night so that it can be understood at least by experts? Who is in a position of being able to regulate "in his/her mind" all the traffic lights in Frankfurt, taking into consideration changing traffic conditions?

However, "artificial intelligence" in its broadest sense has not yet come into existence. Instead, there has been a wide-ranging increase in the use of information technology to carry out what our grandparents considered to be completely inconceivable for the human brain. Human intelligence is still beyond the capabilities of techIPSs because it can cope with a *broad spectrum* of external as well as internal information and with highly complex procedures.

7.3 Is the TechIPS Really a New Level in Mega-Evolution?

My classification of the various IPS-levels is a *phenomenological* one. I identify as levels of IPSs distinct *functional* structures. From this point of view it seems justified to consider the techIPS as a new level which has evolved during the past 3000 years. TechIPSs use physical information processing as well as programs which result from neural information processing in human beings. Thus, techIPSs evolve from the highest level of neural information processing taking socIPSs into consideration as well. However, in contrast to the other steps in the mega-evolution of IPSs, many inventions have more or less been planned by human intention. Thus, "variations" in technical information processing which relate to social "selection mechanisms" seem to be somewhat different from, for example, mutations and selections in the level-evolution of genIPSs or neurIPSs.

But we have to be careful in comparing mechanisms in *level*-evolution with processes responsible for *mega*-evolution. In the latter we just do not know the important facts for the early steps, e.g. from physical to genetic or from genetic to neural information processing. It does not seem unlikely that at those initial steps important "inventions" were made which evolved as special events from the basis of the lower level IPSs. "Invention" here means a set of specific *initial* and *boundary conditions* under which just the new level of IPS is selected from a broad variety of possibilities. Such a "survival" of a given set of conditions must have been the preliminary reason for the genesis of physical, genetic, neural and social information processing systems. Whether these sets arose at that time due to an "act of consciousness" is, however, quite difficult to decide upon, since, from our position as human beings supplied with a "highly advanced" information processing system, we have no idea at all about the standards of "consciousness" on lower levels in mega-evolution.

There is a second difficulty in discussing the techIPS as a new IPS-level: whereas all the older levels have already progressed for a long time in their level-evolution, techIPSs are at their *very* beginning! Therefore it does not seem surprising that techIPSs (as they function today) use only programs that have been specified "in detail" by human beings. This might – and will – change at later stages of the level-evolution of techIPSs. Artificial intelligence is already going along this line, e.g. by designing computers that can "learn" from their environments and thereby develop their "own" data structures and procedures without steady human support.

In the third place, technical information processing does not directly support self-organization of the computer itself. With its input/output relations it is always directly or indirectly "connected" with a human being as the interpreter of its output. However, this is not so different from, for example, the relation of genIPS and neurIPS, since they cooperate very intensively in performing all functions necessary for a cell or a living being to exist.

In total, techIPSs have to be seen as a phenomenological step in mega-evolution, though deeply connected to human information processing.

8 Computers and Society – Sociotechnical Information Processing

8.1 The Present Level in Mega-Evolution of IPSs

Present-day computers do not exist on their own; they "need" human beings. However, vice versa, modern societies do not exist without techIPSs in economy, administration, the military, etc. Social and technical information processing are closely integrated. This process of mixing technical and social information processing is a progressive one. I consider it to be the most recent distinct level in the mega-evolution of IPSs. Systems consisting of complex relationships between social and technical information processing will be called **sociotech**nical **Information Processing Systems** (sotecIPSs).

A more detailed description of this newly emerging level can be found in Chap. VI.1. In the following, some essential features are summarized. (1) In sotecIPSs there is a continuous "sharing of work" between human and technical information processing. In many cases today human beings do the "creative part", whereas techIPSs handle the "routines" (as in, e.g. computer-aided design within computer-integrated manufacturing). (2) There is an increasing number world-wide of installed techIPSs which play an essential and crucial role in the functioning of societies (e.g. the computerized banking system). (3) The way technical IPSs function has an increasing influence on how societal affairs are dealt with (e.g. tax legislation, just-in-time logistics, medicine, operations in the Gulf War). (4) In many cases, sotecIPSs take the form of world-wide structures (e.g. financial systems, early warning systems in the military, data communication within telecommunication networks). Thus, sotecIPSs are – apart from gravitational information processing – the largest integrated information processing structures on earth.

8.2 Basic Features

Since sotecIPSs are at their *very early beginnings* in their level-evolution I can only give *tentative answers* to the questions posed in Sect. 2.5. Nevertheless, it seems quite clear that sotecIPSs differ from other IPSs in *many* respects. However, they indeed form a new and quite distinct level.

(A1) SotecIPSs handle rather formalized, syntactically well-defined internal information. Input into the techIPS has to have a "properly coded" structure, output is organized in strict formats. But these restrictions are continuously being given up in more and more areas; there are, for example, "seeing" and "hearing" robots which today are already able to "behave properly" in complex environments. Very slowly, artificial intelligence is on its way to breaking down the "formal interfaces" between neurIPS, socIPS and techIPS.

(A2) Inside these complex, and often quite large, sotecIPSs all types of new information can be produced. The systems are able to learn, to recognise vast amounts of internal information and to control not only themselves, but also "outside structures" (e.g. the large reservation and booking systems embedded in the sociotechnical structures of tourism).

(A3) Internally the sotecIPSs function in a "symbiotic" way by integrating socIPSs and techIPSs. They are heterogeneous systems. Information is stored in both human brains and technical storage devices ("sociotechnical learning"). Processing is done at high speed in computers (for "simple" procedures) and at low speed in human brains (for "complex" procedures). In many cases there is a complex sharing of memory as well as procedures between socIPS and techIPS.

Up to now computers in the sotecIPS environment have been limited to algorithmic structures. They are able to simulate rational tasks. Mental qualities like feeling, understanding, fear, happiness and consciousness are properties of human beings only which cannot easily be transferred to techIPSs.

(A4) The sotecIPS is essential for its self-organization. There is something like a spiral feedback loop: the existence of a sotecIPS stimulates social activities which steadily increase techIPS structures. This then results in a new social situation which stabilizes this setting and stimultaneously asks for further differentiation and increase of the techIPS, and so on. (Financial systems or just-in-time logistics are good examples of this type of "self-organized growth".)

(A5) There are two essential structures in all sotecIPSs: human information processing on the one hand and physical information processing and tele-communications machinery on the other. Human beings are steadily increasing the technical part of sotecIPSs. In many instances they (already) are dependent on techIPSs for doing this (e.g. chip design, software production for larger systems).

(A6) SotecIPSs are able to perform a wide range of tasks. With increasing use of techIPSs in almost all areas of societal life a rapid increase in the number and the performing qualities of sotecIPSs can be observed. Many areas of declarative and procedural information that were, up to the 1950s, available to human information processing only are now being spread in sotecIPSs (e.g. production of music and videos). In addition, there are new types of informational structures which can only originate in the complexity of sotecIPSs (e.g. computer graphics simulating chemical reactions used by the chemical industry).

There is a broad variety in speed and capacity of sotecIPSs. Real-time systems can be very fast (e.g. the control of electricity and its production at the national or even international level). Capacity, in terms of size and power of integration, varies as well. Whereas early warning systems in the military, for example, operate world-wide (controlling "war or peace"), just-in-time systems may work for one company only.

(A7) As mentioned above, sotecIPSs rely heavily on human as well as technical information processing. The sotecIPS structure, however, cannot be seen as the result of a simple "addition" of those "lower" levels. Rather, a new level emerges which integrates social and technical information processing in a mutual relationship. All "components" change their basic "behaviour" in this integration into the sociotechnical structure.

Table 1. Essential features of the six IPS levels

Level	physIPS	genIPS	neurIPS	socIPS	techIPS	sotecIPS
A1: Messages and internal information	"Exchange particles" Cellular automata	DNA, m-RNA, t-RNA, ribosomes	Electromagnetic messages; molecular and electrical information	Markers, paintings, scripts; "coded information" (language)	Various forms of input converted to internal electrical impulses	More or less formally coded information used internally. All types of input
A2: *New* information produced internally	Only pragmatic information "causing" forces	m-RNA, t-RNA, biomolecules	"Electrical" signals, macromolecules	Markers, paintings, scripts; "coded information" (language)	Data output: electrical signals	"Knowledge" in brains and databases
A3: Internal functions of the system, "learning"	No memory for external information; type of processing unknown	No memory for external information. Biochemical machinery; transcription, translation, regulation	Storage of external information; processing of declarative and procedural information	Common internal procedural and declarative information; distributed processing	Storage of external information. Processing of data according to pre-stored programs	Distributed processing and storage in brains and computers; task-sharing
A4: Contribution of IP to self-organization	IP is the basic principle for controlling matter structures	*The* mechanism for keeping cells and organisms alive	Keeps up internal communication; allows for "active" adjustment to the environment	Essential structure for setting up and running "societies"	No self-organization of the physical structure itself. However, self-organization at the level of software	IP is essential for the existence of sotec structures; continuous reorganization
A5: Material structures forming the IPS	"Elementary particles", quanta, waves	Organic (chain) molecules	Neurons, sensory systems	Living beings, *Homo sapiens*	Integrated circuits, physical components	Human beings, computers, telecommunication systems
A6: Performance (procedural, declarative; speed, capacity)	Long-range (electromagnetic and gravitational), short-range (weak and strong); extremely fast	Strict, quite complex procedures; not very fast	Complex (unknown) procedures, high capacity in the brain; slow	Quite complex procedures, flexible but slow	Large algorithms: large databases; quite fast	Complex mixes of heuristics and algorithms. Fast as well as slow. Distributed processing

A7: Integration of "lower-level" IPSs	Unknown ("cellular automata"?)	PhysIPS as the basis for biochemistry	GenIPS as an "environment"	NeurIPS	PhysIPS as well as human IPS (in form of programs)	NeurIPS, socIPS, techIPS
"Specialities"	Does not need energy for internal information processing; very rapid evolution in early cosmology	There may have been a slow evolution in the very beginning (crystals; microspheres; hypercycle)	"Distributed systems" with task-sharing within one organism	Distributed systems with sharing of common procedures and internal information in *many* organisms	Completely man-made; extremely short evolution giving rise to complex structures	Mega-structure at its very beginning. No "traditional" science available (no "sociotechnology")

At this point, I cannot say much about the level-evolution of the sotecIPSs because this is only at its very early beginning. Still, it might be possible to derive some ideas for the future of sotecIPSs from looking at the level-evolution of the "lower" levels of IPSs. (These aspects are dealt with in Chap. VI.1.)

9 The Heuristic Value of the Information Processing Paradigm

This basic concept, aggregated in the eight hypotheses, shows the possibility of explaining nature and societies, in general terms, as an "information processing universe". Within this the six levels of *phenomenologically* different information processing systems can be clearly distinguished. At every distinct level a specific way of handling external and internal information can be identified. These levels came into existence through evolution.

The results of the analysis of these levels have been summarized in Table 1. In the seven columns a *very brief* answer is given to our central questions (A1) to (A7) that were posed in Sect. 2.5 and answered in more detail in Sects. 3–8. Under the heading "specialities" additional insights are summarized.

Analysing the results, one can state clearly that the evolution of IPSs (in mega-steps as well as within each level) is a basic pattern in the unfolding of the universe. There is a *general mechanism* in the way *all* structures are built: they are formed and stabilized by means of self-organization using a given set of "basic" (internal) information and procedures. Evolution of all structures is characterised by the increasing amount and complexity of the information used internally as well as the "architectures" of the information processing systems and their procedural complexity.

Within the general evolutionary concept the information processing paradigm offers a new understanding of the origin of structure and order of natural systems: through aggregation of the components of lower-level IPSs a new type of information processing system can emerge which is characterised by new properties and functions. In accordance with the "emergence principle" such a new IPS-level is *not* reducible to the lower-level system(s). The emergence of a higher-level IPS is characterised by its ability to cope with new types of external and internal information (e.g., from physIPS to genIPS new types of internal information occur: DNA, m-RNA and t-RNA).

The evolution of information processing systems can be understood as the consequence of continuous internal processing of external information, thus constructing new structures. The emergence of new properties and functions through variations and aggregation is a characteristic phenomenon of mega-evolution. The unfolding of a "new" system depends on the presence of "proper" conditions in the informational environment. Environmental constraints can prevent the unfolding of the system.

In contrast to the traditional point of view of the "empire of sciences" which is characterised by differentiation into a large number of disciplines and special theories the paradigm makes way for a holistic understanding of nature based on just *one* fundamental principle: information processing and its evolution. An important feature is the assumption of *internal* information within all systems. The internal structure of a system allows this information to be processed appropriately. A further consequence of the paradigm is the characterisation of all natural systems as a process of continuous alterations because of the continuous information flow "through" every IPS. Informational interaction between environment and a system is an interaction between the "outside" and the "inside". "Outside" pragmatic information (e.g. sound, behaviour, forces) can be observed selectively by an IPS. The emergence of energy/matter structures is the result of processes going on "inside" the IPS. Last but not least, reproducibility (in its strong meaning) *cannot* exist in nature, because there is a steady growth of "external" and "internal" information which *never* allows the return to a given status. Reproducibility is only "possible" if "weak instruments" are used which do not show the evolution of the "information processing universe".

The total approach is *not* in contrast to the traditional matter/energy paradigm; instead it is the (very) beginning of a complementary approach. This is, first of all, concerned with the "why" questions instead of the "how-does-it-function" questions. I assume that by thoroughly studying information processing within structures we shall get a better understanding of *internal* mechanisms, and thus may be enabled to explain *why* the matter/energy world "behaves" as it does in typical scientific experiments.

I do not at all ignore the huge volume of experimental findings and theories derived from the assumption of the "real existence" of matter and energy. In particular, I am quite aware of the extremely high pragmatic value of this approach – particularly for organising and structuring technology. In contrast to this success, the information processing paradigm is still weak; particularly it has to overcome two handicaps: (1) there is no elaborate set of observer systems and experiments available to get "directly" inside the "information processing world"; (2) there is no quantitative theoretical approach available for information processing in nature which could be used for a "grand unified theory". At present the new paradigm can (at best) offer an alternative *perspective* to the results of the orthodox (matter/energy) disciplines.

One can ask why pieces of such a new "grand unified theory" have not yet been established. Because no methods have been found to investigate the details of information processing systems? Or, because of the present allocation of interests (and resources)? I believe it is the latter; the matter/energy paradigm has "absorbed" all possible contributors.

Understanding of nature often started in history with a "general idea" which later on was proven to be false or true by the setting up of proper methods for its evaluation (e.g. the antique concept of the atom or Maxwell's equations). The concept of the "computing space" argued for by Requardt (Chap. III.1) might be a rudimentary starting point for such a general approach, even though *much more* scientific work has to be invested to get it a little further.

My approach, of an "information processing universe", allows a new understanding of its evolution and its *future*. It can be assumed that by coping with the informational environment available at a given time, new information is generated all the time and fed back to the original IPS. Hence, there is a steady increase in information internally available in the systems which permanently stimulates information processing. Thus a *steady increase* in the *size* and the *complexity* of structures occurs. What can be observed by means of classical approaches is the matter/energy "surface" of this development only. The theoretical approach put forward in our basic concept tries to give an answer to the question of *what processes are "responsible" for the unfolding of the universe.* (See also the Epilogue.) Hopefully, it will be verified or falsified by deeper insights into nature.

II Contribution to the Concept of Information

Information: Course and Recourse

George Kampis

Eadem mutata resurgo

(Although changed,
I rise again the same.
Inscription on the tombstone
of J. Bernoulli, student of
the logarithmic spiral)

1 The Context   49

2 Historical Roots   51

3 D. M. MacKay and the Quantal Aspects of Scientific Information   54

4 Incommensurability   57

5 Information and Topology   58

6 Emergence and Information Processing   59

References   61

The principal aim of this paper is to find guidelines, for the author himself and also possibly for the wider community of information theorists, along which to arrange conceptually the various ideas on information. At the surface level, there seems to be no fundamental unity that links these papers; mine is no exception. Every author assumes this or that particular meaning for the term "information", then, in turn, they discuss an entailed twin evolution through geological and cultural ages: in other words, the genesis and recognition of information. What is this thing "information", after all? We could perhaps say "scientific information" instead. This technical term was introduced by Leon Brillouin (1956), perhaps the first to think systematically about the difference between "information" understood in the narrow sense of a measure, and the other thing, which everybody talks about. In fact, he suggested distinguishing entropic ("bound") and "free" information, and he argued that the latter is just arbitrary. So information is anything we wish. Now I shall reconsider the question in an even wider context to seek another meaning.

1 The Context

My thesis is that today, information assumes a common meaning through a prevailing societal practice, and as such, it is a word that reminds us of a paradigm bound to expire. Yet, on the other hand, it can be expected to re-emerge in a new skin, thereby representing both continuity and interruption in the development of an old idea: the idea is, of course, to get to know the world as it is.

There is nothing more characteristic of a period than the encompassing terms in which fundamental knowledge is expressed. These terms constitute the basic "faith" or "mythology" of a society. This helps us to understand why information is important and why it is not. It is impossible to speak about information without considering this context first.

It is usual to describe our time as the era of information. This era is increasingly dominated by concepts like "information society", "informatics" and "information processing paradigm". Two remarkable aspects of this development are these: first, technology and social life are increasingly information-based (whatever that means); second, it is assumed that the more information and the more technology we have, the better. Implicit in these statements is the belief that promoting such a development will solve problems, both in science and in real life. All we have to do is to encode things properly so as to represent a domain of reality, then turn the wheel, and out comes the result. Information is the philosophers' stone.

More scientifically, what all this rests upon is the claim that everything can be represented by combining symbols, the meanings of which are completely specified, and everything can be changed according to some plan based on manipulations of such representations. Hence, representations and the means by which to store, manipulate, replicate, analyse and realise them gain enormous significance in this doctrine. This reveals the picture of an essentially mechanistic, computable, controllable and transparent universe. We have machine mind, machine society and machine culture; everything accessible to reason.

Now, these linear, expansive thoughts are increasingly recognised as nonsensical. They can be dismissed (and have been dismissed already) at various levels.

At the level of philosophy, the dream of faithful representability, central to the information paradigm, has been refuted repeatedly, by many thinkers of this century, among them Popper, Wittgenstein, and Feyerabend.

Likewise, at the level of practical scientific method, ideas and idols of transparency, controllability and perfect classical rationality have been abandoned already. It is common experience that not every analysis leads to a useful synthesis or the possibility of realisation; in other words, science admits that our understanding is necessarily fairly incomplete. We should live together within the limits imposed by the complexity and uncertainty of real-world processes. Many scientists speak of "bounded rationality" (after Simon), and new scientific paradigms, sought by people from Prigogine to Maturana, Böhm to Penrose, try to offer ways out from the walled-in platonic universe of the freely exchangeable little tokens of information.

In terms of practical informatics, a recent crisis of the market has shown that expert systems are far from being experts yet, and decision-support systems cannot make any autonomous and meaningful decisions. The research community has already realised that some limitations of principle are involved here. The well-foreseeable failures of large and complex computer systems are common knowledge today. Some link is missing. The world is not like a computer and knowledge does not come in pieces.

Also, at the level of sociology the "cult of information" (Roszak 1986) has been sharply criticized. However, mass media and the educational systems promote this cult openly, the message of which is that everything is achievable, and everything is ultimately simple. They suggest that everything should be put in terms of some definite procedures for which there are universal frameworks already at hand.

Thus, despite its limitations, information has been a triumphant concept in the social history of the twentieth century. At the same time, intellectual history proceeded in an essentially opposite direction. For information that can be processed, manipulated, transmitted and realised, assumes **omnipotence** and **omniscience**. But today's best knowledge is about the limits of cognition and action. A feeling of omnipotence was common in modernity and, in general, in what was called the "industrial society". Today we enter the era of the "post-s": "post-industrialism" and "post-modern science". They warn us to be more modest. We are not creators but inhabitants of the world. Not masters but participants in a game bigger than us.

In short, it can be predicted that, due to the spreading of notions of a coming new era (I intentionally do not say New Age, but of course, the New Age movement is one of the important concepts here), with the fall of the myth of simplicity, the notion 'information' will lose the privileged status and significance it has today.

Will it submerge? After all, physics and many other sciences are well off without this word anyway. Is there anything fundamental in nature that necessitates the use of an information concept? Or is it merely a relic of an aggressive age without much scientific interest in the future?

My opinion is the opposite. I believe that the notion of information can be revitalized if the variety of its meanings is recognised, and that is what many papers of this volume endeavour to achieve. I think there is a common content, or commn denominator, to these new efforts. The present essay tries to give an account of this, by reviewing an old concept and relating it to the work currently done in the field, and to the more general ideas we are considering throughout this book.

2 Historical Roots

The word itself comes from the Latin *informatio*, which means illumination, exposition, outline, unfolding, or, boldly: a "body of knowledge". The modern study of information began when scientists started to apply probability theory to the analysis of communication channels. This was done in the Shannon theory of information. The core of this traditional theory is remarkably simple. An excellent description of these foundations is to be found in Atlan (1983).

In a message **x** made of an arbitrary number of symbols, $x_1, x_2, \ldots, x_n$, taken from an alphabet of **N** different symbols, the information content or "message entropy" is given by the formula

$$H(x) = - \sum_{i=1}^{N} p(i) \log_2 p(i)$$

where $p(i)$ is the relative frequency or probability of finding symbol x_i in x. Atlan writes that, "the function H does not characterise so much the given message x as it does an ensemble of messages using the same alphabet of N different symbols, with the same relative frequency for each of the x_i." Thus, information theory is in this narrower sense fairly limited in its scope, and deliberately so: it assumes (i) *ergodicity* (i.e. statistical homogeneity) of the process it considers and (ii) a preliminary choice of all messages that can be sent. In other words, it is assumed that the probability distributions do not change over time or across subsets of messages of the ensemble; in particular, the already received messages do not influence further messages in any way. Furthermore, the language cannot change and we should know this language in advance.

This concept has often been criticized for its restrictiveness. It was criticized already very early (Bar-Hillel and Carnap 1952) for not dealing with anything but *sign statistics*, that is, certain syntactic properties of messages. These allow for designing methods for the efficient and economical handling of messages, but nothing else. Indeed, the symbols and their meanings can be anything, as long as they can be labelled as $x_1, x_2, \ldots, x_n$ as required in the theory.

It is easy to see the origin of these restrictions. Shannon and Weaver were, at the time they published their epoch-making work, employees of the Bell Telephone Company. Now, it is required by law that telephone networks must forward messages regardless of their content – hence, by means independent of their content (this point was kindly brought to my attention by Professor H. Primas, Zürich, for which I am grateful). Hence the statistical approach; all messages should, perforce, get an equal treatment.

Weaver himself made a distinction between three categories of problems related to information: (i) technical problems – how to forward a message so that errors are kept to a minimum in spite of noise; (ii) semantic problems – how to make sure that the message received means the same as the one emitted; (iii) efficiency problems—how to make sure that the response or action of the receiver when receiving a message is the desired one. Clearly, these aspects involve what in semiotics are known as the syntactic, semantic and pragmatic aspects of information (Morris 1946).

The massive number of *independent* communication events mediated by telephone exchanges justifies the assumption of ergodicity. If somebody tells somebody else that he won a billion dollars, this may alter the future conversations between these two, but will not change conversations between everybody else. That is, individual differences are averaged out. The same refers to differences in language. In the end, all words and languages break down to letters, phonemes and morphemes, of which there is but a finite variety. This constitutes a stupid, but in a sense universal, "set of messages" suitable for telephone channels.

This is the specific historical setting that started the information sciences. There are many reasons for the unexpected success and expansion of this theory. When it was born, it was merely a mathematical formalism, as we have seen. The broadening out of the scope of this theory, and, together with this, the change of the meaning of "information" began with a few straightforward genera-lisations. A first step was to change the scope from a statistical measure of messages to a characteristic of any kind of interconnected events. From then on, any sequence of symbols was viewed as a "message". Ultimately, then, any outcome of measurements could be read as symbols and hence subjects of information theory. "It is as if a transmission of messages were taking place between the system and the observer... Thus, the information content of the system is in fact the amount of transmitted information in this channel" (Atlan 1983).

The prototype of a system dealing with information content of this kind is clearly a computer. A computer is characterised by the curious fact that, at a behavioural level, it appears to consist of *nothing but pure information*: the information content of a computer, as a system, equipped with some program, is concentrated in the information content of the program, which can be consi-dered a message in an obvious way. This "concentration" of information is a consequence of Turing's universal simulation theorem, which makes it possible for a present-day computer as a machine to simulate any other machine, that is, to reproduce the behaviour of any machine on the basis of the description, or program, of that machine.

Hence, outputs of such computer systems are determined by their programs alone, as if no system were in between, and so the information content of the output is merely a function of the information content of the program, the input. That is what we express by saying that computers "process" information: they manifest the complete control of a black box by means of input sequences. Computers have revolutionised our lives and technology, as have the information concepts that rode with that tide. Informatics, the art of dealing with information, was thus born.

There is reason to believe that the informational transparency of computers is due to their specific design and is fairly uncommon among system in general. Maybe this is why the study of fundamental questions of information has never stopped; let us only refer to two major collections from the early 1980s (Machlup and Mansfield 1983, and IJTP 1982); the questions discussed range from thermodynamics to algorithm theory, and raise the doubt, whether the concept is universally applicable.

3 D.M. MacKay and the Quantal Aspects
of Scientific Information

Current information technology suggests that there is in fact no real alternative
to the above informational notions on which it is based. But this is not the
case. This will be best seen if we recall the classic work of MacKay (1950). Already
then, in 1948 to 1950, that is, at the time the Shannon theory was first published,
he suggested an alternative system of ideas which in part parallels, but in an
essential way, however, transcends and complements Shannon's theory.

MacKay's basic idea was to distinguish between what he called the *metric
content* and the *logon content* of information. His metric information is in many
respects similar to Shannon information and to what computers can process.
His "logons", on the other hand, express a property completely different in nature.
He defines the logon content as the minimal number of the most fundamental
statements that specify a given phenomenon, expressed on some logical basis.
In other words, he speaks about the number of irreducible statements necessary
in order to specify a given system from a formal viewpoint.

At the surface level, this sounds like a definition that asks for a procedure
for finding a most comprehensive and most compressed (in this sense "minimal")
description of a domain. Now, since domains of reality are scientifically recognised
by means of measurements, and measurements are, in the spirit of the remarks
we made about messages, representable as symbols of some language, this
definition of information might be thought of as a measure of *algorithmic data
compression*.

This is a misunderstanding, however; it is a common mistake, made even by
those who know him, to consider MacKay (just as Gödel) a forerunner of
algorithmic information theory. The latter theory, developed by Solomonoff,
Martin-Löf, Kolmogorov and later by Chaitin (see Kampis 1991a), aspires to
express the least number of logical operations necessary for reconstructing a
sequence of symbols. The two tasks are similar but not identical. Kolmogorov
complexity (as algorithmic information is commonly called) is expressed as the
length of the shortest program that makes a universal Turing machine print out
a given sequence. Since every new program statement specifies one more operation,
their total number characterises the overall complexity, or information content,
of the sequence of operations. Let us note that this construct refers to the program
as a static object. There is, in theoretical computer science, a complementary
concept expressing another aspect of complexity, namely, the difficulty of actually
carrying out the already specified operations. To that end, the number of computational steps of the running time are considered on some reference machine.
This is what *computational complexity theory* deals with (Ausiello 1983).

However, what MacKay has considered was none of these. He did not mean
"specification" in the co-dependent sense of "derivation from program". He
considered the least number of *definitions* instead. To illustrate the difference,
it is like requiring n symbols to write a program in PASCAL, or requiring n
symbols to define PASCAL as a language (for some compiler, or a handbook,

for instance). In the two sorts of constructs the same symbols might appear; for instance, we can use the English alphabet for both. Still, they mean fundamentally different things. The one is posterior, and the other is prior. You cannot write PASCAL programs if there is no PASCAL language.

In fact, MacKay himself called the logon content *a priori*, or structural, information content, and the metron content *a posteriori*, or metric. In the numerical sense, Kolmogorov information turns out to be a version of Shannon information, and this unites them as versions of MacKay's metrc information.

But what does the other, the logical information content stand for? He says this:

"One must devise apparatus and/or prepare some system of classification, such that an adequate number of independent categories can be defined when describing the result [emphasis mine]. For example, if fluctuations varying in frequency between 1 and 100 per second are to be observed, the apparatus must be capable of responding in a time of the order of 1/100 second, i.e. of giving 100 "independent" readings per second. Or again, if ten shades of colour are recognised as distinct modes of description (i.e. coordinate-values on a "colour-axis") of the members of a population requiring to be classified, then ten columns must be provided in the observer's notebook" (p. 291).

Or, more explicitly:

"...the differentiating capacity of the least-discriminating link determines the number of independent categories in the result... There is a sense in which this number, that is to say *the number of independent dimensions or 'degrees of freedom'* [emphasis mine], can be regarded as a measure of the information supplied by the experiment" (p. 291).

The term "logon" for denoting this number is borrowed from Gabor. Now, having arrived at this definition, what characterises the definitions of such observational categories?

"A scientific statement... can be dissected ultimately into a pattern of elementary 'atomic' propositions (Wittgenstein 1922). Each atomic proposition states a fact so simple that it cannot be further decomposed...its existence is its only attribute."

"Ultimately, the latter must make contact with the primitive sense-data in terms of which it acquires meaning."

"Ideally, a scientific statement is based entirely on *observable* evidence, ... [on] elementary propositions relating to observations. We should expect therefore to be able to define a measure of *information-content* corresponding to the number of such propositions substantiated by a given experiment. On this view we are led to define one kind of *unit* of information as *that which decides us to add one elementary proposition* to the "frame" (the pattern of propositions) which is logically sufficient to define the results observed" (p. 292).

I quoted somewhat lengthily from this text in order to prove that it speaks about something which most discussions of science neglect: not about *how much*, but about what is expressed in a scientific description. The final conceptual step, then, is straightforward:

"... the minimum change possible is the addition of one element to the logical pattern ..." (p. 293); that is why *logon* content is quantal. Note that the word "quantal" is used in the same sense as in "quantum mechanics"; it does not refer to quantities as such. Rather, it refers to the fundamental units of knowledge; not token-like, however, but type-like.

There is in this text an implicit reference to Russell's *theory of types*: that the "atomic" statements cannot be defined on common grounds or subsumed under a more general category of which they are cases. This is much the same as what Russell suggested when saying there are classes of objects that cannot belong to the same set because they are of different type. Hence, such classes (and the corresponding propositions) constitute irreducible *qualities*.

Also of interest is the reference to the notion of "frames". A *"frame of description"* is a concept fundamental to science and broadly used in Artificial Intelligence (hence in informatics) since Minsky (1977). A frame in Minsky's sense consists of a scheme of rubrics that can be filled in; it is like a questionnaire or a form sheet with little boxes. Such frames are important because they are very close to the way people intuitively think about objects. Frames provide very natural tools for representing properties. Also it is known that frame-based systems are just as general as any universal computer. That is, we have a tool that is both ultimately simple and extremely powerful. This tool is often embodied in object oriented languages, used throughout informatics.

The frame concept immediately brings to mind MacKay's "man with the note-book". Now, the problem with such frames and notebooks is that they do not always represent those aspects of reality which we want them to. Especially, they cannot change an aspect that is represented for another one that is not, and there seems to be no algorithm for deciding what it is that we want. This "frame problem" came to be known as the *"robot's dilemma"* (Pylyshin 1986), and poses a serious challenge for AI research.

Put in these terms, MacKay's notion deals with a version of the frame problem: the question of how to define the slots of a frame, i.e. how to assign categories to an originally unshaped reality. In other words, it is concerned with how to partition reality into suitably defined chunks. Ultimately, his logon content measures the amount of articulation needed on behalf of the observer in order to bring a system into a tractable form. This is a pre-scientific aspect of model formation, because it has to do with the *distinctions* or the elementary acts of differentiation that define a system's abstract identity. That this aspect has great importance is obvious; Spencer-Brown in his acclaimed book (1969) claims that it is precisely such distinctions that provide the most fundamental material for all sorts of logic, and consequently, for public human thought. Once you have a set of distinctions, the rest is implied.

In a scientific model of some domain, it is always the choice of the various *variables* (numerical or logical) which carries out this kind of articulation. So, to summarize, what MacKay's theory tells us is that there are two complementary aspects of the information conveyed by a scientific statement. The one concerns the terms, that is the variables, in which the statement is formulated, and the other the mathematical properties of the statements made over the field of variables. Speaking of information content, now, we have first to consider this question: what variables are necessary in order to express a given domain of phenomena?

It is clear that MacKay's long-forgotten paper offers much wisdom relevant to the problem of the paradigm shift in information concepts.

4 Incommensurability

By defining a variable we give names to features that we think are of interest. While doing so, we realise an interaction in terms of those features, selected out of a possible infinity of implicit features we might have chosen to focus on.

It is a basic tenet of natural science that the information contents conveyed by the different variables (i.e. the ones related to different features) are, in general, *incommensurable*. If we do not invent the right variables, we are lost. Before Newton's time, force and position were already known to be important variables of mechanics. Velocity as a derived quantity (as change of position per unit time) was also known. Yet, mechanics was a mess largely because velocity (or rather, its conjugate, impulse) was not recognised as one of the *independent variables*. This was first done in Newton's equation which linked force with speed (or actually, with change in speed).

The importance of variables is well known in biology, too. The mute and the deaf will never understand each other; in the evolutionary context this was first formulated by von Uexküll. He said the various species live in different environments ("Umwelt") even if they live in geometrically the same place, because they perceive different aspects of this environment, through their phylogenetically acquired sense organs that mediate different modalities and evaluate, therefore, different variables. In short, the visual, tactile, auditory, olfactory and kinaesthetic modes of perception lead to different semantic universes.

This provides us with reinforcement concerning the correctness (and feasibility) of the logon concept. Information content as the number of variables is an expression of those variables that in some essential way "belong to each other", that is, that are related by the observed system itself. Increasing this number means an enrichment of the system: this can be achieved, for instance, by making a closed system open.

This also allows us to see that information content is something inherently *relative* or situational. Even a simple physical system, such as a pendulum, changes its information content when subjected to external interaction. An isolated pendulum has one degree of freedom. A coupled pendulum has at least two. One can manipulate it externally by setting the initial position of the swinging sphere, as in Eco's book *Foucault's Pendulum*, where the exact initial condition bore significance in encoding and decoding the secret message of the Rosicrucians. Jacopo Belbo interacted with the Pendulum in an unforeseen way, and the *sacrifice humaine* spoiled the code.

The choice of a channel of interaction that fixes certain variables for the interaction can be called, after Buchler (1966), an *integrity*. This is a concept directly related to the notion of "reading frame" discussed in my own research paper in this volume (Chap. II.4). As a result of incommensurability, it becomes important to distinguish between *internal* and external information of a system, depending on who, so to speak, is the "client" who uses a given variable (whether it is the system or an external observer). Internal and external information may depart from each other (Kampis 1991b).

Another important implication of the incommensurability of logons is that if we want to realise some information content by a system, we need at least as many variables as there are irreducible categories to be represented. This observation, a version of Ashby's *law of requisite variety* (1956), leads us to a better understanding of much of the recent work on information.

5 Information and Topology

Classical information theory was only concerned with the "metron" content. The modern theory of dynamical systems provides us with new methodologies, *semi-classical frameworks* which, in some way, can reflect espects of the "logon" content, too.

Professor Haken's dynamic information theory (1991), Professor G. Nicolis's dynamic instability model (1991), or Professor J. Nicolis's theory of chaotic information processing (1985) all reflect the same idea, namely, that dealing with information means dealing with categories. The Shannon theory focussed on an element selected from a set. In the dynamic models an equal emphasis is on the idea that you need the set first. That is, the logon content requires richness on behalf of the processor's or receiver's system. Dynamic information theory represents categories in the static aspects of dynamic systems, whereas chaotic information processors provide a means to embed these categories in the course of dynamic behaviour. The static classes we find in dynamical systems are the various attractors and attractor basins. Accordingly, the number of separate and triggerable basins must equal the number of logons to be distinguished. Ultimately, it is the phase space topology (determining the number and kind of basins) that relates information to dynamics. This idea connects models that use stable or metastable states for representing information to MacKay's theory, and furthermore, to work on classical information systems. The same thought applies to synergetics (Haken 1991), brain-like computing, and connectionism (Anderson and Rosenfeld 1988, Hopfield 1982, Smolensky

1988, Bechtel and Abrahamsen 1991). An interesting application to brain theory and neurophysiology is Kien (1991).

Chaotic systems are characterised by the property that their trajectories continually diverge from each other, the degree of which is measured by the so-called "Lyapunov exponent". As a consequence of divergence, such systems can exhibit a unique richness in their behaviour. In the initial condition of a chaotic system, every digit has a significance that increases with time. Hence, it becomes possible to encode as many independent variables in a chaotic system as we wish (a practical limit being the word length of the system we use). Such a chaotic system can pick up information on-line, that is, while functioning. Other systems with a less complicated dynamic behaviour soon reach saturation and their behaviour becomes monotonous (constant or cyclic, that is). Again, it is the logon content that requires a dynamic on-line information processor to be chaotic. The theory is developed in Nicolis (1985), Tsuda (1991) and elsewhere.

On the other hand, it is clearly the "metron" content that is formulated in the expressions of "uncertainty" about dynamic events related to their event spaces. The formal treatments of semi-classical systems usually deal with this aspect. But in a dynamical system we find both.

6 Emergence and Information Processing

We have briefly commented on how logons appear in dynamic models. Now we go beyond this and consider the case when logons can change.

It is always external information that we base our models and abstractions on, and so when internal information departs from that, we experience that our model breaks down. In terms of traditional information theory, one would speak of "errors" (as if nature could go wrong). In terms of MacKay's logon concept, we should speak about the change in non-metric information content.

This language offers a logical basis for talking about *emergent phenomena*. Emergence is a concept that refers to internal novelty in systems. Such novelty occurs whenever a model can no longer cope with the complexity of the system modelled. From the mathematical point of view, Professor Semenov (1991) shows that there are abstract information processing systems that cannot be modelled computationally, if certain restrictions are applied to the models. This indicates, in the spirit of Gödel's theorem and the "Halting Problem", that, even in purely formal systems, complete certainty and transparency are not achievable. The case is worse in natural systems where there is an unlimited supply of things we do not take into account in a given model. This amounts to saying that emergent phenomena are quite common in the outside world. For discussions, see Kampis (1991c), Cariani (1989), Minch (1987).

Emergence is likely to play a crucial role in many domains, among them evolution, human and animal communication, the self-organisation of systems, and the evolution of information processing in nature. Professor Csányi argues that current evolutionary models (most notably Darwinism and the implied theories) do not account for the emergent and hence self-creating nature of biological and social evolution (Csányi 1989). In this volume (Chap. IV.3), he warns us that communication is never informative without an active contribution of the receiver who (or which) interprets it freely. Again, we can express this is the terms of MacKay: the success of communication depends on *inventing* the right logons on behalf of the receiver. This inventive aspect is what makes interpretations often ambiguous and leads to an incomplete correspondence between sent and received meanings. Similarly, Professor Laszlo (1991) says that a signal or action only becomes information when becoming related to systemic structure. Examining the evolution of natural information processing, Professor Haefner in the basic paper discusses how the subsequent steps of this evolution come along with an enrichment of the interaction potential of information carriers; in other words, how new information carrying and processing capacities appear.

In my opinion, this whole circle of thought might lead to a new paradigm of emergent information processing that operates with logon content rather than metron content.

That such a paradigm is likely to emerge is indicated from various directions. For instance, in computer science there is a recent interest in what they call "emergent computations". One aspect of it has to do with the advent of parallel computations.

From a rigorous point of view, an emergent phenomenon occurs if we link two computers physically rather than symbolically. That is, we link them not by their communication protocols but by plugging an arbitrary wire from the one to the other. A similar situation occurs during the installation phase of large computer networks. It may happen that no one knows exactly where we are. When machines are connected and disconnected, anything can happen, from short circuits to an inadvertent launching of nuclear weapons.

A somewhat more interesting (or, at least, more attractive) emergent phenomenon is the result of *nonlocal interactions*, a notion also entertained in physics (Requardt 1991). In large parallel computer systems nonlocality appears when parts of the system that were previously unrelated begin to interact. This is similar to wiring computers together except that here it is the system itself that establishes the connection. In a large enough parallel system there are always elements that never "met" so far. Hence, they can realise emergent interactions only when two physically separate computers are linked anew. In other words, the local, asynchronous computations of a distributed system can be *independent* or isolated on one time scale but may become co-dependent on another, after a certain time has passed. Now, combining independent information sources is a phenomenon typically involving a shift in the reading frames (Kampis 1991b), and hence of the variables of the system – as we discussed above. Of course, this is only true for systems, components of which are truly independent in the informational sense. This is not the case for parallel computers as they are manufactured today but there seems to be no theoretical obstacle to introducing new architectures with these criteria in mind.

Consequently, there can be expected a future trend in the development of information systems, one that moves towards systems utilizing changes in their logon content, in other words, in their definitions, to achieve their goals. This might lead to a coming *evolutionary technology* of self-modifying systems where the focus will no longer be on design but on autonomous activity of the systems. Of course, to exploit emergent processes is to give up control, that is prediction, of the exact course of system behaviour. We will no longer know exactly what the system does and why it does it. Yet, such systems might be able to produce desirable outcomes. Without doubt, changing our perspectives so radically would produce many as yet unusual problems, and there will be many heretical ideas to swallow. In exchange, we may gain creativity and efficiency. To be sure, this was the course which biological evolution entered upon a few billion years ago and which in the long run produced organisms like humans who behave quite unpredictably (to the extent that they are very dumb in this sense: they are poor at mimicking machines). But humans are, at a higher level, both intelligent and reliable.

After all, maybe the evolution of our information concepts can mimic the pattern presented by the evolution of the living information processing systems. We have many things to learn from them yet.

Acknowledgments. This paper has emerged from a series of discussions I had with several people. I thank them for being both supportive and critical. In particular, I thank Professor V. Csányi, Budapest, Professor K. Haefner, Bremen, and Professor S. Salthe, New York, for their valuable feedback and suggestions. Also my thanks go to Ms Ute Riemann-Kurtz, Bremen, for her continuing interest. I thank Professor Haefner separately for inviting my contribution to this volume. The final version of the paper was written after my return from San Sebastian, Spain, where I participated in a meeting on related topics with Dr Peter Cariani (Boston), Dr Eric Minch (San Francisco), and our hosts, Dr Alvaro Moreno, Julio Fernandez, and others. I thank them for their hospitality and the ideas they shared with me.

References

Anderson JA, Rosenfeld E (ed.) (1988) Neurocomputing. MIT Press, Cambridge
Ashby WR (1956) An Introduction to Cybernetics. Chapman and Hall, London
Atlan H (1983) Information Theory. In Cybernetics: Theory and Applications (ed: Trappl, R), Hamisphere, Washington, pp 9–41
Ausiello G (1983) Complessita di calcolo delle funzioni. Editore Boringhieri, Torino (also in Hungarian)
Bar-Hillel Y, Carnap R (1952) An Outline of a Theory of Semantic Information. Technical Report No 247 of the Research Laboratory of Electronics, MIT; reprinted in Bar-Hillel, Y: Language and Information. Addison-Wesley, Reading, Mass., 1964

Bechtel W, Abrahamsen A (1991) Connectionism and the Mind. Basil Blackwell, New York.
Brillouin L (1956) Science and Information Theory. Academic Press, New York
Buchler J (1966) Metaphysics of Natural Complexes. Columbia Univ Press, New York
Cariani P (1989) On the Design of Devices with Emergent Semantic Functions. Ph.D. thesis, Department of Systems Science, SUNY at Binghamton, New York
Csányi V (1989) Evolutionary Systems and Society: A General Theory. Duke University Press, Durham
Csányi V (1991) Chap IV.3 in this volume
Haken H (1991) Chap III.3 in this volume
Hopfield J (1982) Neural Networks and Physical Systems with Emergent Selective Computational Abilities. Proc Natl Acad Sci USA **79**, 2554
IJTP (1982) Physics of Computation. Papers presented at the conference on Physics and Computation held on May 6–8, 1981 at MIT's conference center at Endicott House, Dedham, Mass. International Journal of Theoretical Physics **21**, no. 3/4
Kampis G (1991a) Self-Modifying Systems in Biology and Cognitive Science: A New Framework for Dynamics, Information, and Complexity. Pergamon, Oxford
Kampis G (1991b) Chap II.4 in this volume
Kampis G (ed.) (1991c) Creativity in Nature, Mind, and Society. Special Issue of World Futures: The Journal of General Evolution, Gordon and Breach (in press)
Kien J (1991) Chap IV.2 in this volume
Laszlo E (1991) Chap II.2 in this volume
Machlup F, Mansfield V (eds) (1983) The Study of Information. Wiley, New York
MacKay DM (1950) Quantal Aspects of Scientific Information. Phil Mag **41**, 289–311
Minch E (1987) The Representation of Hierarchical Structure in Evolving Networks. Ph.D. thesis, Department of Systems Science, SUNY at Binghamton, New York
Minsky M (1977) Frame-System Theory. In; Thinking (ed. Johnson-Laird, PN, Wason P), Cambridge Univ Press, Cambridge, pp 355–376
Morris Ch (1946) Signs, Language, and Behavior. Braziller, New York
Nicolis J (1985) Chaotic Dynamics of Information Processing with Relevance to Cognitive Brain Functions. Kyternetes **14**, 167–172
Pylyshyn ZW (ed.) (1986) The Robot's Dilemma, Ablex Publ, Norwood, NJ
Requardt M (1991) Chap III.1 in this volume
Roszak T (1986) The Cult of Information: The Folklore of Computers and the True Art of Thinking. Pantheon, New York
Semenov A (1991) Chap II.3 in this volume
Smolensky P (1988) On the Proper Treatment of Connectionism. Behav Brain Sci **11**, 1–74
Spencer-Brown G (1969) Laws of Form. Allen and Unwin, London
Tsuda I (1991) Chaotic Itineraries as Dynamical Basis of Hermeneutic Processes in the Brain. In: Kampis 1991c
Wittgenstein L (1922) Tractatus Logico-Philosophicus. Routledge and Kegan Paul, London

Aspects of Information

Ervin Laszlo

1 The Definition of "Internal Information" 63
2 Theoretical Foundations of Third-State Systems 67

1 The Definition of "Internal Information"

This definition of internal information uses the general features of the known and knowable universe as a framework. It is an "ontological" as opposed to "epistemological" (or methodological) definition. Its rationale lies not in the exercise of abstract reason but in the presumed logic of the observable world.

According to current cosmological physics, the particles that make up the observable units of matter are synthesized in the course of time from the potential energies that pervade the universe. Matter synthesis began following Planck-time, as the inflationary universe settled into a Robertson–Walker universe. Any further synthesis of matter requires that a pair of particles be derived from the underlying potential energy field, with isomorphic structure but opposite charge. Black-hole theory as developed by Hawking requires that superdecayed matter in a superdense state ultimately vanish in a process known as black-hole evaporation. It appears, then, that matter both emerges from, and dies back into, a potential energy state. Its maximum span of existence is given by the time-horizon of the final evaporation of galactic-cluster-sized black holes in an open (non-recontracting) universe, estimated at 100^{117} years (100^{122} if protons are not subject to prior decay).

While in an actualized state, matter is theoretically equivalent to energy, according to the Einstein formula $E = mc^2$. Moreover, in quantum physics it is possible to conceptualize particles as standing Schrödinger waves. (Particles always exhibit dual particle/wave properties and indeterminacy of behaviour giving rise to uncertainties in measurement.) In its wave aspect, matter is a disturbance that requires a medium (*in* which it is a disturbance), however, with the disproof of the classical ether concept (in Fresnel's formulation) and in the absence of alternative theories of cosmic fields, the persistence and propagation of matter is ascribed to space-time, conceived as a highly structured vacuum. It can be reasonably argued that matter, and all observable phenomena, derive ultimately from a subquantum energy field (see Chap. III.1). Matter-energy appears to be an actualization of an energetic potential present in space-time (rather than space-time itself).

Independently of the interpretation of the ultimate nature of matter, it is rational to hold that matter-energy is a unitary process arising from, and dying back into, a space-time energy potential field. If so, the nature of the process that constitutes the observable phenomena of the cosmos is best defined as an energy flow phenomenon in space-time, having the form of interfering energy-waves (including standing waves). Consequently our first definition reads:

Existents = matter-energy flows in space-time.   (1)

Matter-energy builds up from the originally synthesized leptons and hadrons to higher-level configurations, using the theoretically postulated varieties of quarks as building blocks. The evolutionary process in space-time conduces from matter in the hydrogen state to the heavier atoms that progressively fill the chemical table of elements. In time molecules build from the already configured atoms, and from molecules supramolecular (i.e. polymeric, macromolecular and cellular) structures are occasionally built. From free-living cells colonial and multicellular organisms and ecologies of organisms can evolve. Consequently definition (1) needs to be completed as follows:

Existents: progressively structured matter-energy flows in space-time.   (2)

In the evolutionary process, already configured flows of matter-energy are "systems in the third state" (far from thermodynamic equilibrium). Third-state systems interact, receiving inputs from (some) other systems and giving rise to outputs that are inputs for the same, or still other, systems. The input flows have varying degrees of impact on the systems, depending on the correlation between the structured matter-energies in the flow and the structure (the already configured matter-energy flows) of the system. Those flows that do have some, potentially observable, degree of impact on a system are said to represent "information" for the latter. This gives us the definition:

Information: effective-impact flow relative to a given system.   (3)

In this sense, in the cosmic evolutionary process the existents are structured flows of matter-energy, and information is an interactive emergent.

Further, it is possible to distinguish between "internal" and "external" information. The definition is relative to given systems. Any (nonstructured) flow that occurs in the time-space interstices of (structured) flows is external information. Since flows other than streams of photons appear to be to some degree structured, external information is a limiting case. Internal information, on the other hand, has wide application. Any flow that effectively impacts on a system and creates modifications in its structure is internal information.

Internal information: effective-impact flow *within* a given system.   (4)

The above definition can be objectively as well as subjectively interpreted. The objective definition is that internal information is any or all effective-impact flow within any (or all) system(s). The subjective definition is that internal information is the effective-impact flow within the reference system and all other flows are external information. The subjective definition has limited usefulness, however, as it divides internal from external information in exclusive reference to one given system.

The (objective or subjective) definition of internal information is applicable to any or all third-state systems. In every such system there are input and output flows. In highly structured systems boundary conditions filter some flows and allow others to penetrate. The penetrating flows are information for the system if they effect some modification in its structure. Non-negligible modifications tend to be reflected in the output flow of the system, so that the information represented by the penetrating flow passes from environing space-time into and through the system and back into environing space-time. Inasmuch as other systems may occupy a given space-time domain, the modification of the flow by the system's structure alters its output and thus its impact on environing systems, and produces correspondingly modified impacts on their particular structures. Consequently internal information is interactively processed between systems.

The statistically probable outcome of interaction is the "conformation" of the interacting systems. Any given system tends to map its environment, including

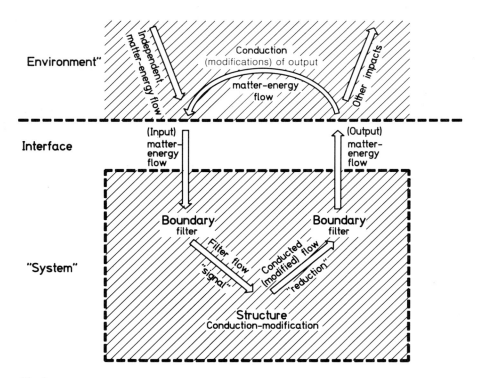

Fig. 1.

the environing systems, into its own structure. The mutual mapping of systems of their environments over time produces progressive inter-system conformation.

In highly structured systems the processing of internal information is already complex and interactive in its first phases. The human nervous system, for example, generates internal signals with which the input flows (the "perceivable array" of light, sound and other sensory data) interact. The receptor organs not only register, but also actively interact with, the perceived elements of this array. The result of the interaction constitutes efferent signals conduced to a decider (the higher analytic centres of the brain). These centres likewise not merely register, but also interact with, the signals, dynamically setting the sensitivities and focus of the receptor organs. The interaction product resulting from the efferent signals and the signals generated by the analytic centres themselves can be further analysed by the latter on progressively higher levels, where the output of one level is input for the next. The limiting case is reflective consciousness of

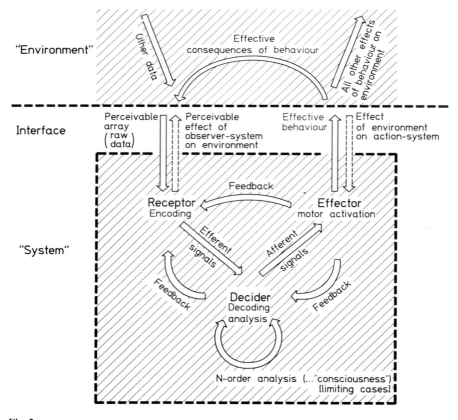

Fig. 2.

the human variety (the final output of the internal analysis of internal analyses). The analytic process may produce a stream of efferent signals conducted to the motor centres (the effector organs) of the system. Here the signals interact with feedback signals from the system's environment and give rise to some form of behaviour.

System behaviour—the output flow of the system – enters the system's environment, and may become part of the input flow of other systems sharing that environment. Depending on the relation between the output flow and the structure of the environing systems, the flow produces varying degrees of impact. Some flows do not produce modifications to the structure of environing systems: they remain without observable effect. Others are likely to create modification in the structure of some environing systems, which in turn produce modifications in their own behaviour. Some of the behavioural changes are likely to produce outputs that fall within the observable and observed range (the perceivable and actually perceived array) of the original system. In these cases there is a continuous loop of information entering the system and becoming internal information; passing through the system and into its environment; becoming internal information in other systems; and feeding back to the reference system.

We should add that in the high-order analyses of highly structured systems the input flows are progressively filtered. Only a small segment of the total input flow of the organism is available for high-order analysis. (A human being, for example, is only conscious of a very small fraction of the flows that impact on his/her organism.) Selection occurs by means of a set of codes by which the analytical centres identify signals for active processing. A mismatch between efferent signal and preset code can lead to the suppression of the signal and loss of its effective information content. Such selectivity is required to distinguish between information of survival value and other types of information, ensuring that the former can be adequately responded to. At the same time, it results in a loss of vast quantities of information, some of which, however, is gradually recovered in learning as the third-state system elaborates and extends its information processing codes.

2 Theoretical Foundations of Third-State Systems

Systems in the real world can exist in one of three kinds of state. Of the three, one is radically different from classical conceptions: it is the state far from thermal and chemical equilibrium. The other two states are those in which the systems are either in equilibrium or *near* it. In a state of equilibrium energy and matter flows have eliminated differences in temperature and concentration; the elements of the system are unordered in a random mix and the system itself is homogeneous and dynamically inert. The second state differs only slightly from the first: in systems near equilibrium there are small differences in temperature and concentration; the internal structure is not random and the system is not inert. Such systems will tend to move toward equilibrium as soon as the constraints that keep them in nonequilibrium are removed. For a system of this kind, equilibrium remains the "attractor" which it reaches when the forward and reverse reactions compensate one another statistically, so that there is no longer any overall variation in the concentrations (Guldberg and Waage's law). The elimination of differences between concentrations corresponds to chemical equilibrium, just as uniformity of temperature corresponds to thermal equilibrium. Thermodynamics further tells us that while in a state of nonequilibrium systems perform work and therefore produce entropy; at equilibrium no further work is performed and entropy production ceases.

In a condition of equilibrium the production of entropy, and forces and fluxes (the rates of irreversible processes) are at zero, while in states near equilibrium entropy production is small, the forces are weak and the fluxes are linear functions of the forces. Thus a state near equilibrium is one of *linear* nonequilibrium, described by linear thermodynamics in terms of statistically predictable behaviours, as the system tends toward the maximum dissipation of free energy and the highest level of entropy. Whatever the initial conditions, the system will ultimately reach a state characterised by the least free energy and the maximum of entropy compatible with its boundary conditions.

The third possible state of real-world systems is the state far from thermal and chemical equilibrium. Systems in this state are nonlinear and occasionally indeterminate. They do not tend toward minimum free energy and maximum entropy but may amplify certain fluctuations and evolve toward a new and more dynamic regime that is radically different from states at or near equilibrium.

How systems can become more dynamic in the course of time is explained by the fact that systems in the third state are open rather than closed or isolated. Consequently the Second Law does not fully describe what takes place in third-state systems, or, more precisely, between third-state systems and their environment. Although internal processes within the systems obey the Second Law (free energy, once expended, is unavailable to perform further work), energy available to perform further work is "imported" by these systems from their environment. They have a transport of free energy – negative entropy – across their boundaries. Change in the entropy of open systems is defined by the well-known Prigogine equation $dS = d_iS + d_eS$. Here dS is the total change of entropy in the system, while d_iS is the entropy change produced by irreversible processes within it and d_eS is the entropy transported across the system boundaries. In an isolated system dS is always positive, for it is uniquely determined by d_iS, which necessarily grows as the system performs work. However, in an open system d_eS can offset the entropy produced within the system and may even exceed it. Thus dS in an open system need not be positive: it can be zero or negative. The open system can be in a stationary state ($dS = 0$), or it can grow and complexify ($dS < 0$). Entropy change in such a system is given by the equation $d_eS = (-d_iS \leq 0)$; that is, the entropy produced by irreversible processes within the system is shifted into the environment. When the free energy within the system and the free energy transported across the system boundaries from the environment balance and offset each other, the systems are in a stationary state.

Open systems in the third state require a specific kind of environment. This is essentially a flow-environment, in which a rich and constant source of energy irradiates the systems. It is known that such a flow passing through complex systems drives them toward states characterised by higher levels of free energy and lower specific entropy. As early as 1968, experiments conducted by biologist Harold Morowitz demonstrated that a flow of energy passing through a complex molecular structure organizes its components to access, use and store increasing quantities of free energy. The explanation of this phenomenon in thermodynamic terms was given by the Israeli thermodynamicist Aharon Katchalsky in 1971: he showed that increasing energy penetration always drives systems consisting of a large number of diffusely coupled nonlinear elements into states of increasing nonequilibrium. Since then, many experiments and observations have demonstrated details of the process. They range from the creation of simple Bénard cells in a liquid, to the emergence of life itself in the biosphere.

Wherever there is an enduring energy flow in a medium, new orders arise spontaneously. This is true of the flow of heat from the centre of a star such as the sun to its outer layers: the flow is self-organizing and produces the typical Bénard cells. It is also true of the flow of warm air from the surface of the earth toward outer space. The earth, warmed by the sun, heats the air from below while outer space, much colder, absorbs heat from the top layers of the atmosphere. As the lower layer of the atmosphere rises and the upper layer drops, circulation vortices are created. They take the shape of Bénard cells. Closely packed hexagonal lattices, such cells leave their imprint on the pattern of sand dunes in the desert and on the pattern of snowfields in the Arctic. They represent a basic form of the kind of order that emerges in open systems when they move into states far from thermodynamic equilibrium.

Chemistry sheds further light on how systems of interacting elements move from states at or near equilibrium to the domains of nonequilibrium characteristic of open systems in the third state. Under controlled laboratory conditions sets of chemical reactions are irradiated and forced to move progressively further from chemical equilibrium. Relatively near chemical equilibrium the reaction system is still successfully described by solutions to the chemical kinetic equations that apply at equilibrium as well as those that correspond to the Brownian motion of the molecules and the random mixing of the components. But, as reaction rates are increased, at some point the system becomes unstable and new solutions are required to explain its state, branching off from those that apply near equilibrium. The modified solutions signify new states of organization in the system of reactions: stationary or dynamic patterns of structure, or chemical clocks of various frequency. Where the equilibrium branch of the solution becomes unstable the reaction system acquires characteristics typical of nonequilibrium systems in general: coherent behaviour appears, and a higher level of autonomy vis-à-vis the environment. The elements cohere into an identifiable unity with a characteristic spatial and temporal order; there is now a dynamic system, whereas near equilibrium there were but sets of reactants.

Life on this planet may have been due to such a process of progressive organization in third-state systems embedded in a rich energy flow environment. It is well known that steady irradiation by energy from the sun was instrumental in catalysing the basic reactions which led to the first protobionts in shallow primeval seas. It is also possible that the energy of the earth itself has played a role, in the form of hot submarine springs in the Archaean oceans. The hypothesis calls for a series of relatively small continuous-flow reactors, similar to the chemical reaction systems responsible for Bénard cells. Nature's reaction systems may have consisted of cracking fronts in the submarine rock, causing sea water to heat rapidly and to react with chemicals in the rock and in the surrounding sea. As the hot fluid rose toward the surface it dissipated its heat. The chemicals that served as the basic building blocks of life were constantly mixed in the continuous flux of energy in shallow seas created by magma erupting into the sea bottom and reinforcing the energy of the sun. Out of these series of reactions could have come the protobionts – the lipid vesicles from which the more complex forms of life evolved in subsequent eons.

When sets of reactants are exposed to a rich and enduring energy flow, systems with ordered structure and behaviour emerge; and if the flow endures, the systems that have already emerged tend to become further structured. Because these systems move further away from equilibrium, they become more unstable. Their continued persistence is usually ensured by the catalytic cycles that evolve among their principal components and subsystems. Such cycles are selected in the course of time in virtue of their remarkable stability under a wide range of conditions: they have great resilience and fast reaction rates. Complex systems in nature almost always exhibit some variety of catalytic cycles.

There are two varieties of catalytic cycle: cycles of autocatalysis, where a product of a reaction catalyses its own synthesis, and cycles of cross-catalysis, where two different products (or groups of products) catalyse each other's synthesis. An example of autocatalysis is the reaction scheme $X + Y \rightarrow 2X$. Starting from one molecule of X and one of Y, two molecules of X are catalysed. The chemical rate equation for this reaction is $dX/dt = k \times Y$. When Y is held at a constant concentration there is an exponential growth in X. Cross-catalytic reaction cycles have been studied in detail by the school of Ilya Prigogine. A model of such reactions, known as the Brusselator, consists of the following four steps:

$$A + A \rightarrow X \tag{1}$$

$$B + X \rightarrow Y + D \tag{2}$$

$$2X + Y \rightarrow 3X \tag{3}$$

$$X \rightarrow E \tag{4}$$

In this reaction model X and Y are intermediate molecules within an overall sequence through which A and B become D and E. In step (2) Y is synthesized from X and B, while in step (3) an additional X is produced through collisions of 2X and Y. Thus while (3) in itself constitutes autocatalysis, (2) and (3) in combination make for cross-catalysis. In relatively simple chemical systems autocatalytic reactions tend to dominate, while in more complex processes, characteristic of living phenomena, entire chains of cross-catalytic cycles appear. The physicists Manfred Eigen and Peter Schuster showed that catalytic cycles underlie the stability of the sequence of nucleic acids that code the structure of living organisms and assure the persistence of the many forms of life on this planet.

Given sufficient time, and an enduring energy flow acting on organized systems within permissible parameters of intensity, temperature and concentration, the primal catalytic cycles tend to interlock in hypercycles. These super-ordinate cycles maintain two or more dynamic systems in a shared environment through coordinated functions. For example, nucleic acid molecules carry the information needed to reproduce themselves as well as an enzyme. The enzyme catalyses the production of another nucleic acid molecule, which in turn repro-duces itself plus another enzyme. The loop may involve a large number of elements; ultimately it closes in on itself, forming a cross-catalytic hypercycle remarkable for its fast reaction rates and stability under a variety of conditions.

The formation of hypercycles allows dynamic systems to emerge on successively higher levels of organization. The shift from level to level of organization through catalytic hypercycles produces the convergent aspect of the evolutionary gigatrend. Convergent systems on successively higher levels of organization have a wider range of possibilities to set forth the process by which they access, use and retain increasing amounts of free energies in correspondingly complex structures. On the new level the amount of complexity that can be developed in a system is greater than on lower levels due to the greater diversity and richness of the components and subsystems. The wider range of structural possibilities offers fresh opportunities for evolution. Molecules built of many atoms and cells built of many molecules can evolve toward the complex polymers that are the basis of life; living organisms based on many cells can evolve toward the higher forms of life, and local ecologies based on many species and populations can build toward mature regional and continental ecosystems. Human societies themselves, built of many populations and levels of organization, tend to evolve toward progres-sively more embracing units: nations and regional communities of nations, and ultimately a global community of all nations and peoples.

Mathematical Aspects of PANINFORMATICA

Alexei Semenor

1 Discretization 73

2 Messages 73

3 Information Machines 74

4 Information Processing 74

5 Turing Thesis 75

6 Programs 75

7 Universal Algorithms and Functions 75

8 Gödelian Functions 76

9 Computational Structures 76

10 Psychological Complexity of Message 76

11 Measurement of Complexity or Entropy of Message 77

12 Accuracy of Physical Measurement. Two Types of Message 78

13 Monotonic Complexity 78

14 Randomness 79

15 Illegal Sequences 79

16 Relative Complexity 79

17 Information in One Object About Another 79

18 Oracles 80

19 Relativization 80

20 Informatics Game 80

21 Computable Processes 81

22 Probability 81

23 Complexity of Computation 82

References 82

Three basic essences of the modern universe – matter, energy, information – correlate to different views and pictures of our world.

A picture called PANINFORMATICA is the subject of our investigation. We shall concentrate on a specific aspect of that picture – the mathematical aspect. Mathematics consitutes the main foundation for scientific "explanation and prediction" in modern science. We face the very challenging question "What is the scientific content of informatics (computer science) beyond mathematics?"

At the same time mathematics can serve as a source of important images and metaphors for informatics. So, in our paper we present major mathematical concepts and facts related to the contemporary mathematical understanding of information processing [1].

1 Discretization

The first general scheme of the informatics view of the universe is **discretization**. When we use it to its full extent, we consider the universe to be a collection of cells, each of them locally interacting with the neighbours. On different levels of organization we can distinguish different kinds of cells – from "space cells" to "humanity cells". The interaction occurs in discrete time. A possible relationship between space cells and time cells can be obtained from consideration of the maximal speed of signal propagation (speed of light). The possible *topologies* of the immediate connections between cells are not evident. Any regular lattice (such as those in crystals) would raise the question of isotropy of different directions in space. From an intuitive view all directions are equivalent. In crystals they are not. Perhaps the interconnection structure could be irregular or variable, as in polymers or liquids, and establishing an immediate connection could be considered a primitive action. Later we shall consider different mathematical versions of information processing substance. For physicists themselves the problem of the local dimension of physical space is not clear.

Now we would like to mention one example of cell structure: a two-dimensional Game of Life. It is remarkable that this structure, with very primitive rules of behaviour, is universal in the sense that it can imitate (model, simulate) the information processing of any other structure. The idea of universality is one of the key ideas of the informatics view of the universe. We will return to universality later in our considerations.

2 Messages

So, the interaction between cells occurs in discrete time and we are interested in the existence or nonexistence of a primitive interaction between cells, without differentiating between different possible kinds of interaction. This leads to the notion of **message** as a binary sequence.

Of course, messages can be in different forms. The first generalization is a message in a nonbinary alphabet. A natural general concept for a message is that of a constructive object; for example, a labelled graph is a constructive object. Constructive objects do not exist independently; they constitute **ensembles** of constructive objects. For example, binary strings constitute such an ensemble. It is not clear whether a necklace of letters is a constructive object or not. Some structures can exist on an ensemble of constructive objects. One example of such a structure, in which one object can be more accurate than another, will be considered later.

3 Information Machines

In our informatics view of the universe, its existence (or life – remember the Game of Life) is considered to be a process of transformations of information. This information processing can be described in different terms.

Usually it involves some kind of machine (computer, or computing device) from some family. We call this family a computational model. A computational model can be constituted, for example, by all three-dimensional cellular automata with different constructions of the elementary automata. Another possibility is to consider all Kolmogorov machines ("self-modifying graphs") as a computational model [2]. The classical model of the Turing (or Post) machine is not so natural and useful now.

A version of Kolmogorov machines proposed by Schonhage is in a strong sense isomorphic to simple abstract computers. In the Kolmogorov definition input and output degrees of graphs for one computational model are limited. In the Schonhage modification input degree is unlimited. In both these models only one controlling centre exists. At the same time, control in cellular automata (or iterative arrays) is distributed. In the 1960s Kolmogorov proposed to his students (Barzdin and others) that they consider graphs with parallel control. He insisted on the importance of this model in the 1980s in his conversations with the author.

4 Information Processing

In many cases we are not interested in the inner structure of a machine, but in its behaviour. In general, behaviour is determined when you know all possible processes of informational interaction between the machine and the environment. In the simplest case it is enough to distinguish in the whole environment input and output and two moments of time – beginning and end. In that case we obtain a function: input → output. Such functions are called computable. Some computable functions are partial (not defined for some arguments) and cannot be extended to total functions.

5 Turing Thesis

By reducing the whole behaviour of a machine to the computable function given by it, we get a class of functions computable by machines from one computational model. This class depends *a priori* on the model. Indeed we can describe natural models with a rather small class of computable functions. But several classes have been constructed starting with the idea of obtaining the furthest reaching class of functions which are computable in any sense.

It happens that for all such classes the set of computable functions is the same. We can suppose that each of these classes actually coincides with the set of all computable functions in our intuitive sense. This hypothesis is called the **Turing thesis**.

One possible description, in a very simple way, could start with a small set of primitive algorithmic functions and a small set of primitive operations on functions which construct new functions.

A natural strengthening of the Turing thesis would be a formulation of some computational model which can "imitate" in some sense any machine. Kolmogorov's initial formulation of his algorithms depended on this. A formalization of this idea is the concept of real-time simulation of one machine (and model) by another. Real-time simulation tells us about the input–output interaction of both machines but in conditions when the input is given in parts. In this condition the output of one device has to be represented by another (the simulating machine with a constant delay).

It is not known whether (and improbable that) Kolmogorov machines can simulate Schonhage machines in real time.

6 Programs

A specific machine from a computational model can be given by its description. This description can be called the program of the given machine. Note that this program itself is a piece of information which can be processed.

We can now propose a mathematical understanding of what was referred to in [3] as **syntactic**, **semantic** and **pragmatic** aspects of information: if we have a binary string (i.e. a syntactic object) it can be interpreted as a program (its semantics) with some behaviour, or input–output function (i.e. pragmatic).

7 Universal Algorithms and Functions

The usual description of a computational model actually gives a general algorithm for applying any specific machine from the model to any argument. This algorithm is called a **universal** computational **algorithm** (for a given computational model).

The corresponding function U(p, x) has two arguments: p is the description or program of a specific machine, and x is its argument. So, an abstract computable function V(y, x) is called **universal** if, for any computable function f(x), we have y such that f(x) = V(y, x).

8 Gödelian Functions

So, for different computational models, we have the same set of computable functions. In other words, the corresponding general algorithms give us universal functions.

But all universal functions U(Y, x) which are given by existing computational models have an interesting property: for any computable function V(y, x) (universal or not) there exists a computable function T, for which

$$V(y, x) = U(T(y), x)$$

So T translates programs for V into a program for U. A function U which has this property is called Gödelian.

9 Computational Structures

We know that universal functions U(y, x) have to be partial (i.e. not defined for some pairs of y, x). We can define a **computational structure** to be an algebraic structure with the domain consisting of all binary words (or any other ensemble which could be used) and a (partial) function with two arguments.

Of course, we could define in the same way a computational structure for any Gödelian universal function. The remarkable fact is that all these structures are algebraically isomorphic [4].

10 Psychological Complexity of Message

We are now ready to present the classical notion of the quantity of information in the Kolmogorov sense, or Kolmogorov complexity, or (Kolmogorov) entropy of a message or any finite object. But to make it clearer and more natural we prefer to start with some psychological considerations.

What is the psychological meaning of the complexity of a message? Of course it depends on our individual histories. It is much easier to memorize a list of the names of your friends or relatives than an arbitrary list of names in a foreign language. But at the same time there are some common laws. For example, what is easier to memorize:

– one page of the word "Yes";
– one page of poetry;
– one page of random letters?

We believe that the (psychological) complexities of the three memorizing tasks are ordered in the same way for all human beings.

A possible interpretation of this psychological phenomenon is that texts which are simple permit shorter *descriptions*. This leads us to the following sequence of definitions.

11 Measurement of Complexity or Entropy of Message

Let us return to our basic concept of information. It is usual to say that the meaning of a signal depends on the state of a subject who sends or receives the signal. In other words, it depends on the history or it can be included in a very long chain of signals. This supports the view that it is wrong to define information (or its quantity) for an individual message, but that one should define it only for a message from some class (or sequence).

A **mode of description** M is any computable function which transforms descriptions p into objects x:

$$M: p \rightarrow x$$

The complexity K for a given mode M is defined by

$$K_M(x) = \min\{l(y)\,|\,M(y) = x\}$$

Here and below l denotes the length of a binary string. As we can see the complexity depends on the mode of description. A simple example of a useful mode is given by Shannon coding which uses frequencies of occurrence of different letters. Morse code is an application of this idea. So these modes give us specific complexities. But Kolmogorov's idea is much broader. In a philosophical and mathematical sense, it is universal.

Of course, for any fixed object x we can construct a mode of description which transforms a given string (say 00) into this object. So, the complexity of any given object can be arbitrarily low. But if we consider all objects then we can reach some kind of objectivity: there exists a mode of description U such that for any M we have

$$C \times K_U(x) < K_M(x) + C$$

The function K is called the **Kolmogorov complexity** or **Kolmogorov entropy**.

So, U gives "almost" the shortest description. Could we improve this "almost"? The answer is "no", if we formalize the question in the following way.

It is evident that the number of objects with complexity $< n$ is not more than 2^n. Let us try to construct any family of functions which (i) satisfies the same condition on cardinality of objects of limited complexity and (ii) is invariant (stable over isomorphisms of the computable structure). In that case the class does not give us better estimates for complexity.

12 Accuracy of Physical Measurement.
Two Types of Message

Up to now we have not considered structure on our set of messages, i.e. binary strings. But in our observations we can meet at least two types of message. In some conditions (for example, in physical measurements) one message can be considered as more accurate *or* precise than another, but not contradictory to it: the binary string 1.1011 can be a more accurate measurement of a physical value than 1.10. In this case a finite string is actually an approximation of an infinite sequence. This is one type of message. The structure on them is the **extension** structure. For another type which we have considered before, any two different messages are in contradiction: if we have 110 (six) apples we cannot have 1101 (thirteen) apples.

13 Monotonic Complexity

The definition of complexity which was given earlier corresponds to objects which are not related to each other. To consider the other possibility we have to modify this definition. The modification consists of the introduction of an additional argument, time, to the description mode. Now a mode is a computable monotonic function $M(y, t)$. Monotonic means that for a later moment t' of time t and extensions y' of description y we can obtain from M an object which has to be an extension of $M(y, t)$.

In this circumstance we can define a complexity for M similar to the previous case:

$$K_M(x) = \min\{l(y)\,|\,tM(y, t) \text{ extends } x\}$$

We can also prove the existence of a mode U with "almost" the shortest description. The complexity (or entropy) corresponding to this mode U is called the **monotonic complexity**.

14 Randomness

The notion of monotonic complexity is closely related to another important notion of the informatics universe. This is the notion of randomness. An individual infinite binary sequence is **random** if the monotonic complexity of its initial segments of length n grows almost maximally, i.e. is greater than $n - c$ for some c. More information on randomness can be found in [5].

15 Illegal Sequences

Random sequences constitute only one philosophically important class. Another is constituted by sequences which we can call **illegal**. Let us for a moment call a computable function f a law if, for any finite sequence x, function f gives us some object f(x). Let us say that an infinite sequence satisfies law f if, for any of its initial segments, an object which is not the next object in the sequence is produced. A sequence is illegal if it satisfies no law. Illegal sequences do exist and any random sequence is illegal.

Are all real sequences (of events) some combination of random and algorithmic (= computable) sequences or are there other sequences?

16 Relative Complexity

We can add one more argument z to a mode of description, of the same type as argument y and result x. In this case z serves as an additional source of information. A theorem on the existence of the optimal (universal) mode $U(x|z)$ – the entropy of x relative to z – can be proved in the cases of both independent strings and approximations. So, for some U, M, C, x, z (an optimal mode of description),

$$K_U(x|z) < K_M(x|z) + C$$

The corresponding function K is called the **relative complexity**.

17 Information in One Object About Another

This leads us to a definition of the **quantity of information** contained in one object x about another object y:

$$I(y:x) = K(y) - K(y|x)$$

One nontrivial property of the information function I is its "almost" computativity:

$$I(y:x) \sim I(x:y)$$

"Almost" means here "up to addition of $\log(I(y:x))$". This log cannot be reduced to an additive constant.

An important philosophical question is "Can the common information of size $I(x, y)$ be represented as an object of (almost) this size?" The answer is "no."

18 Oracles

In the classical theory of algorithms an important role is played by so-called **oracle algorithms**. In the case of such an algorithm, the next state of the computational process is determined not just by the previous state and the program of the algorithm but also by an answer to the question "Does the previous state belong to the fixed oracle set A?"

19 Relativization

By substituting algorithms with fixed oracle A instead of the usual algorithms (without an oracle) we can obtain a new theory of algorithms. An empirical fact about research in the theory of algorithms is that all theorems which were proved without oracle can be proved for the **relativized** (this means "with oracle"; it has no connection with the relative complexity discussed before) case. This phenomenon is called relativization.

So, it is very hard for a human being to recognise whether there is a nontrivial oracle at the base of the universe.

At the same time computational structures with oracles are not isomorphic to computational structures without oracles.

The whole situation will be explained from some point of view in the next section.

20 Informatics Game

Let us consider a **game** of two players, *Mathematician* (M) and *Nature* (N). Mathematician wants to prove a mathematical fact Φ, Nature wants to refute it. The game consists of two tables for two players. These tables contain finite pieces of tables for two-argument functions (as in our definition of computational structure).

The players make their moves alternately. Each move consists of making a finite extension of the corresponding table. Nature wins if by the end of the game the fact Φ is not valid in its table and the table of Mathematician can be translated into its table. Mathematician wins in the opposite case.

What is the connection between our simple game and the previous discussion? It is very natural! Let us consider only "relativizable" facts Φ (that means facts whose validity does not depend on an oracle).

Such a fact is true if and only if Mathematician wins the game.

21 Computable Processes

We have mentioned different kinds of processes: deterministic (or algorithmic or computable), random (or stochastic), etc. In the model of the cellular universe mentioned above, the universal world process is computable, starting with some initial (perhaps very small) configuration at the moment zero. We can also consider local or global reversibility of this process, etc.

Independent of the general answer about the determinism of the universe, computable processes are very important.

So, let us consider a **computable process** – a computable sequence of objects. Each object is a binary string (for simplicity we do not consider the case of the extension relation on strings). For such processes there exists a *universal* process which contains any other computable process with some linear slowdown. It is interesting that the universal process can be chosen to be a total function (being universal even for partial functions).

22 Probability

For any object we can measure the frequency of its occurrence in the initial segments of a process and then take the lower limit of all frequencies. This will give us the (guaranteed) probability of meeting the object in the process. In the universal process this probability is maximal up to a multiplicative constant. The probability for this process is called the *a priori* probability.

A priori probability can be considered a version of complexity. Actually it differs by not more than an additive log from the Kolmogorov complexity function relativized by oracle O. (This oracle tells us, for a given universal function, whether this function is defined for any given argument.)

23 Complexity of Computation

In speaking about the complexity of objects, we have never mentioned the complexity of computations. This is a very large, important field, but we shall limit ourselves to a brief comment. The major kind of complexity is the so-called time complexity – the number of steps in the computation of a result from some input. To speak honestly and realistically about the complexity of objects or a simulation, etc., we have to take into account the computational complexity of the object or simulation, etc. and also the computational complexity. Some fragments of the theory we have described can be developed within this framework.

References

1. Haefner K (1988) The evolution of information processing. Draft December University of Bremen, Bremen
2. Uspensky V, Semenov A (1987) Theory of algorithms: major discoveries and applications (in Russian). Moscow, Nauka. First version was published in Springer Lecture Notes in Computer Science **122**, Springer, Berlin, Heidelberg
3. Riemann-Kurtz U (1990) Aspects of Information and Information Processing According to the Project EIP, University of Bremen, Bremen
4. Muchnik A (1985) On the basic structures of the descriptive theory of algorithms. Soviet Math. Dokl. **32** (3), 115
5. Semenov A, Uspensky V, Shen A (1990) In Soviet Math Surveys No. 1, 34

Process, Information Theory and the Creation of Systems

George Kampis

 1 The Relevance of Theory   84

 2 Information Theories and Reading Frames   85

 3 How much Information Can Be Stored in a Piece of Matter?   88

 4 Whitehead's "Process Philosophy"   90

 5 Philosophy and Natural Science   92

 6 Component-Systems   93

 7 Referential and Nonreferential Information   95

 8 Where is Information Stored in Systems?   96

 9 Computation, Construction and Self-Reference   98

10 Conclusion   100

References and Notes   101

This paper can be but a fragment or an introduction. I set forth the view that *information* has two essentially different aspects, and I try to analyse the consequences. I think information involves, on the one hand, the knowledge and the possession of some records, and on the other hand, a process and an action that brings forward the former, that is, a physical agent that makes it possible to have knowledge about something.

These aspects complement each other and cannot be reduced to one common quality. The second, that is, the processual aspect, corresponds to a *creative action* in which the information is *de novo* produced through the actualization of potentialities, and this information content is only definable as bound to the respective action and relative to it; information ceases when the action stops. The first, the possessional aspect of information, corresponds to a distilled result of some past interactions, something that already exists on its own, can be stored, recalled and transformed mechanistically, without having to fear losing it.

Technical information theories, initiated and motivated by a cybernetic set-up, deal with this latter kind of information only. They restrict their attention to techniques of transmitting and processing already acquired information and to questions of efficiency and resource requirements that are closely related. In the Shannon theory the amount of information can be used to characterise, through several transmutations, the required bandwidth of real-time high-frequency transmission channels. Also related to this is Ashby's classic notion of *requisite variety* that tells us how desired controllers can be linked with systems to be controlled.

Many people think that with such formalizations and axiomatizations, the concept of information has assumed a precise and advanced form, with the aid of which we have arrived at the highest levels of abstraction with respect to information. This is, however, not true. We can risk the statement that the mistaken notion that *information theory* is a *theory of information* has caused significant damage to the universal knowledge process. It is remarkable that Shannon himself warned us of the danger in this assumption [1] when he said that of the many possible properties of information, mathematical information theories deal solely with a *measure that is determinable under given specific conditions*. Shannon's "communication theory" is only syntactic and analyses the combinatorial properties of typographical symbols – and so do all other formal theories, as we shall soon see. The conclusion is that instead of – or rather, besides – the technical characterisations of the atomistic, machine-like, thing-like, direct-representation-like pieces of information, the development of a process-specific, system-theoretical information conceptualization is necessary and the question "What is information?" must also be answered.

1 The Relevance of Theory

But why should we mind what is meant by "information"? What will be changed if we define a word differently?

Let us begin with an almost-triviality. There is good reason to assume that information, like complexity (viz. "intricacy"), about which there has been much discussion recently in scientific circles both in the natural sciences and in the humanities, is a derived concept which cannot be identified with some basic physical quantity. Now physical quantities such as energy have the characteristic that one has absolutely no freedom in giving their definition. Namely, energy is not just a number, but feeds back to the system where it performs work, and therefore on the basis of this unequivocal effect it can be unequivocally determined. This line of thought can be continued by noting that today it is generally believed that the basic physical quantities are already known – at least to the degree the epistemological difficulties of physics enable us to know them at all. So there is not much space left for further concepts.

The question is whether, despite this lack of "non-fundamental" nature with respect to physical law, it is possible to define information in a similarly un-equivocal sense.

My opinion is that this is in a sense indeed possible. When defining the information concept, not only the nominal rules for using a word are fixed. I agree with Hegel: definitions are *shorthand descriptions*, signs, behind them being an entire universe. This universe can be unfolded and shown up, and its rules can be analysed. Herein lies the answer to the question as to what the 'mental gymnastics', the seeking for the definition, is good for. Information concepts are used in a variety of fields: in informatics (the European term for computer sciences), which is often wrongly equated with practical computer science and is subsumed under mathematics departments; in biology, where related notions occur in the popular metaphor of "genetic information" and in the problems of evolutionary growth; in cognitive science, which aims at cartographing the abilities of mind to acquire, represent and manipulate information; or in the social sciences, where societal information flows, coordinating self-maintenance of material production systems and the idea networks of cultural systems, are in the foreground of interest. For researchers in these fields the structure of the underlying world of the theory of information – and, through that, the structure of our own world – provides a theoretical framework that marks a direction of research. Therefore, *by shaping a theory of information, a research program for several fields of science is shaped at the same time.*

I try to show that information can be freed from its present-day mechanistic connotations. Systemic information can feed back to systems and to their modes of functioning in the same way as physical quantities do, and, therefore, information **is** a fundamental concept for a qualitative and conceptual, if not a physics-like quantitative and symbolic, characterisation of systems.

In particular, elaborating the theory of creative (causative) information and passive (epistemic) information, which I will call, as I did earlier [2–4], *referential information* and *nonreferential information*, respectively, we may arrive at the source of a radically new view in theoretical biology and cognitive science. It is noteworthy that theoreticians like to imagine biology as *complicated physics*, and despise as unscientific all holistic theories that wish to dispense with this myth; in cognitive science, in turn, a *computational paradigm* is favoured, the basic concept of which – maybe after several twists – is invariably the notorious idea of the direct representability of the level-specific mental events as computer algorithms. The theory of information, if developed properly, will invalidate both.

Starting from a closely related basis it has been possible to work out a new theoretical framework for evolution and cognition. This is described in Professor Vilmos Csányi's [5] and the author's [6] monographs.

2 Information Theories and Reading Frames

When "information theory" is mentioned, most people think of the theory of Shannon and Weaver. It is known, however, that several rival conceptualizations exist, some of which are quite different in spirit. It is beyond the purposes of this paper to give a detailed account of the numerous information theories that have been developed in the past. Such a review would be practically impossible anyway. Partial reviews are found in [6–8]. Yet it could be useful to have a glance at some of the most characteristic ideas which enjoy general popularity.

The Shannonian theory is based on the assumption of repeatable mass events and on an **a priori** assignment of symbols to events that fixes in advance both an invariable encoding and a universe of symbolic structures called "messages". After these preliminaries, dealing with information reduces to dealing with expectable statistical properties of selection and identification operations that pick out elements of the universe at random. Outside the validity of these hypotheses, Shannonian information (or "metric entropy") is not applicable.

There is a particularly useful distinction between *uncertainty information* and *description information*. Shannonian information corresponds to the former, in other words, to an uncertainty of artificially defined outcomes. Description information, on the other hand, refers to an informational concept as applied to natural systems on their own. The two coincide only in the case of *micro-physical entropy*, which expresses, because of the microscopic-macroscopic interface involved, both a most informative description of a statistical physical system and, at the same time, an uncertainty about the exact microstate of the same system. This point has been correctly elaborated by the renowned physicist E. T. Jaynes [9]. Unjustified extensions of this "Clausius-Shannon identity" were at one time quite popular. There was a time when it seemed obligatory to think that information was just another name for entropy. This mis-understanding was due to a neglect of both the existence of organizational levels of phenomena and the restricted validity of the statistical assumptions. Also related to this is the *entropy-negentropy* problem, cf. [10], which consists in the seeming paradox that maximum information content means maximum Shannon entropy but is often associated with a minimum physical entropy in realised systems. The solution is that things can be both "complex" (in the naive sense of being entropic) and organised (in the sense of being negentropic); these are simply two viewpoints that cannot be represented on the same axis.

Communication and evaluation of symbolic information are accompanied by a *classification* of possible events into a set of discrete categories. There are information theories that hope to characterise the physical requirements of the classification process directly [11, 12]. R. Thom considered "information" to be an expression of forms, and developed a theory for the emergence of physical forms by means of successive instabilities. His basic observation was that a message (or any information carrier) is a physical structure whose constituents are arranged specifically so as to realise one of its possible *alternative* confi-gurations. Forms have, in general, the same property: they have no unique stabi-lity domains and can exist in many (but pairwise incompatible) states – they are, therefore, suitable carriers of information. In Thom's model a dynamic form-shaping process is like the movement of a rolling ball along the ridge of a mountain surface. The ball tends to roll down into one of the isolated valleys (i.e. stability domains or "basins") that surround the mountain (the boundary). Now, to have an intricate classification structure one needs a very complicated dynamic stability picture and a wrinkled basin boundary that makes processes sensitively dependent on initial conditions. Thus the closer target of dynamic information theories is the mathematics of multistable and/or chaotic systems capable of exhibiting such "nontrivial phenomena".

"Nontriviality" of dynamic behaviour also plays a role in another informational concept, that of "dynamic entropy", which measures the convergence or divergence of "neighbouring" trajectories (ultimately, the relationships between different time functions that correspond to different initial conditions).

A completely different approach to information theory is exemplified by the *algorithmic information theory* (also known as the "Kolmogorov–Chaitin complexity theory"). The basic idea is that the information content is taken to be expressed as a minimal computer program that represents a given set of events (a message, a system or a process) in a symbolic code. Here *description* (d-) and *interpretation* (i-) *complexities* (viz. information contents) have to be distinguished. D-complexity is the difficulty of specifying the minimal program (i.e. a code) that prints out the original code. I-complexity is the difficulty of interpreting a given code.

Of interest is *"semantic information theory"* [13, 14], utilised in cognitive science and artificial intelligence. It measures the informativeness of utterances on the basis of their different abilities to exclude possible states of affairs. Given a set of logical primitives, the observation is that the possible statements that can be made up from the primitives by using logical connectives are not equally specific. For instance, the statement that "It is either raining or not" is always true whereas the statement "If it is raining and the sun is shining then there is a rainbow" is only true if there is a rainbow, which is a rare event. This difference can be formalized with the aid of the truth tables of the system.

No doubt these information definitions can be very useful in a number of situations, but a common characteristic becomes striking if we take a closer look – and that is the real reason I included them. Notably, we can observe that they all involve the use of *externally fixed categories* of events and, with this, something else, which can adequately be called a fixed *"reading frame"*, is also introduced. A reading frame will be defined as a mode of selection of basic primitives and their meaning. For instance, when a symbolic alphabet is fixed, this canonizes a way of interaction with the system which, without this categorisation, is "shapeless" to itself; who can tell whether the physical forms "*a*" and "*A*" are "really" different in respect of their information content? They are different if we (or the system) distinguish between them and are identical if we do not. Such a choice is prior to all mathematical "information theories" but is a proper subject when studying "the theory of information". I claim that because of the neglect of such questions the role and the significance of the reading frames has not been recognised and it has not been satisfactorily discussed so far. For instance, what about those systems that *can themselves form their own categories*?

A work that directed my attention to what I now suggest we recognise as the problem of reading frames was that of V. Csányi, who in his 1982 book

[15] referred to a concept of information as one of the basic categories for the interesting evolutionary model he developed. He suggested we distinguish between "structural" and "functional" aspects of information. This distinction indicated the possibility of a new theory. Since then it has turned out that several authors have formulated related views, among them, most notably, Robert Rosen [16], who discussed the role of "observables" in a spirit very close to that of my concept of "reading frames". Another author is J.-P. Ryan [17], one of those who formulated the thesis of relativity of information very clearly. General studies by L. Löfgren, e.g. [18], on the relativity of descriptions have also been very important. They together form the historical basis for my research. My purpose was to clarify the consequences of the idea of relativity in detail and to work out a conceptual system in which it can be grasped.

3 How Much Information Can Be Stored in a Piece of Matter?

The first step towards this end is the discussion of the question [19]: "How much information can be stored in a given piece of matter?" In particular, we can ask "How many bits can be stored on a magnetic tape?"

Let us consider a magnetic tape n squares long in some suitable convention, and let us use it for storing binary digits. This assumes a tape recorder that reads and writes the successive squares of the tape according to some rule; the similarity of this scenario to the operating principles of Turing machines should be obvious. Now the first answer anyone would give is that exactly n bits can be stored under the given circumstances.

All right, but why exactly n? The naive answer is that it's because there are 2^n possible binary sequences n squares long and because to represent any of the 2^n strings we need $\log_2(2^n) = n$ bits – in fact, this is exactly how bits are defined. However, I claim that there exists a somewhat better answer which, unlike the naive answer, can be significantly generalised.

The point is that if I fix *one particular* binary sequence then there are at most 2^n different ways of interpreting it, that is, there are so many unequivalent ways to read it out using some tape recorder. Thus, if I have a given information carrier and an arbitrary set of various tape recorders, this number gives the number of equivalence classes to which the tape recorders can belong. For instance, I can have a recorder that masks out position 1 and only reads the rest, and so on; by taking all combinations into account we get, as can be checked, the number 2^n. Now whereas the idea of having tape recorders that are insensitive to certain squares may sound somewhat odd, let us note that from the viewpoint of *one given* tape recorder it is just natural that there are qualities of the tape to which it has no access but to which an *extended* tape recorder does. That is, there is no such thing as a "universal tape recorder" and thus we have to deal with a number of possible readout methods, *reading frames*, one by one. So, in the case of the above magnetic tape, a selection of one of the 2^n tape recorders has to be accomplished in order for us to be able to speak about the information content unequivocally. We are justified in saying that it is this selection that is represented in the quantity of information.

But, then, nothing prevents us from increasing the set of tape recorders beyond any limit. Not only the sensitivity to different squares of the tape can be considered. For instance, I can tie a knot in the tape, or I can cut it into pieces; then I can introduce tape recorders that consider the existence/nonexistence of a knot or a fragmenting as one additional bit – and we can use as many knots and fragments as we wish. (We may assume for simplicity that these "tricks" do not influence the reading ability of the original "pedestrian" machines that have access to the magnetised squares only. Think of Turing machines which are small enough to follow the tape whatever happens with it. This is not important but aids understanding.)

There is no reason to stop at any point. We can glue ends of the tape together to form a Möbius strip or any other topological object; membership of topological equivalence class can bear numerous further bits. Geometrical shape, dielectric properties, colour, or the strength of the material *or any material properties whatsoever* are just as good.

Thus, no matter how many information-carrying properties are already defined, we can always add to the list and we can always introduce a tape recorder that has access to still further properties *through an altered reading frame*.

So, in full generality, the information content of any piece of matter is infinite, or, more precisely, indeterminate. It cannot be named without naming a specific system of reading frames (interaction modes) to which it is relative.

In particular, if we can manage to build a tape recorder that can *internally shift its reading frame* and that has access, with the passing of time, to previously unused material properties of its 'tape' (viz. information carrying components), then it can continually produce new information through its "processing" steps. Thereby we can arrive at systems that can *define and alter their information content autonomously*. This idea has significant consequences for the concept of information – and for the chances of science being able to model systems.

There exists a direct correspondence between these "reading frames" and those "form sheet representations" called "*frames*" in AI and cognitive science, the conception of which comes from M. Minsky and others. This relationship is discussed in [6]. Thereby, we arrive at a concrete point where we can establish a link between the theory of information and the claims of strong AI for representing "intelligence" by means of computable representations. Frame representations are known to be as powerful as universal Turing machines and thus our results shed light on the limitations of both, a question to which we return at the end of this paper.

Two subtle remarks are in order here. The first remark is that, physically speaking, in order to shift the reading frame (that is, to alter the given interaction mode that dominates the system) we need an extension of the phase space of the system. Then we deal literally with a *different system* that has different variables.

Let us file away this remark for future reference. The other remark is that, according to the usual view, there is a distant but nevertheless finite and well-definable limit for the amount of information that can be stored in any material structure. Although this number is so high that it would not influence our conclusions seriously, it may be interesting to stop here for a while. The said limit, known as the *Bremerman limit*, has to do with physical fundamentals. A given piece of matter has a given mass, and it is therefore equivalent to a given energy content, according to Einstein's law $E = mc^2$. This energy content can be used for information storage. With the assumption that every bit occupies a different energy level, the Heisenberg uncertainty relation between energy and time comes in, which when applied here says that if we want to read out the stored information in reasonable time then we need well separated energy levels. In this way, by estimating the necessary spacing between energy levels, we can give an estimate of the number of different available levels and therefore the maximum number of bits. However, this attractive train of thought suffers from the obvious error that energetic differences are taken to be necessary. In other words, a particular reading frame, which only distinguishes energy levels, is assumed implicitly. But in reality different topological or geometrical shapes need not belong to different energy levels, and they can still be distinguished by applying complementary forms [key-and-hole recognition). So, in principle, even Bremerman's limit could be transcended. But, as I said, this is not necessary since it does not really matter whether a potential number is actually infinite or just larger than any intelligible number.

4 Whitehead's "Process Philosophy"

The idea of having access to an increasingly large number of information-carrying properties during processes is closely related to the philosophy of A.N. Whitehead, which has strikingly similar elements.

Whitehead is one of the most interesting representatives of modern thought. A great mathematician, Russell's co-author in the giant work *Principia Mathematica*, a work in the spirit of logical monism, was also a great philosopher, who in his philosophy rejects not only logical monism but the direct applicability of mathematics to reality wholesale. It is this seemingly contradictory pair of facts that makes his work most important: it lies on the crossroads of two competing developments of scientific ideas.

A basic category in Whitehead's thinking is that of *process*, by which he means something unusual that cannot be inferred from the name. To get an idea of what is involved, consider a few quotations:

—"[static] notions belong to the fable of modern science – a very useful fable when understood for what it is." [20, p. 90]
—"Nothing in realized matter-of-fact retains complete identity with its antece-

dent self. ... [identity] dominates certain kinds of process. But in other sorts of process, the differences are important, and the self-identity is an interesting fable." [20, p. 94]

—"[...] In so far as identities decay, these laws are subject to modification. But the modification itself may be lawful. [...] laws of change are themselves liable to change." [20, p. 95]

To fix things, Whitehead mentions examples of mechanical gadgets that, due to internal corrosion processes, collapse after a time or start to function differently, changing their identity from one mode of interactions to another. In other words, the point is that there are many more degrees of freedom in matter than those manifested in any given interaction scheme, and that sooner or later these implicit properties will (or might) come to the surface and change things.

Whitehead also developed the basic notions of a *processual* interpretation of mathematical operations. This interpretation goes beyond the idea of algorithmic computation – and therefore, also beyond the ordinary concept of a process. He wrote this:

"The statement "twice-three is six" is referent to an unspecified principle of sustenance of character which is supposed to be maintained during the process of fusion. The phase "twice-three" refers to a form of a process of fusion sustaining this principle of individuation." [20, p. 91]

"A prevalent modern doctrine is that the phrase "twice-three is six" is a tautology. This means that "twice-three" says the same thing as "six"; so that no new truth is arrived at in the sentence. My contention is that the sentence considers a process and its issue. Of course, the issue of the process is part of the material for processes beyond itself. But in respect to the abstraction "twice-three is six", the phrase "twice-three" indicates a form of fluent process and "six" indicates a characterization of the completed fact. [etc]" [20, p. 92.]

Whitehead was aware of the fact that it is the tautological version of "twice-three is six" that would correspond to ordinarily conceived processes. But his idea was that such processes do not in fact exist; even the machines are not machines. I believe the quotations clearly reflect his view that, although in some systems for a shorter or longer period of time the prevailing interaction modes can be "frozen out", in reality nothing remains the same and the interactions and their defining qualities necessarily change. It is this encompassing, limit-transcending process that he calls "process'; and that is why he says realised mathematical operations are not tautological. He claimed that in order to execute mathematical operations one needs the general form that *he* called process. (In his time there were no computers and thus he could not discuss the possibility of realising operations by machines, which are in his terminology non-processual, static notions. More about this can be found in [6].)

We can conclude that Whitehead's philosophy anticipated the existence of systems that change access to their dominant properties; in fact he said all systems are of this kind if we watch them for a sufficiently long time.

One could say, on this basis, that the question of information is settled; here is a general philosophy that gives justification to a "processual" conception. But my opinion is that this philosophy says both too much and too little. It speaks about a basic property of the whole universe, but says nothing about the existence of concrete systems of interest that would actually utilize self-transcending modes in their normal behaviour, that is, that would have the property that in their understanding the advanced Whiteheadian notions are practically important. From the selected point of view of science we can understand mechanical systems very well even though at a closer look they will indeed cease to remain mechanical systems after, say, one billion years. But who cares? So, we are not yet finished.

5 Philosophy and Natural Science

Our task can be better illuminated if we consider the difference between philosophical explanations and the explanations offered by the natural sciences.

Well, on the surface level, there is no such difference. Everything is philosophy. Both historically and in a logical reconstruction it can be shown that natural science is not "something else" – in fact it is a kind of philosophy. Procedurally, however, that is, in terms of the methods applied, a distinction can be made.

Philosophies are often said to be incommensurable, because they create entire new conceptual worlds that in general have no common denominator, and for which there exists no *independent* method of selection – any method is itself philosophical and then the circle is closed.

The case is somewhat different with scientific theories. At the cost of being somewhat less fundamental, we can relate them to each other. Of course, it would be far too naive to think that scientific theories can be evaluated with no problems, but it is right to expect a tight relationship and a degree of harmony between theories.

I developed the standpoint [6, 21] that one basis of selection between competing scientific theories can be what I called their "*relevance*" understood in terms of *adequacy* and *interpretability*. This conception of "relevance" preserves much of the common sense of the word. The idea is the following: a theory is adequate if it answers our questions correctly (for instance, if it can predict outcomes, etc.). A theory is interpretable essentially if there is a *model* behind it, the defining properties of which can be publicly demonstrated. Public demonstration can take place either immediately or by means of other, already accepted and at the moment unchallenged, theories.

In line with these remarks, my next purpose is to outline a model, the model of *component-systems*, which can be positively shown to exhibit the kind of advanced phenomena we have spoken about so far.

6 Component-Systems

As with so many other things, the very concept stems from our discussions with Mr Csányi. The most complete development of the topic is found in [6].

A component-system is defined by the properties that

(1) the set of the different types of its components is open-ended; and
(2) the system produces and destroys its own components during its typical activities.

It can be seen from the definition that such systems are in fact quite common. At this point it is not yet obvious that anything unusual is involved. Chemical systems (especially systems of macromolecules), minds, languages and societal information networks all provide examples. That chemicals produce each other hardly needs discussion and it is also clear that the number of possible macromolecules is as indeterminate as it is expected to be in the above definition; by means of combinatorics the number of molecules can be increased beyond any limit. There is evidence showing that mental ideas have in the mind but a limited life-span, and that they are discarded and produced anew. Thinking is an activity that actually constructs the ideas that are being thought. They are not simultaneously stored in any ready-made form. Those who have doubts will find our later discussion of "immensity" illuminating. That nontrivial languages are open-ended is again a triviality; that linguistic expressions are made directly from other linguistic expressions (and, insofar as verbal thinking is concerned, *by* other linguistic expressions) seems to me also an acceptable statement. So they are component-systems, too. Likewise, since societies are closed for their specific information flow, the carriers of societal information are themselves products of a general societal production process and therefore also conform to our model.

After having claimed that such systems are quite common, what follows will be perhaps amazing. I claim nothing less than that component-systems are not physics-like, not mechanistic by any standard, and that they are systems that factually realise a continual shift of the reading frame defining their mode of functioning and, through that, their systemic information content.

I first show that these systems indeed *do* produce new variables. To understand how this is done and how this relates to the shift of reading frames, it will be instructive to recall a concept of O.E. Rössler [22, 23]. In discussing chemical evolution he characterised chemical systems as "privileged zero systems". The "privileged zero" property means that if we set the values of the variables to zero, the system disappears; this stands in contrast with all conventional dynamic systems for which a zero-crossing has no special significance (and can in fact be transformed out by a proper rescaling of coordinate systems). For an electrical system, the same number of variables is necessary to specify the states when all voltages are zero and when they are not. In a chemical system (and, by definition, in all other component-systems) we need zero variables for the "switched-off" state. Related is the unusual "economy" of component-systems: if we buy five chemicals (i.e. five variables) we can effectively bootstrap a system which, according to the equilibrium Boltzmann distribution, may have some $10^{10,000}$ accessible variables. Or, on a small scale, to start a popular Belousov–Zhabotinsky system we need five different molecules, and the "mature" oscillating system has twenty, which the system produces by itself.

The Physicist W. Elsässer introduced the concept of *immensity* for characterising situations when "too large" numbers similar to the open-ended variety of components or to the number $10^{10,000}$ occur [24]. An "immense" number is defined as a number whose logarithm is still a large number. So, the set of components that are actually reachable by ordinary processes of component-systems is immense.

Now, immensity has the obvious consequence that an immense number of states or variables cannot be simultaneously realised even *in principle*. If we consider modest proteins of length 100 amino acids, there are 20^{100} of them. Contrast this with the estimated number (10^{80}) of atoms in the universe.

We recognise that component-systems realise a *significant combinatorial explosion*. The phenomenon of combinatorial explosion is well known from several fields of mathematics and natural science, but in those cases that have received serious attention so far, its consequences appear to be less significant. For instance, the "chess problem" is thought to involve 10^{120} different moves. Much as a component-system, the game of chess is not directly representable either, but *in "ordinary" systems such as chess the consequences of immensity do not feed back to the rules of the system.*

In a component-system, they do. Because the system produces its own components, it is the very (non-simultaneously realised) set of components that determines the transformations which the components undergo.

As a consequence of immensity we deal with a high degree of *individuality* of the components. Of the total number of components that can play a role in a given system's processes, only a small fraction can be co-extant at any moment in time. It follows that most components are not accessible most of the time, and the individual appearance/disappearance of a component becomes a significant event. *A typical event in a component-system is the "de novo" production of individual components that have never been present in the system so far.*

Now since component–component interactions can be as heterogeneous as the components themselves, with the production of any new component, potentially a new "window" is opened through which we can peep into other components new properties. Needless to say, we deal with a new reading frame.

Furthermore, *in a "de novo" interaction we can never know in advance which reading frame will be opened and what property it will activate.* The only way of getting this information is to let the components interact.

In other words, we cannot map all the "relevant" properties of the components in advance. Metaphorically speaking, we can open any window but because of immensity we cannot open all of them, and so the new window (and the new property) is only definable if the new component that utilizes it is produced – that is, it is only definable *a posteriori*. That means that any concrete and detailed *prediction or computation* of such a system's behaviour is strictly impossible. In component-systems we deal with a *creative causation* that brings forth new information (i.e. surprise, unexpected event, miracle) according to the scheme envisioned in our "magnetic tape" example.

That there are "de novo" events, and that they are creative and unpredictable, is not denied, I think, by anyone on the metaphysical level. But the usual reaction is that of ignorance, based on the belief that such events are rare and atypical and that they are therefore not proper subjects of scientific study. The model of component-systems shows the falsity of this expectation. Creation and information production is just *the* normal way of functioning for this large and important class of systems.

How biological systems, minds and other systems utilise the unique properties of component-systems for their purposes is discussed elsewhere [5, 6]. We can relate F. Jacob's ideas on "*evolutionary tinkering*" [25], Rosen's *principle of function change* [26], Levins's "*fitness set*" theory [27], etc. A good independent discussion of some of these ideas is found in [28].

7 Referential and Nonreferential Information

Now we are prepared to discuss the outline of a new theory of information in the terms introduced at the beginning of the paper.

On the grounds of our extensive discussion of reading frames and the systems that utilise relativity of information when shifting their reading frames, we are to understand that there is in general no way of circumventing this relativity and reducing it to a fixed and well-specified process-independent frame.

It follows that *the frames through which we interact with a system and gain information about it, and the ones that are used by the very system to define its internal interaction modes, can be different, and are in fact different, for class component-systems.* Due to the different reading frames the systemic information content will be different along the two axes. Information to be acquired by the observer and information to act in the system are no longer the same; it cannot always be ensured that we use the same windows. The primary information content that refers to the system is manifested in its internal, causal, creative unfolding. This is what I suggested we call *referential information*. A secondary information content that I proposed to call *nonreferential information* is what we can gain and store about the system. Due to the difference of the frameworks the two cannot be directly converted into each other. Conversion is only possible by observation of "realised matter-of-fact", that is to say, only when the referential information content becomes publicly available through its end product.

8 Where is Information Stored in Systems?

After the preceding discussions this should be a straightforward question. And yet, it is interesting because the answer is so different from what is generally thought.

Of course, the answer is that information is stored *nowhere and everywhere* in systems; it is in the selection of the reading frames, and the latter are defined by the system as a whole. This is very well visible in component-systems where the temporarily available set of components fixes the temporarily valid frames, and this happens by a "synergetic" effect to which every individual component can contribute in an uncontrollable way modulated by the presence of other components as well. I think this is the closest to what von Bertalanffy once called "*constitutive characteristics*" [29].

We can go further and consider a few known problems of information theory that can be handled in a novel way once in possession of our results.

First let us reflect on the "letter paradox". The paradox (insofar as it is a paradox at all – and not just an unintelligible side product of an unintelligible information definition. .) is that a letter written by Einstein can contain as much information as there is in a random letter tapped out by a monkey [30] (we know that the "monkey typist" is a favourite idea in evolutionary studies). Now several ways were suggested to remedy this fact – but, for heaven's sake, if two letters have the same Shannon information content then *they can be coded into each other*. In fact the most economic Shannon code is always a random code. So the conclusion is that it is not a difference of the letters but again a difference of the reading frames that is involved here, and a random letter becomes perfectly meaningful if we know how to decode it meaningfully.

In an extreme form, this point can be illustrated by recalling a sci-fi story from the renowned Polish author S. Lem's book *Kyberiad*. In a notable scene an "Information Burgler" (a greedy creature that has to be fed on new information) forces our heroes, the robots Trurl and Clapantius, to write texts for him forever. To escape, they build a "Second-Order Maxwell Demon", which differs from its relative, the "ordinary" Maxwell Demon, which can only open and close doors according to the motion of molecules, in that it converts thermal motion, present everywhere, into meaningful messages. The "happy ending" (happy for our heroes, at least) is that the Information Burgler gets drowned in the resulting flood of information.

Now of course this story has a weak point but there is a significant idea in it. The idea is that it is *indeed* always possible to convert randomness to meaning by some specific transformation. The weak point is that we do not know which is this transformation and there is no way of inventing it, other than trial and error, and the number of possible trials (i.e. the different reading frames that can be applied) is infinite. This is the same situation as with component-systems and their "windows".

At this point an unexpected conclusion about secret military codes follows (I hope the enemy is not listing – but who is now the enemy? [31]). Much research is done in the direction of unbreakable codes. In World War II the secret codes were broken several times and their safety was not satisfactory. Paradoxically, however, a fact implied by our results is that *a coded message cannot be decoded ever, on the basis of information internal to the message*. Therefore, if we don't know the code anyway, it cannot be identified – the message itself is of no help. The reason why this statement seems to be false (although it is not) is that we can have in several cases *good guesses* about the nature of the code applied. This is true for the ancient secret code systems which worked on the basis of a few universal keys (often they even kept the original alphabet), and it is also true for the present-day military codes because we use the same electronics and the same technology. This defines "natural" readout frames that have a less-then-immense variety. But if one could introduce radically new technology, indeed there would be no more need for coding. For instance, a CD disk would have been strictly unreadable for both CIA and KGB 15 years ago. The respective reading frames were not yet defined. You need lasers, digitization methods, etc. which, when they do not exist, are not felt to be missing. It's like the blind spot. If you don't have *any* knowledge of the system in which a CD is written (for instance that it will produce sound, or a colour TV picture with three colours and 625 rows), there would be no chance of restoring the original information. Why not use smell or written messages? There are, again, infinitely many significantly different ways of decoding – why apply this one and not that one?

The romantic topic of *communication with alien civilizations* is another case in point. C. Sagan, F. Drake and a number of other people keep spending lots of money on what should now be understood as a hopeless project. Whereas it is not possible to decide about the question of aliens factually (cf. "the proof of the pudding is in the eating"), in the theoretical sense there is absolutely no reason to expect that we can decode a mesage which was coded by someone else. Yes, one could say that cosmic messages are meant to be understandable and not to be secret, but again, what counts as a natural or typical mode (reading frame) for one system can be unintelligible for another. If there are intelligent beings and we can really communicate with them, that will be a real miracle (or a proof of parallel evolution, see below). But, alas, to count on miracles is not a very scientific attitude.

We can go even further: *we* humans cannot communicate with each other either. We are all aliens [32]. Our communication is fragmented and imperfect and is only possible, to the limited degree to which it is possible, by the culturally and evolutionally, partially standardized (and therefore shared by all (or most) humans) reference frames we apply. That is, due to our common history there is a guaranteed partial overlap from which a workable system can be bootstrapped by means of iterative processes. Yet, since every individual has a slightly different genetic background and life history, and therefore a different person-specific system of interpretation frames, the maximum that can be achieved is a *reproducibility* of the communicated information but not its *transfer*. We have to abandon the view that knowledge can be directly "handed over". If that were so there would be no problem with education and socialization. Information is not a thing that could be transferred. In communication only a physical action is taking place that evokes different referential informational components in the speakers' minds and builds up, in the course of cognitive "processing", a different non-referential information content. Yet, this content can be fed back to the communicative domain again and thereby we can achieve a stage when the different conceptions evoke actions that in the communicational domain (but only there) coincide. This is what we can call the process of *understanding* – it should be realised now how limited it is. A grand *theory of communication* that utilises similar ideas has been developed by the English cybernetician G. Pask [33].

In short, it is time to forget about the classic notions of "information processing" that assume a rigid and transparent universe in which codes can be freely exchanged and meaning is unequivocal. Instead, we live inside a fluid system that produces its own meaning according to its own laws. If we wish to characterise this situation with one word, I suggest we say that *information is hermeneutical*: it is only definable by context and participation – as has always been suspected (but now there is a model behind it).

9 Computation, Construction and Self-Reference

We can now go deeper into the discussion of mechanistic and creative systems, or, in other words, a discussion of *the formal and the material*.

This is the more timely since discussing computational systems amounts to discussing what is usually called "information processing systems". Computers are prototypes for systems that manipulate token-like, atomistic, symbolic (non-referential) information.

Computers and other information processing devices can be but *subsystems* of real information-using systems. Before and after "processing", their information content should be interpreted by someone else. Computers have no real internal semantics.

Internally, they only perform *mappings* that transform certain sets into certain other sets. Such an operation is in itself not particularly thrilling. Whatever is done by a computer is done in an entirely *syntactical* domain; no algorithm can interpret its own symbols. This view was most sharply expressed by Wittgenstein [34] and recently by Searle [35]. Searle's example of the "Chinese Room" illustrates the point that one can translate Chinese texts without understanding a word of Chinese if the procedure of how to do that is given – this is exactly the situation with computers.

Essentially, all mechanistic systems are information processing systems in this restricted sense. Their processing consists in the transformation of state information by means of a computation rule: the state transition function.

That this ultimately simple and uniform information processing model is not applicable to real systems can be seen without theory if we consider the difficulties with the 'semantics' of genes, neural networks, or written texts. For instance, it is clear that a DNA is only functional in the cell and if we isolate it, it will do nothing and its information content cannot be interpreted.

The semantics of computer programs is, in comparison, a simple and unequivocal matter. But we have just said they have no semantics. Well, when we said computers have no semantics, that was not quite true. What was meant is that they have no internal, system-specific semantics, and this is true. But programs do have a kind of semantics that is external to the system and the system can have no access to it. Namely, from the statement that no formal system can interpret its own symbols (that is, no formal system has internal semantics) it also follows that *no formal system can execute itself.* One needs an external interpretation to that end (additional to the interpretation frame in which the symbols are determined).

The kind of semantics that plays a role here is indeed simple. It is a known fact that every algorithm can be rewritten as a *production system.* A production system is defined as a set of conditional *IF ... THEN* rules. Now the semantics of a production system is simply that *the production rules are interpreted externally as productions, in other words, as derivations to be carried out.*

In contrast, component-systems and all closed information systems do have internal semantics, an internal interpretation. They are semantically closed.

This fact can be well illustrated by the notion of the *Uroboros* [36]; this gives us another interesting parallel.

A well-known problem of the early evolution of life can be put in the following strange form: in order to have proteins one needs proteins. That is, the synthesis of specific proteins is a thermodynamically most unlikely process that goes with reasonable speed only if there are enzymes (that is, proteins) to catalyse it. This is clearly a chicken-and-egg problem. A protein system is an analogue of the legendary animal called "Uroboros", a favorite subject of medieval bestiaries, one that was believed to eat its own tail.

It is easy to understand at this point that internalization of meaning in component-systems through production of meaning-giving reading frames by the very system is like having a Uroboros. In other words, there is a *self-referential* nature to component-systems; every external point of reference is lost.

Now I think there are two basic ways to cope with this self-referentiality. The two are different enough but agree in that they are incompatible with the computational paradigm.

One approach is to define self-referentiality as a *new ontic primitive* of description. Such a definition can be consistent with traditional mathematics and science but goes outside it. This question was delightfully discussed by L. Löfgren [37]. Self-reference has received much attention recently; more general works are [38, 39]. As is well known, the assumption of self-reference as a 'third logic state' is exactly the metaphysics behind *autopoiesis* [40] which is, therefore, to be acknowledged as a possible answer to the needs of post-modern (i.e. anti-representational) science and to the problems we meet in component-systems. There is no space here to go into a discussion of autopoiesis; I only mention it as an alternative approach for both biology and thinking.

I do not see, however, how the theory of autopoiesis could be interpreted with the aid of a simple model. Our model of component-systems suggests a different solution. In a component-system *self-reference can be opened up*. In this model the internalization of the semantic relation is due not to some new basic quality but to the frame-shifting properties of material construction that arise because of the reasons discussed.

The essential point in the comparison of computation and the material construction processes of component-systems is this: every process produces new properties as it proceeds but computation *excludes* the new properties artificially from future consideration. If we execute an algorithm, not only the deduced results but also a number of side effects are produced. This is just inevitable. For example, if I take a cellular automata space, the evolution of the configurations leads to curious forms of activated regions therein. These forms have properties that are defined by their arrangement as a whole and arise therefore as "new" when the arrangement is realised. The reason why cellular automata are notwithstanding "boring" – as mere executors of previously fixed algorithms – is that these new properties cannot be "seen" by the system's reading frames. There is no hand to reach out for them. Only the same good old interpretations are repeated monotonically, over and over again.

In this respect, *computation can be conceived as degenerate construction*, and computationally describable systems as devastated, impoverished systems. The material is infinitely richer than the formal.

The Church-Turing Hypothesis, which says (in its strongest form) that every process can be reduced to computation, is positively false: computation is a very particular mode of process. It is synonymous with *lambda-definability*. Lambda-calculus fixes through its inevitable *a priori* variable assignment a single invariant reading frame. Component-systems create and define their own variables and thus their own laws 'in the process of doing'.

10 Conclusion

The new theory of information we have presented and its intimate relationship to creativity and non-mechanistic principles of system behaviour invalidate a number of old assumptions.

"Genetic information" is just a metaphor and so are the Darwinistic speculations based on the unique role of genes. There is a system in living organisms the logic of which is different from the logic used in traditional science and is, to a great extent, yet to be understood.

Linguistic and logical information also has to be rethought. New ways are necessary for cognitive science. In the light of the double nature of information, under "representation" something else should be understood. Instead of determinism, a co-determinism and a non-symbolic semiosis have to be considered. How a new science of life and mind can be developed is a question for the future.

Acknowledgment. This paper was written during the author's stay at the Department of Theoretical Chemistry, University of Tübingen, Germany. It is an edited version of a lecture he gave in Bremen on March 5th, 1990. Financial support of the Alexander von Humboldt Foundation and the Hungarian Academy of Sciences (OTKA grant no. 1-600-2-88-1-616) is gratefully acknowledged. The author warmly thanks Professor V. Csányi (Budapest) for his encouragement and continual cooperation, and Professor O.E. Rössler (Tübingen) for his hospitality and several deep discussions. He thanks Professor K. Haefner (Bremen) for stimulation. Furthermore, he thanks Professor J. Nicolis (Patras), Professor F.J. Varela (Paris), Dr. M. Requart (Göttingen) and Dr. P. Cariani (Boston) for discussions.

References and Notes

1. Shannon CE, Weaver W (1949) The Mathematical Theory of Communication, Univ of Urbana Press, Urbana
2. Kampis G (1986) Biological Information as a System Description, In: Cybernetics and Systems '86 (ed. Trappl R), Reidel D, Dordrecht, pp 36–42
3. Kampis G (1987) Some Problems of System Descriptions I. Function, II. Information, Int J General Systems **13**, 143–156; 157–171
4. Kampis G (1987) Elements to the Systems Modeling of Evolution, Ph.D. Thesis, Budapest
5. Csányi V (1989) Evolutionary Systems and Society: A General Theory, Duke University Press, Durham
6. Kampis G (1990) Self-Modifying Systems in Biology and Cognitive Science: A New Framework for Dynamics, Information, and Complexity, Pergamon, Oxford, to be published
7. Atlan H (1983) Information Theory, In: Cybernetics: Theory and Applications (ed. Trappl R), Hemisphere, Washington, pp 9–41
8. Klir GJ (1985) Architecture of General Systems Problem Solving, Plenum, New York
9. Jaynes ET (1979) Where Do We Stand On Maximum Entropy? In: The Maximum Entropy Formalism (eds Tribus M, Levine RD), MIT Press, Cambridge, pp 15–118. This is an extensive summary of work done by its author in the fifties and by others since then
10. Schrödinger E (1944) What is Life? Cambridge Univ Press, Cambridge
11. Thom R (1983) Mathematical Models of Morphogenesis, E Horwood/Wiley, Chichester
12. Nicolis J (1986) Dynamics of Hierarchical Systems, Springer, Berlin
13. Bar-Hiller Y, Carnap R (1952) An Outline of a Theory of Semantic Information, Technical Report No. 247 of the Research Laboratory of Electronics, MIT; reprinted in Bar-Hillel Y: Language and Information, Addison-Wesley, Reading, Mass, 1964
14. Johnson-Laird PN (1983) Mental Models, Cambridge Univ Press, Cambridge
15. Csányi V (1982) General Theory of Evolution, Publ House of the Hung Acad Sci, Budapest
16. Rosen R (1978) Fundamentals of Measurement and Representation of Natural Systems, North-Holland, New York

17. Ryan J-P (1975) Aspects of the Clasuius-Shannon Identity: Emphasis on the Components of Transitive Information in Linear, Branched, and Composite Physical Systems, Bull Math Biol **37**, 223–253

18. Löfgren L (1972) Relative Explanations of Systems, In: Trends in General Systems Theory (ed. Klir GJ), Wiley, New York, pp 340–407

19. Kampis G, Csányi V (1988) A systems Approach to the Creating Process, IFSR Newsletter No. 20, 2–4. The example of the "tape recorder" that changes its interaction mode is from V Csányi

20. Whitehead AN (1966) Modes of Thought, MacMillan, New York. Another work in which Whitehead's views are exposed is Whitehead AN (1929): Process and Reality: An Essay in Cosmology, Cambridge Univ Press, Cambridge. A volume devoted to discussion is Holz H, Wolf-Gazo E (eds) (1984) Whitehead and the Idea of Process. Verlag Karl Aber, Freiburg München

21. Kampis G (1988) On the Modelling Relation, Systems Research **5**, 131–144

22. Rössler OE (1981) Chaos and Chemistry, In: Nonlinear Phenomena in Chemical Dynamics (eds Vidal C, Pacault A), Springer, Berlin, pp. 79–87

23. Rössler OE (1984) Deductive Prebiology, In: Molecular Evolution and Protobiology (eds Matsuno K, Dose K, Harada K, Rohlfing DL), Plenum, New York, pp. 375–385

24. Elsässer W (1975) The Chief Abstractions of Biology, North-Holland, Amsterdam. Starting in the fifties, Elsässer wrote five or six books about this topic

25. Jacob F (1981) Le jeu des possibles, Fayard, Paris

26. Rosen R (1973) On the Generation of Metabolic Novelties in Evolution, In: Biogenesis, Evolution, Homeostasis (ed. Locker A), Springer, Berlin

27. Levins R (1968) Evolution in Changing Environments, Princeton Univ Press, Princeton, NJ

28. Cariani P (1989) On the Design of Devices with Emergent Semantic Functions Ph.D. dissertation, Dept of Systems Sci, SUNY at Binghamton

29. Bertalanffy L von (1968) General System Theory, Braziller, New York

30. Lwoff A (1968) Biological Order, MIT Press, Cambridge. Quoted and discussed in Riedl R (1979) Order in Living Organisms, Reidel, Dordrecht

31. The Unavoidable conclusion is that we are our own enemies. This is true on a global scale, because there is but one Earth, and also true on a national scale in Hungary where the author comes from. Hungary has been a member of the Warsaw Pact but is likely to orient itself towards NATO, the good old enemy, in the future

32. Csányi V, Kampis G (1988) Can We Communicate With Aliens?, In: Bioastronomy: The Next Steps (ed. Marx G), Kluwer, Dordrecht

33. Pask B (1975) Conversation Theory, Elsevier, New York

34. Wittgenstein L (1922) Tractatus Logico-Philosophicus, Routledge and Kegan Paul, London

35. Searle JR (1980) Minds, Brains, and Programs, The Behavioral and Brain Sciences **3**, 417–424. A very famous article, reprinted several times

36. Fox RF (1989) Energy and the Evolution of Life, Freeman, New York. The author is S. Fox's son

37. Löfgren L (1968) An Axiomatic Explanation of Complete Self-Reproduction, Bull Math Biophys **30**, 415–425

38. Hofstadter D (1979) Gödel, Escher, Bach, Basic Books, New York

39. Bartlett SJ, Suber P (eds) (1987) Self-Reference. Reflections on Reflectivity, M. Nijhoff, Dordrecht

40. Maturana HR, Varela FJ (1980) Autopoiesis and Cognition, Reidel, Dordrecht; Varela FJ (1979) Principles of Biological Autonomy, North-Holland, New York

Mega-Evolution of Information Processing Systems

Erhard Oeser

The project "Evolution of Information Processing" is not the first attempt to describe evolution in an interdisciplinary context, since the extension of the biological theory of evolution to inorganic nature on the one hand and to human society on the other is a logical consequence of this theory already recognised by Darwin. He explicitly said that "the principle of life will be recognised as part or sequel of a universal law" [1]. His contemporary Spencer even tried to formulate such a general law and turned it into the basis of a synthetic evolutionary philosophy, which was based, however, to a large extent on speculations not empirically verified. In this general law, development and evolution are conceived of as an increase in complexity, i.e. as a process leading from unspecified and incoherent homogeneity to specified coherent heterogeneity [2]. With the formulation of such a law Spencer went a step beyond Darwin, who had considered the origin both of life from inorganic matter and of the human mind to be unanswerable questions.

In our century a fundamental distinction for the whole domain of evolution has been made by Julian Huxley [3]. He started from the idea that reality clearly consists of a simple universal evolutionary process. But this process is divided into several levels or phases leading to different products according to their individual speed of development with each of them being based on a special developing mechanism: cosmic, inorganic and prebiotic evolution is controlled by physical or simple chemical processes, whereas organico-genetic evolution is determined by the mechanisms of recombination, mutation and selection, and the so-called sociocultural evolution is based on the transfer mechanisms of direct communication and tradition, which are independent of hereditary mechanisms and thus of the alterations within generations.

Today, all scientific disciplines, from physics and chemistry to the social sciences and humanities, provide empirical facts and special theories able to justify the search for an integrated theory of universal evolution. But the strict and consistent transfer of the information processing paradigm to all three levels of universal evolution, the prebiotic, biotic and postbiotic levels, is a new element in the considerations of the concept of universal evolution and leads to a new and more differentiated concept of evolution. This new concept of evolution covers also the physico-chemical domain of prebiotic evolution and the new domain of technical information processing systems. In contrast to the biological concept of macro-evolution, which means the evolution on the level of the higher categories of living systems, this concept of universal evolution, which covers all levels of information processing systems from the physical information processing systems (physIPSs) to the sociotechnical information processing systems (soctech IPSs), is to be called "*mega-evolution*".

Whereas in a universal model of the evolution of information processing the concept "evolution" is transferred from the biological domain where it originated to the more complex level of human society and technology, the concept "information" is used at a "lower" level.

This is due to the fact that the concept "information" stems from the human domain. It originated as a single concept of classical epistemology [4]. It was soon used in military technology as "message" and in the pedagogical and didactic domain as "instruction" [5]. In our century the concept of information, originally a semantico-pragmatic concept, was specified in the theory of communication engineering and in the theory of technical observation apparatus. This specification required, however, the restriction of the concept "information" to the syntactic level. In the beginning the classical symbol processing systems were confined to this level as simple algorithmic mechanisms. Whether today's knowledge processing systems of AI have put an end to this restriction is still a controversial question.

It is certain, however, that the concept "information" must be specified and expanded beyond its vague meaning in common language and its restricted meaning in communications engineering in order to link it with the model of mega-evolution. The basis for an extension of the concept "information" is provided by the theories of information technology.

Basically there are three theories [6]:

- the theory of information transmission (represented by Shannon) oriented towards communications engineering [7],
- the algorithmic theory of information processing (represented by Turing, Kolmogorov, Chaitin etc.) [8],
- the theory of representation (represented by MacKay) [9].

According to the three dimensions of the concepts of information, the syntactic, semantic and the pragmatic dimensions, these three theories are located at different levels.

Shannon's theory of information is in fact a fully syntactic theory of signal transmission. The meaning of signals is unimportant during the transmission process, since it is determined before hand by encoding a joint sign repertory which has to be decoded again after the process. The statistical content of a message is therefore related in this case to the encoding effort before and after the transmission of signals. Thus, as is well known, there is no access from this theory alone to semantics, because not even a measure for the statistical information content of individual signal sequences can be defined.

The algorithmic theory of information is also primarily a syntactic theory. But it already permits an individual measure for information in symbol sequences, because the concept of random sequence becomes more constructive [10]. A random symbol sequence cannot be shortened, whereas a non-random, i.e. rule-determined symbol sequence can be abridged by indicating only the program of the constant production (= the algorithm) of the symbol chains. This results in a reduction of redundancies and thus in a condensation of information.

The transition to the semantic level of information processing results at this level by not only assigning meanings to the regularly produced individual syntactic symbol sequences, which can be implemented physically and technically in different ways, but – at the same time – by interpreting semantically the operations leading to the production of symbol sequences.

It is true that this semantics, which is realised by a techIPS and which is necessary for information processing, is a formal semantics artificially constructed "from outside" by syntactic rules. But we have already reached the second dimension of the concept "information". From this semantic dimension of the concept "information" derived from physico-technical theories, further connections to other types of IPS can be constructed. The way a genetic IPS functions, for instance, cannot be interpreted only in the sense of the theory of information transmission in Shannon's tradition. The model derived from it of the genetic code, transferring genetic information from one generation to the next, does not yet take into account the fact that genetic information also contains general instructions for individual development, i.e. the ontogeny or morphogeny of the organism in the next generation. This process of the regular structure of an organism and its modes of behaviour can be seen, according to the algorithmic approach, as a computer *program* determining the processing of genetic information transferred from one generation to the next [11]. In addition, this model of algorithmic information processing implies a formal equivalence between computer programs and theories, due to the formal provability of the equivalence of computability and derivability. This allows us to reach an at least intuitive, but conceptually consistent, integration, with the genetic program of a genIPS also having the same meaning as the theories at the level of a humIPS. For the living organism is basically something like a living theory of its environment which it anticipates in principle through its (onto)genetic development program, although the changes and details of its development are determined by the influences of its environment.

But in order to be able to precisely describe the influence of the environment on the IPS, another theory is needed which also includes the pragmatic dimension, and which precedes the technical theory of communication as a theory of information transmission. This theory is the theory of representation of information, as it has been established by Donald MacKay, supplementing Shannon's theory. Based on the theory of technical observation apparatus it provides for the important insight that semantics and pragmatics cannot be separated, because the latter determines the former. For the observer or the user decides what relevant information in the semantic sense is and what is not information. MacKay says that "the observation-process results in the appearance of a representation in the representation-space of the receiver or observer" [12]. The representation model does not consist of two fixed instances, one sending signals and the other passively receiving them, but rather of a fixed, i.e. structured instance, the observer, and a variety of possible sources, developing the possible representation-space with its borderlines being determined *a priori* through the structure of the observer or the observing apparatus. Thus the theory of representation of information describes how information originates as representation, while the theory of communication describes how already existing finished information (= representation) is transferred, and the algorithmic theory of information describes how this transferred information is processed into new information. On the basis of this semantic-pragmatic theory of representation the fundamental meaning of the concept of information arises.

Information is what effects an alteration in a system. This means that information does not exist everywhere in the universe, but rather only for a certain system. Information is therefore *ex definitione* a concept relative to a system. "Information" and "system" mutually define each other [13]. This is also valid for the level of physical systems: "All organized structures contain information" [14], because an individual physical body can only be regarded as existent if it is informationally coupled with its nearer or farther environment. This means that it is only real in existence if it develops inner states in connection with its outer reactions and that the other real bodies that act upon it should be represented by these inner states.

Thus we can see the difference the theory of representation makes between structural (*a priori*) and fluctuating (*a posteriori*) information in an extended dynamic sense approximating the evolutionary model of information processing. In the theory of technical observation apparatus the structural-material conditions, to which in this case the structural content of information is restricted, do not change because of the "real" fluctuating information. In organismic information processing systems and in "higher" knowledge processing systems the structural content of information, at least the so-called "knowledge basis", changes every time information is acquired. But in the evolution and ontogenetic development of organisms even the processing mechanism changes, since knowledge basis and inference mechanism are not separated here.

Summing up, we can say that first at the level of the analysis of basic concepts for a model of the mega-evolution of information processing covering all IPS, the two basic concepts "evolution" and "information" can be linked to a consistent terminology in such a way that the concept "system", originally a static concept in the theories of information technology, becomes dynamic.

"Evolution", "information" and "system" are the three basic concepts which cannot be deduced from one another and which have to be linked in a consistent way with a theory of the mega-evolution of information processing.

The unreflected, sometimes uncritical and even careless use of the word "evolution" is a characteristic deficiency of the ubiquitous application of the theory of evolution. Although these applications in dynamic, diachronic and historic contexts are no longer confined to biology, the scientific differentiation and specification of their fundamental concept has taken place in biology. Thus the analysis of the concept "evolution" and its use has to start there.

Even a short glance at the different meanings or definitions assigned to the concept "evolution" since 1800 shows that this term has always been polysemous in biology, too. This can be attributed to the complexity of the real phenomenon of the evolutionary process itself [15].

Thus it is easier to begin with what evolution is not. "Evolution" cannot be defined only as "the process of change in time", since such a definition does not explain either the causal mechanism, the speed and the mode, or the products and results of this process. Only the integration of all these components results in the complete concept of "evolution". Thus the question "what is evolution?" cannot be answered by a simple definition, but rather with a theory comprising all these components of the concept "evolution" in their logical context. Such a complete theory of evolution does not yet exist in biology. Instead there are just several evolutionary theories with different foci. Even the historical origin of the theory of evolution was characterised by such a pluralism. Lamarck stressed the "vertical" component of change in time. Therefore he considered evolution mainly a process of transformation. Although Darwin accepted this transformatory process of species as a fundamental aspect of the concept "evolution", he was interested much more, within the context of his selection theory, in the horizontal level of evolution, i.e. in the diversification or the diversity of different individual specimens of a species and the populations as new emerging species [16]. This is due to the fact that selection can only be considered the basic principle of phylogeny, if there is a large diversity of variations. The recognition of two different components was as important as the recognition of the fact that they are independent of each other to a large extent. This independence of the diachronic-vertical component and the synchronic-horizontal component can also be seen in the causal mechanism of evolution, since the synchronic and horizontal diversity of organisms as individuals and populations is the result of mutation, while selection causes the gradual transformation of species surviving in the course of the alteration of generations with their predecessors becoming extinct.

The differentiation of the concept "evolution" at the biological level of genetic information processing systems already implies several levels of explanation [17]:

— At the level of traditional genetics, with the gene, the hereditary unit, still being a mental construction, probability laws are concerned describing, in the framework of gradual micro-evolution, evolutionary steps as adaptive alterations of genetic frequencies in defined populations under given conditions of mutation, recombination and selection.
— "Genetic drift", independent of this mechanism, occurring as a casual alteration especially in numerically small populations, is added to these processes.
— At the level of molecular genetics, with the gene not intended to be a construct anymore, i.e. not a theoretical, but an empirical, concept, representing a section of DNA, the chemical hereditary substance, a third evolutionary process which is called "molecular drift" is added, and can led to quick pheno-

typical alterations in a population, independent of selection and genetic drift. In addition it has been stressed, especially at this explanatory level in the sense of a neutral theory of molecular evolution, that most mutations are neutral in selection. As a consequence evolutionary change is even more isolated from selection than has been assumed in the Darwinist synthetic theory of evolution.

This complexity of the concept "evolution", which at present can only be understood by several differing biological theories, with a uniform definition of the concept being impossible, has to be borne in mind when we talk about the mega-evolution of information processing.

In addition, the conception of evolution is changing, especially concerning the definition of the causal mechanism at the various levels of a certain type of information processing system. For there is not only one evolution of information processing systems, but also an evolution in the sense of an increasing linkage of the separated factors of mutation and selection on the different levels of information processing systems.

For it is clear that both factors, mutation and selection, acquire a different meaning, when the evolution of techIPSs is concerned.

Mutation means planned, not casual, change. In technical domains this is nothing other than an "invention". Selection is the conscious "choice" of a technical product by the user. Here, too, both instances are usually separated in different persons. The inventor is the designer of the system and the user eventually chooses one of the products offered by the system designer. This shows that the evolution of techIPSs is linked, directly and not only metaphorically, with sociocultural and especially with socioeconomic evolution.

Even before Darwin various conceptions had been set up concerning a first differentiation of sociocultural evolution and biotic or organic evolution. This fact has been pointed out especially by Friedrich von Hayek, who consequently linked the theory of sociocultural evolution with the theory of complex ordering systems [18].

So when we talk about universal evolution or mega-evolution or just want to use one and the same concept "evolution" in the cosmological-physical-chemical domain, in the biological and in the sociocultural fields, we have to prove a real "kinship" between evolutionary mechanisms in the sense that one originated from the other. This proof must be provided at those critical points of evolution which for Darwin were unanswerable questions: the origin of life and of the human mind. In both cases the proof has already been provided to a large extent by recognising in every single domain a self-increasing development of the mechanisms in question: in the domain of prebiotic, chemical evolution, progress in the explanation of the origin of life was achieved by describing synergetic co-operation phenomena [19], fluctuations far away from thermodynamic equilibrium [2], catalytic cycles and autocatalytic hypercycles [21]. In the biological domain, on the other hand, the idea of an evolution of the hereditary mechanisms became the important issue. As early as 1936 the geneticist A. Shull coined the expression "evolution of evolution" [22].

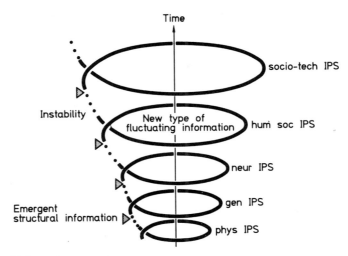

Fig. 1. Mega-evolution of information processing systems

Thus we can only speak in a logically consistent manner of the mega-evolution of information processing systems if a change of systems in their organizational functional structure is permitted, leading to new so-called emergent properties not having existed before.

The mechanism of the mega-evolution transcending the individual emergent levels of information processing systems is nothing other than an iterative process of structural and fluctuating information, which also represents a process of the economical condensation of information, where the lower ordering structures are used at a higher level. In this connection both the quality of information and its basis for quantification are changing at every level of complexity, increasing in a process of self-organization (see Fig. 1).

- At the lowest level of inorganic matter, a thermodynamic system develops through fluctuation of molecular states of order as structural information. This information is the basis for the emergence of a new type of flowing information in the sense of a primitive replicating mechanism, which is the origin of genetic information.
- At the level of genIPS, changes in the sense of flowing information occur by chance and/or by external influence on the genetic system. These changes consolidate organically as neural systems in the sense of new structural information.
- At the level of neurIPS, a new type of flowing information emerges through their states of excitation. These excitation states effect as electrochemical processes new structural patterns forming the mental system or consciousness.
- At the level of individual human consciousness, which – in contrast to other higher organisms – has developed into a stable system with constant self-

identity, knowledge finally emerges as a new type of *a posteriori* flowing information. The acquisition, storage and processing of this information depend, however, primarily on the structurally consolidated *a priori* structures of consciousness and secondarily on the material structures, i.e. the micro- and macro-anatomical structures of the brain. These structures in turn are based on molecular structures of inorganic matter.

– At the level of sociotechnical information systems, collective systems finally emerge through communication and cooperative information processing. The tools (techIPSs) are an important part of these sociotechnical information systems as media of the strictly organized information flow. But these collective systems are not an independent organism of their own, but are totally dependent on the interaction of individual instances contained in these systems.

The following demands on a theory of the mega-evolution of information processing systems result from this general description:

1. There must be a set of principles for the evolution of IPSs that should be regarded as axioms of a universal theory. From this theory concretely applicable partial theories can be deduced.
2. The theory has to unite a maximum of logical unity with a minimum of basic concepts in the sense of the principle of economy of Mach and Einstein. This demand refers above all to the differentiation of the concept of information.
3. The theory of the evolution of IPSs must indicate the origin of information.
4. This theory can not only be a classification or typology of IPSs and a description of principles of the information processing that are valid for all levels of IPS. The theory should also explain the transitions from one level of IPS to another level.
5. It must distinguish as well between a mechanism of evolution that indicates the differentiations of IPS on one level (level evolution), and another mechanism indicating the transitions from one level of IPS to another level (mega-evolution).
6. The theory of the mega-evolution of IPSs should explain the mechanism of evolution itself with the help of the information processing paradigm. Thus there must be a distinction between the evolution **of** information processing and the evolution **by** information processing.
7. This theory must indicate the criteria of a new level of IPS. This is the only way that the number and kinds of the levels of IPSs that can be distinguished can be defined. Possible criteria are:

    – the grade of complexity, i.e. the number and kinds of elements, the number and kinds of relations
    – the kind and sort of information processing that are defined by this grade of complexity, i.e. the functional mechanisms
    – the different performances that are produced on the various levels of IPS.

8. Although a theory of the evolution of IPS does not permit deterministic predictions, at least possible tendencies of development should derive from it.

References

1. The Life and Letters of Charles Darwin, Vol. 3, F Darwin, (ed.) (1988) London
2. Spencer H (1962) First Principles, London, Sect 138
3. Huxley J (1966) Die Zukunft des Menschen – Aspekte der Evolution. In: Das umstrittene Experiment: Der Mensch (Man and his Future) München
4. Der Informationsbegriff in der Philosophie und Wissenschaftstheorie, In: Folbert OG, Hackl, (Hrsg) (1986) Der Informationsbegriff in Technik und Wissenschaft, München, Wien
5. Oeser E (1976) Wissenschaft und Information, Bd. 2 Erkenntnis als Informationsprozeß, Wien – München
6. Oeser E (1989) Der Informationsbegriff und die technologische Wende der Wissenschaftstheorie, in: Wiener Studien zur Wissenschaftstheorie Bd. 3, Wien
7. Shannon CE, Weaver W (1949) A Mathematical Theory of Communication, Urbana
8. Chaitin GJ (1966) On the length of programs for computing finite binary sequences, Journal of the Association for Computing Machinery 13: 547
9. MacKay DM (1969) Information, Mechanism and Meaning, M.I.T. Press Cambridge, Mass, London
10. Hotz G, Der Begriff der Information in der Informatik: In: Folbert OG, Hackl, (Hrsg) (1986) Der Informationsbegriff in Technik und Wissenschaft, München, Wien
11. Schuster P (1989) Informationsverarbeitung in der Molekularbiologie: In: Wiener Studien zur Wissenschaftstheorie, Bd. 3, Wien
12. MacKay DM (1969) Information, Mechanism and Meaning, M.I.T. Press Cambridge, Mass London, p 42
13. Holt AW quote from Petri CA (1976) Interpretation of Net Theory. Bericht der GMD 75-07, Bonn, p 8
14. Stonier T (1990) Information and the internal structure of the universe, An Exploration into Information Physics, Springer-Verlag, London, Berlin, Heidelberg, New York: Paris, Tokyo, Hongkong, p 25; Pawlow T (1973) Die Widerspiegelungstheorie, Berlin, p 79.
    Stock WG: Georg Klaus über Kybernetik und Information: In: Studies in Soviet Thought 38, p 219
15. Oeser E (1974) System, Klassifikation, Evolution, Wien-Stuttgart
16 Mayr E, The Growth of Biological Thought, Harvard University Press, Cambridge Mass, London
17. Oeser E (1987) Psychozoikum, Evolution und Mechanismus der menschlichen Erkenntnis-fähigkeit, Berlin-Hamburg
18. Hayek Fv, Die überschätzte Vernunft: In: Riedl RJ, Kreuzer F (eds) (1983) Evolution und Menschenbild, Hamburg
19. Haken H (1981) Erfolgsgeheimnisse der Natur, Stuttgart
20. Glansdorff P, Prigogine I (1971) Thermodynamic Theory of Structure, Stability and Fluctuations, New York
21. Eigen M, Schuster P (1979) The Hypercycle. A Principle of Natural Self-organization, Berlin, Heidelberg, New York
22. Oeser E, Informationsverdichtung als universelles Ökonomieprinzip der Evolution: In: Ott JA, Wagner GP, Wuketits FM (Hrsg) (1985) Evolution, Ordnung und Erkenntnis, Berlin–Hamburg

III Information Processing Systems
at the Physical Level

From "Matter-Energy" to "Irreducible Information Processing": Arguments for a Paradigm Shift in Fundamental Physics

Manfred Requardt

1 Exposition 115
2 Connectivities 118
3 Philosophical Prelude 119
4 Heretical Views about some Topics of Modern Physics 120
5 Some Points of Departure 122
6 Information 124
7 On the Way to a "Subquantum Theory" (Intermediate Magnification) 126
8 Probing into the Substructure of our Physical World (Extreme Magnification) 130
9 Unfolding and "Emergence" 133
References 134

1 Exposition

> The universe begins rather to resemble
> a great idea than a huge machine.
> (Sir J. Jeans)

The following represents an abridged version of a more detailed investigation I made for the project EIP. The Study was devoted to the question of whether concepts of information theory and cybernetics play or will play a deeper role in the foundation of the epistemological framework on which the regime of physics has been based.

While orthodox physics would probably admit that these concepts may be of some use in those branches of physics that deal (in some sense or another) with complex or 'self-organizing' systems (a typical catchword being, e.g., 'synergetics', cf. [1, 2]), the answer will almost surely be a radical 'no' as far as the inner core of this science is concerned.

We regard it as one of our aims to show that this attitude has not always been the prevailing one in natural science and to emphasize that it must be related to a paradigm shift within meta science as a whole, i.e., a change in the very way of registering and analysing phenomena, forming concepts and ideas and even asking questions.

We want to argue in the following that, in order that, e.g., Weizsäckers "Unity of Nature" [3] will again become a core principle of modern physics, another paradigm shift has to take place in meta science. But before embarking on this intricate subject matter I would like to set the stage by formulating (in admittedly highly condensed form) my view of the central dogma of established modern science.

Paradigm of orthodox physics (P-Me). At the fundamental level the physical reality outside of us (i.e., the observers) is organized by what I would like to call the matter energy principle (for simplicity).

Remark: I hope that this notion will not lead to great confusion and that everyone aquainted with the methods of modern physics will at least be intuitively aware of the epistemological content of that sloppy characterisation.

From (P-ME) and the conjecture that the description of nature given by it is basically complete is derived another paradigm.

Paradigm of Reductionism (P-R). Observation: It is an empirical fact, which may ultimately be made more precise via an (as yet hypothetical) universal and all-embracing theory "of everything", that "nature", at least as far as our anthropomorphic point of view is concerned, appears to be layered in a sequence of certain strata, both in reality and in our theoretical description; these strata can, at least pragmatically, be told apart by us as observers (e.g., physics of elementary systems, many particle physics, chemistry, biology, brain and mental states, culture as a whole etc.).

Strong Version of (P-R). The content of the model theory of each stratum (i.e., notions, theoretical constructs etc.) can, at least in principle, be explained and derived via a certain (algorithmic) process of reasoning (which, somewhat suspiciously, is usually never fully specified) from the model theory of the next lower stratum. This possibility of transition from one level of the hierarchy to the next one never terminates, from fundamental physics to, e.g., human consciousness and the mental world.

Weak Version of (P-R). The above holds at least for the so-called material, or more properly, physical world. It is left open whether the content of, e.g., human consciousness and its very existence can really be reduced to certain (material) "brain states".

Remark: This principle is not really openly debated in physics proper but is rather taken as a matter of fact. As to the wider context, e.g., in the philosophy of science one may mention the compilation of classical essays by Küppers, [4], with respect to the field of mind-brain identity [5, 6].

If one undertakes to scrutinize the basement of this hierarchy of levels of theoretical description (i.e., the stratum, governed by (P-ME)), one soon becomes aware that, in a sense, if one starts to replace a few bricks by others, they will not smoothly fit into the rigid framework and that, ultimately, one will have to rebuild the whole building upon new principles. In other words, our innocent-looking starting point unfolds to a major question of a fundamental paradigm shift from, e.g., (P-ME) to a paradigm where some sort of (possibly rudimentary) information processing, that is, something inherently immaterial is built into the very foundation of the above hierarchy of science. What might be at stake is therefore the question of a paradigm shift from (P-ME) to a new paradigm, which we call, somewhat sloppily, the Paradigm of Universal Information Processing.

Paradigm of Universal Information Processing (P-I). At all levels, most notably on the very fundamental level, a certain amount of information processing is a crucial part of every really complete description by a model theory and cannot be eliminated or replaced by processes described within the old (P-ME) and (P-R)-paradigms.

Remark: What would make this difficult to accept by, e.g., physicists is the prospect that this might bring something inherently "immaterial" into play, which may nevertheless act upon matter and vice versa, and this taking place at the core of the scientific hierarchy.

   Given the complexity of the whole subject matter, our enterprise has inevitably to unfold into a sequence of various subtopics:

 (i) A compilation and careful scrutinization of conspicious-looking phenomena and features which have (in our view) only rather insufficiently been explained within the orthodox (P-ME) framework.
 (ii) An inspection of the "waste basket" of (natural) philosophy in order to find some "suppressed" material concerning alternative and deviating concepts of nature which, in one way or another, may have anticipated (possibly in a rather embryonic form) some aspects and facets of our new paradigm (P-I).
(iii) Assuming then that we can gather sufficient evidence that a paradigm shift to, e.g., (P-I) is advisable, we have to give at least a preliminary draft of this more refined and extended epistemological scheme.
(iv) In a last step we have to reorganize the whole hierarchy of levels of description of nature according to (P-I). In persuing this task we introduce a sort of order parameter called "complexity" which will be of considerable explanatory and conceptual value. One of the more aesthetic reasons why we consider (P-I) to be superior as compared to, e.g., (P-ME) is its enormous power of unification, such that epistemological subsystems which have led a life of their own seem to become amenable to a unified description within (P-I). To give an example: the question of "emergence" and "downward causation", notions completely foreign to an interpretational scheme based on (P-ME) and (P-R), will come into the focus of our attention.

Remark: A paradigm shift from a more biological point of view was also advocated by Küppers in [7].

2 Connectivities

In this short interlude we will try to give the reader at least a faint idea of the amount of material and range of topics which in principle had to be condensed into our study.

Besides fundamental physics proper, which will be our main theme in the following, there exist various other branches of natural science and the humanities in general where similar phenomena and epistemological issues may be observed.

As to (natural) philosophy, it has been one of the major issues for several thousand years whether everything can ultimately be reduced to material processes or 'cause and effect', or whether there is something "immaterial" acting behind the scene.

Another important subject matter belonging to the wider context of our investigation consists of the deep and subtle questions concerning the so-called mind-brain debate. As we are searching for indications of some (possibly rudimentary) form of (perhaps immaterial) information processing on the most fundamental levels of nature, it is advisable to look into that regime where probably the dominant role of it is not even denied by the supporters of (P-ME). This topic is, however, of great relevance for our project in yet another respect which may be circumscribed by the catchwords "mind-brain (non-)identity" or "interactionism", i.e., are all mental states brain states or does an immaterial world exist which can nevertheless act upon the objects of the physical "matter-energy world" (cf., e.g., [19] or [20]).

If one is willing to accept that some interaction between these two worlds exists, the concept of cause and effect also has to be extended to embrace these new modes of "causation", a question which is evidently of the highest interest for our subject matter as well.

At the moment, the brain appears to be the system of the highest "localized" complexity in our universe where it has to be admitted that this concept is up to now not very sharply defined. It would be one of the tasks of (P-I) to give this notion a less ambiguous and quantitative meaning. "Complexity" is linked with the levels of information processing. Part of our approach will be the developement of a sub-theory which makes the observed (almost discontinuous) jumps of "complexity" between the strata making up the hierarchical order of our universe, accessible to a really scientific and quantitative treatment. In doing this we will give the deep ideas of M. Polanyi (see, e.g., [24]) about "emergence" and "downward causation" an exact physico-mathematical content.

The above topic leads us to another bundle of concepts and theories having some bearing on our endeavour, i.e., the theory of "formal languages", "proof theory", Gödels deep observations, "self-referential structures", etc. To give an example, the "Radical Constructivism" of Maturana, Varla et al. (see, e.g., [25]) just defines the brain as a self-referential system.

Other tools we will exploit do belong more to the hard core of our subject matter, i.e., fundamental physics, cybernetics and some advanced concepts of modern mathematics which are perhaps not very common among mathematical physicists such as the "theory of cellular automata", "non-standard analysis" and "random geometry". We think that these topics will play a major role in this business in the future.

In closing this section we would like to mention two intimately related branches from the tree of human subject–object relations which are not usually considered to belong to science at all within the "scientific community. On the other hand, they are of an almost paradigmatical character if one wants to transcend the matter-energy or cause-effect world. The first is the theory of the collective subconscious by C.G. Jung, in particular the so-called theory of (acausal) synchronistic phenomena (see [26], similar ideas can already be found in Schopenhauers work, see, e.g., [27] "...the synchronicity of the causal non-related..."); the other is what is called somewhat misleadingly "parapsychology".

As to the former, we consider it to be a futile attempt to think about the mind-brain identity or interactionism without even mentioning the huge body of material supplied by the theory of the subconscious "terra incognita" as is the usual habit (cf., e.g., [19, 20]). As to parapsychology, none other than the eminent physicist W. Pauli made various contributions in this field and recognised its possible prospects also for physics proper (cf. [26] or the essay by him in [4]; also, the remarks by M. Fierz in [23] are worth mentioning).

Evidently these disciplines would provide us with examples of seemingly matterless information exchange, phenomena which in our view belong to, as we like to call it, the "analogical" and "parallel" behaviour of nature, in contrast to the "causal" (vertical) hierarchical structure of the foreground of nature. However, in order not to discredit our investigation, we will not make any further mention of these still somewhat ambiguous disciplines in the following.

3 Philosophical Prelude

To supply some philosophical underpinning for our "new" paradigm we mention the utterances of two eminent thinkers of the past, their remarks pointing also in the direction of, to some extent, an organismic organization of the depth structure of our universe. It would, however, be easy to fill a whole monograph with related ideas (cf., e.g., the work of Schelling, most notably [12], in which context his considerable influence upon his contemporaries Oerstedt and Faraday as well as, e.g., Whitehead should be mentioned, cf. [13]).

As one source of inspiration we cite from Newton's "Query 28" of "Opticks" [8] which tells its own tale (utterances of the same tenor can also be found in, e.g., [9,10]):

> "... that there exists an immaterial, living, intelligent being, which, as if the infinite space were its sensorium, sees the objects our world from the inside."

In this context it may be noted that Newton studied the works of Jakob Böhme, the German mystic, and that this view was quite common at that time. For example, the philosopher Henry Moore, with whom Newton had personal contact, called the "Weltseele" (absolute spirit) the "great quartermaster general" of the universe. Evidently one may see in these concepts the roots of the "aether theory".

Another highly non-materialistic attitude was adopted by Leibniz. Even his "Monadology" is, in our view and in the light of modern findings, by no means so outdated as it is usually considered to be (in which context it may be noted that the judgements are the more pronounced the less is actually known about his philosophy). One of his fundamental axioms was that "matter" (whatever that is to mean) is only one aspect of the appearance of a more fundamental entity or process. For example, in [11]:

> "... *in order to find these real entities, I had to go back to some sort of real and animated point, i.e., to a substantial atom.... which have a completely different origin and being entirely **different** from the "mathematical points"... and which, without doubt, **cannot** yield via combination the continuum.*"

(These extremely deep ideas we will transform into a more physico-mathematical concept in the following, cf. also the statement of R. Feynman in the next section).

As to further influences and traces of this more organismic attitude see, e.g., [15, 16, 17, 18, 21, 22]. It is one of our aims to give these mostly rather vague and metaphoric concepts a more solid and quantiative foundation.

4 Heretical Views about Some Topics of Modern Physics

In order to show that even in modern physics the general attitude is by no means unanimous, we present in this chapter a brief selection of "heretical" views. We want, however, to express a warning. It would be entirely wrong to suspect that the sources being mentioned are in any way typical of the attitude modern physics has since adopted concerning the fundamental questions of natural philosophy. Quite the contrary, they serve instead as exceptions to the rule which is still overwhelmingly "mechanistic" and orthodox.

We start by citing a remark of P.A. Dirac (one of the founding fathers of quantum theory):

> "*Present quantum mechanics is not in its final form! Further changes, about as drastic as the changes which one made in passing from Bohr's orbits to quantum theory proper, will be necessary. It is very likely that in the long run Einstein will turn out to be correct!*"[28].

Another fundamental concept of the older natural philosophy (and which was alluded to in Sect. 3) has been what was called in nineteenth century physics the "aether". One of the dogmas of modern orthodox physics is that this notion has been proven to be devoid of any objective content by Einstein's theory. Our personal opinion is quite different and we want to cite three independent sources which support our view.

The first one is of a more general character (but, we think, nevertheless of extreme importance) and was originally coined by E. Zermelo in connection with what we would call today "Non-standard Analysis", i.e., the mathematics of the infinitely-small numbers:

> "*The non-existence of actual infinitely-small numbers is not provable as is the non-existence of Cantor's transfinite numbers, and the logical mistake is exactly the same in both cases. Namely it is the implicit habit of attributing certain a priori properties to these objects which they cannot possess. In this particular case the so-called "Non-Archimedean" number systems are concerned, the existence of which is a well-established fact today*" [29].

In our view this is an observation of almost universal relevance and is one of the obstacles to scientific progress, i.e., the anthropomorphic habit of endowing our epistemological systems with certain a priori properties and concepts usually taken from our everyday experience. As to the "aether" concept, what is actually proven is only that such a medium cannot have the properties orthodox and classical physics have been aquainted with and nothing more! In this respect we want to cite Einstein himself:

> "*... in particular the law of inertia seems to force us, to attribute physical and **objective** properties to the space-time continuum as such...*" [30].

A more modern source is [31] where it is asked whether the common feature of spontaneous symmetry-breaking may not be an indication that the "vacuum" is actually a medium which "absorbs" the missing quantum numbers.

We want to close this section by mentioning three other sources which serve to support our view being developed in the following and touch upon particularly relevant cornerstones of our approach. The first is taken from [32], going in fact back to R. Feynman, and talks about the internal complexity of the individual points of the space-time continuum, relating them to something like the microscopic sites in a computer array. In the same direction goes [33]:

> "*... On the other hand, I believe, the theory that space is continuous is wrong, because we get these infinities and other difficulties, and we are left with questions on what determines the size of all the particles. I rather suspect that the simple ideas of geometry, extended down into infinitesimally small space, are wrong.*

A related point of view is adopted by J.A. Wheeler in [34]:

> "*... time cannot be a primordial category in the description of nature but secondary, approximative and derived... Laws of physics came into being by a higgledy-piggledy mechanism... It is difficult to defend the view that existence is built at bottom upon particles, fields of force or space and time*".

I think one does not say too much when one sees certain connections with Leibniz's monadology.

That last source we want to mention is K. Zuse [35]:

> "... *Perhaps it is even so that only by using the term information, laws such as those of the conservation of energy, momentum, etc. can be fully explained... Each of the grid points contains a small computer and that an exchange of information takes place between these computers.*"

We will come back to the very interesting ideas of Zuse in the following in more detail, since they are in some ways rather close to our own approach.

5 Some Points of Departure

When one attempts to replace an old paradigm by a perhaps more refined one, it is of considerable importance to isolate the crucial subcomplexes of the epistemological system in current use which have been constant stumbling blocks for quite a few people (whereas orthodox philosophy is permanently persuading the few sceptics that there is absolutely nothing to worry about). As to this we want to mention an observation of C.G. Jung, which is, we think, just to the point:

> "... *Why has this material been brushed under the carpet so that one has to search for it at exactly those sites, and will in fact find it there, where all authorities have assured us that absolutely nothing will be found?*" [26].

There is, e.g., the broad field of the (correct) interpretation of quantum theory (cf., e.g., [36–38]), typical candidates (representing a whole "universality class") being:

(i) The Double Slit Experiment. To this universality class do belong a lot of other subcomplexes such as the perhaps more famous EPR-(thought)-experiment. In a sense the essential aspects of quantum theory can be condensed into this paradigmatic experiment (super-position principle, phase field, interference). Some of the facets of particular relevance being embraced by this key word are the notion of "locality", "holism", "reduction of the wave packet" and the quantum mechanical "measuring process". The same topic in slight disguise runs also under more fashionable names like "the story of Wigner's friend", "... of Bertelmann's socks", "d'Broglie's paradox", etc. "Bell's inequality" and the experiments of aspect also belong to this thematic complex. Some references (of review character) are [36–39]. Very readable accounts are also [40–42].

The seeming paradox in this widely known experiment is that while the quantum object passes exactly through one slit at a time (at least when tested so), the other slit influences the behavior of the "particles" behind the screen in a strange and characteristic manner (being made visible in the form of a pronounced interference pattern. This is even so when the slits are very far apart so that no influence in the classical sense should persist.

On the other hand, the interference pattern can only be observed if the location of the micro object passing through one of the slits remains sufficiently undetermined. If, however, the position of the particle has been fixed (e.g., by means of an act of observation which is at the same time a perturbation) no influence of the second slit can be detected.

Our interest in these bizarre phenomena stems from the possibility that something "immaterial",[1] perhaps of the character of pure "information," may play a hitherto hidden role. This is related to, in our view, the not yet fully understood nature of the q.m. "phase field" being incorporated in the wave function.

(ii) The Aharonov-Bohm-Effect. Much of what has been said in (i) applies here also. What gives, however, the screw another turn is that the vector potential A, the reality of which is at stake in this experiment, was considered to be only an auxiliary field, being devoid of any ontological content in traditional physics, while at least a certain restricted observable existence was granted to the "phase field" of the Schrödinger wave function. Nevertheless an observable change of the interference pattern of the electron wave function is effected in a region of space where all measures have been taken such that the "observable magnetic field B" is equal to zero. On the other hand, the experimental set up is assumed to guarantee that the particle can never enter the region of space where B is generated. Again it seems to be that something immaterial may act on matter!

The literature about this effect is enormous, see, e.g., the list of references compiled in [43]. Therefore we mention only the original paper of Aharonov and Bohm, [44], and the experiment of Möllenstedt et al., [45].

(iii) Gauge Theory. Both in (i) and (ii) the physical quantity which seems to play a crucial but mysterious role is of a somewhat ephemeral character, namely a phase or gauge field. Gauge theories (cf. [46]) are very fashionable in modern quantum field theory. It seems that they are presently in a probably not fully understood sense more fundamental than the model theories which have been in use in the past. On the other hand, the deeper ontological nature of the gauge principle is, in our view, not yet adequately incorporated into the general concept, having at present more the status of a heuristic principle. We hope to be able to indicate in the following what its proper role might be.

(iv) The Aether Concept. As was already mentioned above, the aether concept was expelled from modern physics as sort of a medieval, scholastic entity. On the other hand, there are in our view a lot of indications (at least for one who is willing to see) that matters cannot be so simple.

---

[1] or, rather, "pre-material". This was suggested by E. Lazslo.

If one goes, e.g., into the mathematical details of quantum field theory (not to mention quantum gravity) one soon observes that the so-called "vacuum state" is an extremely animated object of enormous internal complexity. In a certain sense one can say with J.A. Wheeler that ultimately "vacuum physics" lies at the core of everything [46]. But I think that one can even observe its footprints within the regime of classical macroscopic physics. To give an example, the well-known fact that a moving clock goes slower as compared to a clock being at rest, which, in special relativity, is somehow attributed to the nature of space-time as such, is in our opinion rather a dynamical effect of the surrounding vacuum on the internal structure of the clock.

A lot of other examples could be given at this time if we had unlimited space at our disposal, some of which will be interwoven with our treatment in the following. Still a great mystery is, e.g., the so-called "undistinguishability" of micro-objects, belonging to the same class. It drove J.A. Wheeler to such rather extreme ideas as to claim that *there is only one electron in the world* (this being, by the way, almost the doctrine of the Upanishades, see, e.g., [48]).

The so-called "Schrödinger Cat Paradox" may be mentioned, (for a recent account see [49]), the origin of the "arrow of time", the "nature of matter", the question of how and where conservation laws are implemented or encoded in the microworld, etc. It is our aim to develop a systematic approach which leads right into the inner core of this bundle of problems.

6 Information

The concept of "information" has been in use in, e.g., physics for quite some time, but never played a really fundamental role (cf. the book of L. Brillouin, [50]). One of the reasons may be its too narrow definition, which was more or less based on the so-called "Shannon entropy". Intuitive feeling tells us that this cannot be the whole story.

Meanwhile, wider approaches do in fact exist, one running under the heading "semantic information" (introduced, as far as we know, by Carnap and Bar-Hillel, [51]) and which is designed for the analysis of formal languages. Another, in our view very promising concept is the "algorithmic information" which is closely related to another fundamental concept, which may play an ever increasing role in the future, namely the notion of "Complexity". One of its advantages is that it enables one to tackle in a rather natural way a variety of topics, such as the problems raised by "Gödel's incompleteness theorems" (a very readable account is [52] or [53]).

Nevertheless, we are not entirely sure whether one has already isolated the really crucial part of this (perhaps very deep) concept in the above notions. We instead have the impression that they represent at most some aspects of a still not particularly well defined core concept. It may well be that a whole hierarchy of levels of information exists which parallels the observed levels of "onto-logical complexity" in nature. That is, we would, for the time being, prefer to call them partial implementations of an underlying and possibly universal "super-concept".

To prevent, however, endless debates about whether something happening in nature is rightly considered to be a case of information processing, or could, on the other hand, be ultimately dissolved into simpler pieces via some sort of "reductionism", we will try to circumscribe in a tentative way what we consider to be the crucial and irreducible ingredients of the idea of "information", at least in the way we want to exploit it in the following.

(i) Ontologically Reducible Information (R-I): Science is presently mostly dealing with what we would like to call the ontologically reducible aspect of information. From this point of view it belongs to the same class as the other concepts mentioned above. That is, these notions are to a large extent a reflex of what Mach called "Denkökonomie" (economy of thinking).

In other words, while being of considerable scientific and conceptual value, they are, so to speak, purely mental (man-made) entities. At least in principle they could be resolved into a sequence of more elementary building blocks of a different character (depending on the degree of "compression" which has been necessary in order to arrive at these usually highly integrated ideas). Some examples: "irreversibility" (at least in the classical sense) does not exist in the microworld, "chaotic" systems usually behave deterministic-ally on a scale of sufficiently high resolution, etc. To put it in a nutshell: (R-I), while having a "pragmatic" quality, has no really ontological quality. The concept is "reducible".

(ii) Ontologically Irreducible Information (IR-I): This is a really subtle and delicate concept, and its possible existence may, for the time being, be considered as a mere working hypothesis. While positivistic and reduction-istic science will probably deny its very existence or try to eliminate it, we will advocate this concept. That its conception does not remain in a comple-tely shadowy state will, we hope, become clear in the following sections. For the moment it will suffice to give a rough outline of its semantic content.

Its crucial attribute will be that it is, in a sense which will be made more precise in the following, standing in opposition to, or complementing, the notion of "matter-energy (ME)", i.e., it will be an "*immaterial*" but nevertheless "*ontic*" entity. In one of its appearances it has the capacity to act upon (ME) and vice versa, being, however, in a sense more precise, of a more primordial character than (ME). In another, perhaps less fundamental manifestation, it makes use of the objects of the (ME)-world as vehicles of transportation.

In the latter sense "elementary particles" may, e.g., be viewed as certain compartments, carrying and exchanging pieces of information, being condensed in symmetries, quantum numbers, etc. On a much higher and more unfolded level the same applies to the "genetic code" of living organisms. While it will become self-evident in the following that on the fundamental level of physics, "information" (if it exists at all) is an irreducible entity, this is much less obvious in the higher storeys of evolution.

Remark: In various respects we see connections with the "world-3-concept" of K. Popper (see [19]).

7 On the Way to a "Subquantum Theory" (Intermediate Magnification)

We now come to the two central sections of this investigation. We plan to transcend the world of ordinary physical phenomena in two steps which correspond to two levels of resolution of space-time. But before we embark on this endeavour, we would like to make a personal remark. What will be presented in the following is for the most part not a review or resumé of what is presently thought about this subject matter within the scientific community.

Unless otherwise stated, it is more or less an outgrowth of a roughly 12-year long research process of the author. While some of the material has already been made public in a preliminary report, which was circulated a couple of years ago [54], or can be found scattered over [55], most of it has not yet appeared in a coherent form but will be published in the near future, or is in preparation. We are mentioning this as a measure of precaution, in order to reserve, so to speak, the copyright for the to a large extent still unpublished results and ideas being expounded in the following (unless we tool them from other sources).

As a first step we want to argue that ordinary quantum theory (being presented, for simplicity reasons, largely in its non-relativistic version) has to be understood as a "coarse-grained" "effective" theory of a more fundamental and largely hidden substructure, the erratic details of which have been averaged out and integrated over on the scale of resolution of our ordinary continuous space-time. (At this time, we want to remind the reader of the various sources, taken both from philosophy and physics proper, we cited in the previous sections, and which belong to the wider context of what we have in mind).

On this intermediate level of magnification the concept of information is still lurking over the horizon. In the next step we will go over to a larger magnification which puts us in the regime where the discreteness of space-time becomes visible and where this concept acquires its primordial and irreducible character.

As a last remark we want to make a comment on the mode of representation of our ideas. There is no question that they are to a large extent highly speculative and go far beyond the regime which is presently within experimental reach. Nevertheless, our personal point of view is, that they represent a legitimate extrapolation into the still hidden substructure of our quantum world. Furthermore, taking into account the enormous technical complexity of the subject matter and that the "audience" is of an inter-disciplinary character, the unfolding of the texture of our ideas, hypotheses and arguments can only be rather informal.

We start with non-relativistic quantum mechanics. Examples (i) and (ii) of Sect. 5 suggest the idea that there may be a hidden actor in the game who is not of matter-energy type since, otherwise, he would not have escaped complete notice for such a long time.

An important role in this respect may, furthermore, be played by the preoccupation of modern physics with energetic and material processes (the psychological reasons for which we have alluded to in the introductory sections). As a general saying goes (it goes, in fact, back to Heraclitus): *"One will only find what one is already expecting."* Present-day physics lacks completely the concepts to adequately deal with "immaterial" processes, in contrast to which we want to mention the interesting characterisation of the concept of "reality" introduced by Popper in [5], the definition of which also explicitly embraces immaterial "objects". (As to the problem of "quantum ghost fields", see [56]).

It has been common for quite some time in certain "unorthodox" approaches to quantum mechanics to split the "Schrödinger equation" into two coupled equations (mainly in order to support some "realistic" and "hydrodynamical" picture). Our personal rationale will however be rather different. For references see, e.g., [54] or more recently [57]. With

$$R, S \text{ given by } \phi = R \cdot \exp(iS/h) \tag{7.1}$$

the (for simplicity one-particle) Schrödinger equation

$$ih \, \partial_t \phi = - h^2/m \nabla^2 \phi + V\phi \tag{7.2}$$

(∇ denoting the gradient, V the external potential, ϕ the wave function) becomes

$$\partial_t S + 1/2m \cdot (\nabla S)^2 + V - h^2/2m \cdot \nabla^2 R/R = 0$$
$$\partial_t R^2 + \nabla(R^2 \cdot \nabla S/m) = 0. \tag{7.3}$$

The structure of (7.3) is interesting and suggestive in several respects. The second equation is nothing but a "continuity equation" for "matter", the density of which is R^2, the corresponding "current" $R^2 \cdot \nabla S/m$. In this representation ∇S represents some "external" driving force which the "environment" exerts on the "micro-object" (entering via R^2).

Perhaps more revealing is the structure of the first equation. With $h \to 0$ (i.e., in the "classical limit"), one arrives at the "classical Hamilton-Jacobi-equation". Therefore the name "quantum potential" was coined by (as far as I know) Bohm et al. for the quantity

$$Q = - h^2/2m \nabla^2 R/R. \tag{7.4}$$

This led Bohm et al. to develop a model theory based on some sort of "statistical hydrodynamics". Also, in this context should be mentioned the approaches of d'Broglie ("Double Solution") and E. Nelson and others ("Stochastic Mechanics") (see, e.g., [58] or [59–61]), the latter based on a picture borrowed from "Brownian Motion". Without going into the subtle and intricate details, we think our approach deviates on this intermediate level of resolution in at least one (however, crucial) respect.

As far as we can see, in all the above mentioned model theories the environment, in which the particle is somehow floating, is only a "passive medium" (if it is granted a more than formal existence at all). That is to say, it either guides the moving particle or it acts like a "Brownian medium" in that it imposes some, however completely uncoherent, uncorrelated, stochastic "force" on the micro-object. In mathematical terms: the motion of the particle is of a "Markovian type".

In this respect our point of view is radically different. Since in our treatment this level of "Stochastic Quantum Mechanics" serves only as an intermediate step into a regime of even higher resolution of space-time, more interesting aspects of this "medium" come into focus. A main point, which is missing in the above approaches, we consider to be the dynamic, non-linear nature of the vaccum. What, in our view, Eq. (7.3) does already exhibit on this coarse-grained level is a non-linear interplay of the micro-object (which we regard to be, as d'Broglie, some sort of singularity) and the surrounding medium.

In this process the particle is driven by the "deformation pattern" of the vacuum and, on the other hand, deforms the medium via its erratic motion (this has been developed in considerably more detail in [54]). As the "back effect" upon the medium is considered to be of principal importance by us, the motion of the micro-object itself will no longer be "Markov" but rather "Non-Markov" (!), which is mathematically a quite profound change and makes the technical analysis very complicated. On the other hand we conjecture that, e.g., the intricate interference patterns, which are typical for quantum theory, can only be understood in an ontological sense if quantum theory is ultimately regarded as a "projection" of a more fundamental theory in the above sense, i.e., a **non-linear interaction** of a "stochastic field" with a "stochastic singularity", the reflex of this field being encoded on this intermediate level in the "phase field" S or VS.

Reminding the reader of what we have said in the previous sections, e.g., about the "aether", the following conjectures suggest themselves:

(i) Our idea of a "dynamically active vacuum or medium" and the importance of the "back-reaction" effected upon it by the distribution of matter strongly suggests that a "subterranean linkage" to the general theory of relativity should exist, a point we will promote in the following.

(ii) The role played by S or VS in our treatment has some resemblances to the (still quite formal) "gauge concept" of modern gauge theories. It has been known for some time that gauge fields are structurally equivalent to what is called a "connection on a vector bundle" in differential geometry (as to this mathematical background cf., e.g., [60, 61]). That is, its geometrical implications are apparent in a rather formal and not yet fully understood ontological sense. We have a lot to say about these "connections" which shall, however, be done elsewhere. In any case, we think that ordinary quantum theory is already a "proto-gauge theory" and that a gauge-like significance can be attributed to the phase field.

(iii) As an illustration that our concept of an active vacuum is not a mere ad hoc hypothesis, we want to argue that, in our view, already on the coarse scale of ordinary quantum mechanics some of its characteristic attributes can be inferred via a specific "thought experiment" which was already expounded by us in Sect. 2 of [54]. It acquires its power through a combination of the "principle of relativity" and an application of the "uncertainty relations". It roughly runs as follows: the general saying is that "virtual quantum processes or particles" cannot be observed since (and as long as)

$$\Delta E \cdot \Delta t < h, \quad \Delta p \cdot \Delta x < h \tag{7.5}$$

with ΔE, Δp the corresponding energy-momentum fluctuation, measured, somehow, against the background of the vacuum as such, Δt, Δx being "lifetime" and "extension" of the fluctuation in space. Assuming then that such "virtual excitations" already exist in the pure vacuum we showed that, as the above relations are not "Lorentz-invariant", an observer moving sufficiently fast such that

$$\Delta E' \cdot \Delta t' > h, \quad \Delta p' \cdot \Delta x' > h \tag{7.6}$$

will be able to detect and locate this virtual excitation, i.e., for him it acquires some sort of *real existence*.

On the other hand, this would violate the very idea of special relativity, namely the equivalence of all inertial (Lorentz) frames. Furthermore there has been, up to now, no experimental, empirical evidence for this vagueness of the notion of "reality".

Our solution of the problem was the following: the so-called vacuum is already densely packed with all sorts of virtual excitations, the fluctuation pattern being such that the principle of "Lorentz invariance" is matched, i.e., the distribution of excitations with respect to E, p is a Lorentz invariant function. That is, each observer sees the same vacuum structure (for more details see [54]).

Conclusion. Via a thought experiment we infer that the vacuum is in fact a Lorentz invariant medium with (quite erratic) energy-momentum fluctuations on a sufficiently small scale.

Putting these pieces of evidence together (and quite a few more which we cannot mention at this point) we are able to develop a certain model theory which has the features alluded to in (i) to (iii) as *natural attributes*. As the technical details are, however, quite intricate we will only give a rough idea of what we have in mind. Furthermore, the model serves as a preparatory step on the way to a really fundamental and irreducible theory with "information" as an ontological core concept.

In forming our model theory, two strands of ideas will be intertwined. (i) The first strand consists of the features, in a certain respect paradigmatic, of general relativity, namely that it is the only (even classical) theory which is built around the principle of interaction between matter and space-time itself, and which is boiled down to the harmless-looking equation:

$$G_{ik} = -\kappa T_{ik} \qquad (7.7)$$

The lhs is related to the dynamical and geometric properties of space-time, the rhs to the distribution of matter-energy.

(ii) The other line of ideas could, at first glance, hardly be farther apart from (i). It is the "theory of plasticity of deformable materials" and the so-called "continuum theory of dislocations in solids". As the work of, e.g., Kröner and others has shown, these concepts have their natural expression within the mathematical framework of "Riemannian Geometry" (see [64, 65] or also [66, 67]).

In a quite telegraphic style our ideas are the following (to be published in more detail elsewhere):

– at this level of magnification or resolution of space-time (anti-)particles appear to be regions of *stable* plastic deformation of the medium (called vacuum), these regions probably having a complicated internal pattern.
– like dislocations, these in some respects singular regions are the centers of specific (*non-plastic*) *elastic* deformation structures in the outer space. Via these deformation patterns (which are in some sense elastic deformations of space-time itself) the centers of plasticity interact with one another.
– as argued in point (iii) above, the whole structure behaves in a pronouncedly stochastic way with strong fluctuations on small scales.
– we have reason to believe that seemingly irreducible phenomena such as special relativity as a whole can be naturally derived from this model!
– the same holds for the "gauge concept" and its "geometric" implications.

Remark: To develop this model theory in full a whole bundle of aspects and consequences, most notably the numerous implications as to "ordinary" quantum theory, have to be analysed, which, however, has to be done elsewhere as it goes far beyond the scope of this study.

8 Probing into the Substructure of our Physical World (Extreme Magnification)

Switching now to a magnification which is supposed to be several orders higher, the picture will change radically (this is at least our personal point of view). While on the one hand our usual concept of *continuous space-time* will break down to some extent, more and more of its rich internal structure will on the other hand be revealed.

Whether this fundamental length scale will in fact be given by the so-called

$$\begin{aligned}
&\text{Planck-length} \quad l_p := (hG/c^3)^{1/2} \quad 1.6 \ 10^{-33} \, \text{cm} \\
&\text{Planck-time} \quad \ t_p := (hG/c^5)^{1/2} \quad 10^{-43} \, \text{sec}
\end{aligned} \qquad (8.1)$$

is, in our view, not entirely clear. In any case, the resolution of space-time will be (and has to be) so extreme that particles (in Wheeler's words) have the extension and appearance of clouds in the sky.

On this level of magnification space-time will (probably) appear as a highly interconnected and facetted, complicated structure of grains, and their mutual connectivities as "elementary" building blocks.

These grains, however, will be endowed with their own (probably also quite complex) internal structure, i.e., they are able to take on various internal states, depending on the states of their "neighbours" and (possibly) also on grains which are much farther away. In a sense we would not be extremely surprised if this complex hypernetted organism has in fact some resemblances to the human brain, which seems to be placed just on the opposite side of evolution!

Before we go further, some remarks are in order as to topics which we cannot deal with in the following but which have nevertheless some bearing on this subject matter.

(i) We are not convinced that the process of probing into the deeper layers of "nature" really stops here. In the same sense as in the higher storeys of evolution there may exist a whole hierarchy of levels of "complexity" beneath this layer. What seem to be grains on this level of magnification may actually be the harbingers of an even finer and more intricate substructure. By the way, this substructure will not even be 4-dimensional! The observed 4-dimensionality of our ordinary continuous space-time should instead be viewed as sort of a "collective" and "coherent" effect in the same way as quantum observables become approximately classical under appropriate conditions.

(ii) In a sense we can rightly regard these grains as "monads" in Leibniz's terminology. We are presently developing a mathematical theory which tries to combine concepts of "non-standard analysis" and advanced geometry in order to deal with such complex and erratic structures.

In order to have, however, some model at our disposal with the help of which we can get hold of our ideas in a preliminary manner, while neglecting the perhaps finer but largely unknown details, we would like to consider it, for the time being, as some sort of "cellular automaton". (As to this relatively recent field of research – at least as far as physics proper is concerned – see, e.g., [68, 69]. Also, the far-sighted remarks of K. Zuse in [35] are of interest. There seem to be even some applications in "ordinary" quantum field theory, [70], and "cosmology" – cellular automata are also called "mini-universes"! What makes them rather fascinating is their capacity to create highly complicated patterns by means of extremely simple mechanisms from virtually "nothing".)

As to our purpose, they have the great disadvantage of being, in principle, entirely "deterministic" machines. That is, they consist of a certain grid with a "minicomputer" at each site of the grid which can adopt a (usually finite) number of internal states. There exists a fixed algorithmic procedure by which after each elementary "time step" every grid point condenses the pattern of internal states of its neighbours into a simple piece of information which allows it to adapt its own internal state.

The cellular automata in current use have only a small number of admissible internal states (sometimes only two), very simple processing rules and interactions between grid points of more or less "nearest-neighbour type". These features and the "rigidity" of the grid structure as such have to be considerably extended and generalized in order to be able to serve as realistic model theories of fluctuating space-time. On the other hand, several concepts can already be approximately implemented within this framework.

Metaconcept: the World as a "Cellular Automaton"
  (i) Points of the space-time continuum actually have a rich internal structure like the "infinitely small neighbourhoods" of "non-standard analysis".
 (ii) Certain specific and stable extended structures within this (to give it a fancy name) "hyle" or vacuum can be interpreted and are macroscopically experienced as "matter". That is, they are regions or compartments where a specific overall collective state of the individual cells is maintained.
(iii) These regions are surrounded by and embedded in a fluctuating and dynamic "deformation pattern" which moves and spreads out in space, thus affecting the behaviour of the other compartments (stability is, as in, elementary particle physics only a relative concept!). The stable regions correspond to what we called "centers of plasticity" in the previous section.
 (iv) "Information" is, however, processed *both within the particles and in the exterior space surrounding them*. The information transport in the outer space may be linked to the concept of "phase-" or "gauge fields".
  (v) We conjecture that some (if not all) of the so-called elementary constants of nature such as h and c will play a natural role in this game or reveal their true nature on this level.
 (vi) It would not come as a surprise to us (we have in fact rather concrete ideas as to this point) if some portion of the information processing would "violate" the axioms of special relativity, i.e., what is called somewhat misleadingly "causality"! As to this, what is really embraced by this notion has to be analysed. Evidently, modifications in this direction will have a strong impact on many of the so-called paradoxes of quantum theory.
(vii) This possibility of some "faster than light" information processing may be attributed to certain "non-local connectivities" between regions of space-time as is, e.g., also the case between certain neurons in the brain.
(viii) One of the reasons why this possibility does not openly come into conflict with "Einstein causality" lies in the quite restricted framework of "matter-energy processes" as it is conceived in orthodox physics. As compared to this an entirely different picture emerges from our considerations.

Hypothesis about "Monism" and "(Pseudo-)Dualism"
  (i) Matter-energy is *not primary but rather secondary and derived*. Matter is a phenomenological attribute of regions of space which cannot be resolved

on the level of ordinary space-time physics and appear in that framework more or less as singularities, being embedded in space while they have actually a peculiar internal structure maintained by a collective coherent interaction of the grains belonging to the region itself and to a certain neighbourhood.

(ii) What appears to be really *primary* on this level of high resolution is the processing of information. This happens both *within the "particles" and in the vacuum itself.* In this sense it has both *immaterial* (purely structural) and *material* aspects. In the former manifestation it is not necessarily governed by the principle of "Einstein causality"! It should, however, be noted that this *"subterranean"* information transport is not necessarily available to *macroscopic* human, beings.

(iii) In this sense something *"premateriaI"* may act upon **material** objects and vice versa while it is, on the other hand, more fundamental.

Remarks: (i) Some of the possible consequences of these ideas for our "ordinary" quantum world shall be addressed in the last section.

(iv) Other hypothetical systems of a more fundamental theory of space-time-matter also exist. Apart from the various approaches of, e.g., "quantum gravity", we want to mention the ideas of v. Weizsäcker (as expounded in [3] p. 264ff.) and the "twistor concept" of R. Penrose ([7.1]). At the moment however it is not clear to us in what sense (if there is any) they are related to our own approach.

9 Unfolding and "Emergence"

What we have discussed so far concerns the inner core of fundamental physics or natural philosophy in general. One has to make a further great endeavour to study and analyse the consequences of our "first principles" in the more developed and unfolded layers of the scientific hierarchy, which are, typically, many orders of magnitude apart from each other. To mention only a few topics:

  (i) fundamental constants
 (ii) causality.
(iii) the geometric nature of the 'gauge principle'.
(iv) the "indistinguishability" of elementary particles,
 (v) "quantum numbers" and
(vi) forces and interactions.

Another extremely important topic in its own right is the metaconcept of 'emergence', the classification of "universality classes" of complexity in nature and the notion of "downward causation". It is exactly here where, e.g., Gödel's and Turing's results about the (non)reductionistic nature of certain epistemological systems enter. More about these concepts together with an attempt to develop a quantitative theory can be found in [72].

References

1. Haken H (1978) Synergetics, Springer, Heidelberg
2. Küppers BO (1986) Der Ursprung biologischer Information, Piper, München
3. Weizsäcker CFv (1971) Die Einheit der Natur, Hanser, München
4. Küppers BO (1987) Leben = Physik + Chemie Piper, München
5. Popper K, Eccles J (1985) The Self and its Brain, Springer, New York
6. Feigl H, Scriver M, Maxwell Gl (eds.) (1958) Concepts, theories and the mind-body problem, Minnesota Studies in the Philosophy of Science, vol 2, Minn Univ Pr
7. Küppers BO (1988) On a fundamental Paradigm Shift (Preprint)
8. Newton I (1952) 'Opticks', Dover, New York
9. Newton I (1963) Mathematische Prinzipien der Naturlehre, Wiss Buchges, Darmstadt
10. Leibniz GW (1966) Leibniz-Clarke-Korrespondenz (Cassirer Buchenau) Meiner, Hamburg
11. Leibniz GW "Neues System der Natur", and "Monadology, loc cit
12. Schelling FWJ (1857) Ideen zu einer Philosophie der Natur, Sämtl Werke I Abt Bd 2 Cottá, Stuttgart
13. Whitehead AN (1929) Process and Reality, Macmillan, London (Suhrkamp, Frankfurt 1979)
14. Plotin (1958) "Plotin", Fischer, Frankfurt
15. Requardt M (1982) Goethe und die "Anschauende Urteilskraft", in Goethe, Text u. Kritik, München
16. Bergson H (1964) Materie und Gedächtnis, Fischer, Frankfurt
17. Portman A (1963) Biologie und Geist, Herder, Freiburg
18. Chargaff E (1979) Das Feuer des Heraklit, Klett, Stuttgart
19. see [5]
20. Borst CV (ed.) (1983) The "Mind-Brain Identity", Macmillan, London
21. Mach E (1976) Erkenntnis und Irrtum, Wiss Buchg, Darmstadt
22. Glaserfeld Ev et al., in [25]
23. Fierz M (1975) C.G. Jung zum 100 Geb, Walter, Olten
24. Polanyi M (1966) The Tacit Dimension, Doubleday, New York (Implizites Wissen, Suhrkamp, Frankfurt 1985)
25. Schmidt SJ (ed.) (1988) Der Diskurs des Radikalen Konstruktivismus, Suhrkamp, Frankfurt
26. Jung CG, Pauli W (1952) Naturerklärung und Psyche, Rascher, Zürich
27. Schopenhauer A (1977) Züricher Ausgabe, Parerga und Pavalpomena, Diogenes, Zürich
28. A remark by Dirac PA (1979) made at the Jerusalem Einstein Centennial Symposium
29. Zermelo E (1968) Cantors Werke, Hildesheim
30. Einstein A (1969) Grundzüge der Relativitätstheorie., Vieweg, Braunschweig
31. Lee TD (1979) Is the Physical Vacuum a Medium?, Preprint Columbia Univ Press, New York
32. Finkelstein D (1969) "Space-time code", Phys Rev **184**, pp 1261–1262
33. Feynman R (1965) The Character of Physical Law, Brit Broadcasting Co, London
34. Wheeler JA (1981) "The Computer and the Universe", Conf on the Physics of Comp, MIT (and Int J of Theor Phys **21** (1982) p 557)
35. Zuse K The computing Universe, loc cit p 589
36. Selleri F (1983) Die Debatte um die Quantentheorie, Vieweg, Braunschweig
37. Neumann Jv (1932) Mathematische Grundlagen der Quantenmechanik, Springer, Berlin
38. Jammer M (1974) The philosophy of quantum mechanics, Wiley, New York
39. Wheeler JA, Zurek WH (eds.) (1987) Quantum theory and measurement, Princeton Univ Press, New Jersey
40. Bell JS (1982) On the impossible pilot wave, Found Phys **12**, p 989
41. Bell JS (1981) Bertelmann's socks and the Nature of Reality, J Phys (Paris) Coll **41**
42. Feynman R, Leighlon RB, Sands M (1965) Feynman Lectures vol III, Addison and Wesley, New York
43. Gieres F (1983) über den Aharonov-Bohm Effekt, Diploma thesis, Göttingen
44. Aharonov Y, Bohm D (1959) Phys Rev **115**, p 485
45. Möllenstedt G, Bayh W (1962) Naturwiss **49**, p 81
46. Ta-Pei Cheng, Ling-Fong Li (1988) Gauge theory of elementary particles, Clarendon, Oxford
47. Wheeler JA (1987) in "Quantum Cosmology" (Fang LZ, Ruffini R, eds.), World Science, Singapore

48. Schrödinger E (1989) as cited in Phys B1 **43**, 8, p 333ff
49. Requardt M (1989) The Schrödinger Cat Paradox, Preprint, Göttingen
50. Brillouin L (1953) Science and Information Theory, Academic Press, New York
51. Carnap R, Bar-Hillel Y (1953) Semantic Information, Brit J Phil Sci **4**, 147
52. Chaitin GJ (1982) Gödel's theorem Int J Theor Phys **21**, 941
53. Chaitin GJ "Randomness and Mathematical Proof", Sci Am May 1975
54. Requardt M (1981) The underlying substructure of Quantum Theory, unpublished report, Göttingen
55. Requardt M (1984) The quantum mechanical measuring process, Zeitschr Naturf **39a**, p 1147
56. Selleri F (1982) Generalized EPR-Paradox, Found Phys **12**, p 1987
57. Bohm D, Hiley BJ (1988) Non-locality and locality in the interpolation of quantum mechanics Phys Rep **172**, p 93
58. Broglie de L (1960) Non-linear Wave Mechanics, Elsevier, Amsterdam
59. Nelson E (1972) Dynamical Theories of Brownian Motion, Princeton Univ Press, New Jersey
60. Nelson E (1985) Quantum Fluctuations, Princeton Univ Press, New Jersey
61. Blanchard P, Combe Ph, Zheng W (1987) Mathematical and Physical Aspects of Stochastic Mechanics, Springer, New York
62. Choquet-Bruhat Y, Demitt-Moirelle C, Dillard-Blerck M (1982) Analysis, Manifolds and Physics North-Holland, New York
63. Drechsler W, Meyer ME (1977) Fiber Bundle Techniques in Gauge Theory, Springer, New York
64. Kröner E (1978) in Sommerfeld A, Mechanik der deformierbaren Medien, Chap 9, Harry Deutsch, Thun
65. Kröner E in "Les Houches" 1980, session XXXV (Balian R, ed.)
66. Nabarro FR (ed.) (1979) Dislocations in Solids, North-Holland, Amsterdam
67. Landau L, Lischitz EM (1975) Lehrbücher der Theoretischen Physik vol VII, Akademie Verl, Berlin
68. Wolfram S (1983) Statistical Mechanics of Cellular Automata, Rev Mod Phys **55**, p 601
69. Toffoli T, Margolis N (1988) Cellular Automata, MIT Press, Cambridge MA
70. t'Hoft G (1988) Deterministic and quantum mechanical systems. J Stat Phys **53**, 323
71. Penrose R (1972) "On the Nature of Quantum Geometry" in Magic without Magic, Freeman, San Francisco
72. Requardt M (1991) Gödel, Turing, Chaitin and the Question of Emergence, World Futures, the Journal of General Evolution, **31**, p 123

Inorganic Matter As One of Four Levels of Natural Information Processing

Hans-Werner Klement

1 Introduction 136

2 Four Levels of True Entity 136

3 Genetic and Atomic Codes 139

4 The Hydrogen Atom As an Information Processing System 144

5 Structures and Processes, Codes, Reduction and Emergence 147

6 Information Processing and Information Processing Systems 150

7 Conclusion 151

References 152

1 Introduction

This paper introduces an interdisciplinary model for the evolution of information processing. Due to the topical range associated with its interdisciplinary character, there are interfaces to a number of specialist scientific fields and disciplines. Although desirable, it naturally is quite beyond the scope of this paper to deal in depth with all these fields and disciplines; however, in many cases, it can provide food for thought which may lead to further investigation.

To this extent, one can regard this paper as a programme for further interdisciplinary research. Where the boundary line into the area of speculation is quite deliberately overstepped, even to a minor degree, this is done in the conviction that such vision is necessary in order to arrive at a theory for information processing, encompassing all the levels of evolution.

In this paper, the author refers to previous publications on the same subject since 1974 [1].

2 Four Levels of True Entity

A crucial point of this paper is the question whether "information" occurs in inorganic energy and matter, i.e., in the realm of physics. We are, of course, well aware these days that energy and matter are carriers of, for example, linguistic information in the field of intercommunication between people. However, this is not the subject of our investigation. The question under review here is whether information is also intrinsic in the inorganic world, in total independence of man, having already discovered its occurrence in the field of biology, in genetic code, and the associated processes of life itself. Statistical physics offers a starting point for such contemplation. The similarities between the Boltzmann equation for entropy and the Shannon definition of information have led to entropy being widely conceived of as negative information, and information as negative entropy (negentropy). Information would therefore occur during each process where entropy diminishes, i.e., where increased order occurs. We must keep this statement in mind. However, processes of statistical physics are not the actual subject of this investigation.

At a deeper level, there are also attempts to describe physical processes generally as information processing procedures, like the description put forward by Karl Goser in his article entitled "Das Gravitationsgesetz und das Coulomb Gesetz aus der Sicht der Informationstheorie" [2]. The concept implied in Goser's article is that information has been part of the evolutionary process since time began, e.g., like force, energy and effect. Here, it is not intended to take issue with this, although the starting point for our investigation is quite different.

We regard evolution as a process in which systems containing information have formed on four clearly distinguishable levels, namely on the levels of inorganic matter (atomic level), organic matter (cell level), on the level of living beings and on the level of the societies of living beings. Here, a lower level system always becomes an element for the formation of systems on the next higher level; the cell, for example, becomes the basic element for the formation of living creatures. This means that new orders are formed from existing orders. Therefore, as regards the formation of the real world in layers or levels, we hold similar views to those of Hartmann [3], Lorenz [4], Riedl [5] and others. The lowest level – that of the atoms – already contains information: this is our central statement in answer to the question posed at the beginning. Before we substantiate this statement, Fig. 1 will illustrate our model of the four levels of true entity.

On the first level we find energy quanta as elements. Initially, they form into the "elementary particles" of matter, namely lepta and quarks; and, in analogy with higher levels, it is possible to refer to groups of elements when these elementary particles are seen as wave packets. From lepta and quarks, on the one hand, the nucleus of the atom with protons and neutrons is formed, and on the other hand, the electron shell. We consider the nucleus of the atom as a carrier of information in the same way that, one level up, the nucleus of a cell carries information, whilst the electron shell is seen as the functional carrier responsible for the specific chemical characteristic of the atom. The decisive factor now is that there is a code operating between the nucleus of the atom and the electron shell which is based upon the Pauli principle. To each atomic number Z belongs a determinant $\psi_i(Z)$ of possible quantum conditions, which

LEVEL	ELEMENTS	GROUPS OF ELEMENTS	INFORMATION VEHICLES	FUNCTION VEHICLES	CODES	SYSTEMS (STRUCTURES)	SPECIFIC QUALITIES
1	ENERGY	QUARKS, ELECTRONS	ATOMIC NUCLEI (PROTONS, NEUTRONS)	ELECTRON SHELLS	PAULI-PRINCIPLE, ATOMIC CODE	ATOMS	QUALITIES OF MATTER
2	ATOMS	MOLECULES	CELL NUCLEI (DNS, RNS)	AMINO ACIDS, PROTEIN	GENETIC CODE	CELLS	LIFE
3	CELLS	GROUPS OF CELLS	BRAINS	ORGANS	NEURO-PHYSIOL. CODE	LIVING BEINGS	CONSCIOUSNESS
4	LIVING BEINGS	GROUPS OF LIVING BEINGS	LEADING INDIVIDUALS	POPULATIONS	LANGUAGES	SOCIETIES	COMMUNICATION

Fig. 1. The four levels of evolution and its characteristics (DNS = DNA, RNS = RNA)

the electrons can take up. The nucleus and the electron shell form the first level "information processing system" called atom, which simultaneously is also a building block element for second level systems.

It should be noted here that we reserve the term "matter" for the levels from atoms or chemical elements upwards. Elementary particles are, in this context, not yet matter; they are just that, elementary particles: building blocks of what we refer to as matter. This is to provide terminological clarity which we will return to later.

Our illustration shows what happens once atoms have formed. Among other things, one can see that a certain pattern is repeated, clearly demonstrating a universal principle of formation in nature: from individual elements, like atoms, organisations of these elements are formed, such as molecules. From these organisations, information carriers are developed, such as the cell nuclei, as well as functional carriers; together, these constitute a system at the next higher level. Systems formed in this manner then become building block elements in their turn for the next level up; the cell, for example, becomes a building block element for the level of living beings.

On each of these levels, there is an operative code. At the level of society this is language; in the case of human society, it is human language in all its expressions and manifestations. At the level of individual living beings, it is the still largely unknown neurophysiological code. At the cell level, it is the genetic code, and at the atomic level, it is the atomic code, as already mentioned. The latter is to be examined more closely here.

As far as the relations are concerned between the laws governing the higher relative level and the lower relative level, we are adopting the theory of emergence as our basis. In this book entitled "Physik und Evolution" [5], Krueger wrote on this subject: "The laws G' of each higher relative level of entity (e.g., of biology), in other words, of entities of a greater complexity, emerge entirely from the laws G of the previous lower relative level of entity (e.g., of physics), without any reduction to that lower level being possible. ... The laws G still continue unchanged in G', but G' now generally imposes on the system a narrow classification of marginal conditions (qua parameters of order), which leads to an enslaving (slaving principle according to H. Haken 1979 [7]) of the so-called micro-variables determined by G. Due to this slavery, i.e., imposition of narrow marginal conditions, the effects of the laws G are hardly recognisable; in this "higher" system of G' they lose their empirical significance. Rather, the G' now determine the event until a higher order reduces the relevance of G' in favour of G"."

Emergence of such new laws, new orders, is based on the occurrence of branching, i.e., at so-called bifurcation points branches in the possibilities of system development occur. It is not possible to predict which of the given possibilities will be realised; to a large degree, this is determined by secondary factors, even pure chance and on-line phenomena. Prigogine [8] has demonstrated that such possibilities of "branching" can occur far from thermodynamic equilibrium. He explained why, despite the second law of thermodynamics, higher orders can be formed, i.e., why evolution generally is possible and our world does not tend unrestrictedly towards the highest level of entropy, maximum chaos. Detailed discussion of these aspects is beyond the scope of this paper. One should refer to the relevant literature. I should like to mention the book "Evolution" by Laszlo [9]), which also covers bifurcation at higher levels and the evolution of society.

3 Genetic and Atomic Codes

Figure 1 illustrated that on four levels of evolution, similar patterns can be recognised. For example, we consider the atom and the cell to be systems subject to a common principle of formation. Both systems have a nucleus and, as far as the nucleus of the cell is concerned, we know that genetic information in the form of DNA is stored there. This information controls the biological "behaviour" of the cell and the structure of the system on the next higher level, the level of living beings. This involves complex processes, and when we speak of the genetic code, it does not just concern a table in which amino acids are allocated to the so-called codons. To the extent that these processes have been understood, they are assumed to be generally known; in the following, therefore, we will remind ourselves only of the steps central to what is being considered in this paper (see also Fig. 2).

1. By means of an enzyme, the RNA messengers (m-RNA) and the RNA transfers (t-RNA) are synthesised on matrices of the DNA stored in the nucleus of

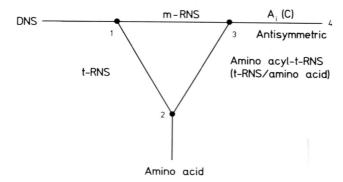

Fig. 2. Genetic information processing, finally producing proteins

the cell. DNA represents the legislative language of the genetic code by the "letters" A (adenine), G (guanine), C (cytosine) and U (uracil) or T (thymine). For a given code, there are always two languages that can be distinguished and which translate into one another. It is possible for one of the languages to set the rules, whilst the other contains the relevent implementation instructions. They are therefore referred to as "legislative" and "executive" languages. The former is the one under review here. The m-RNA is present in the form of a chain in which "words" of three separate letters are strung together, which are denoted as codons. The t-RNA consists of individual molecules which, at a certain position, contain the anti-codons also consisting of three letters.

2. By means of a catalytic process one in each case of the t-RNA molecules and one of 20 amino acids are combined to form aminoacyl-t-RNA molecules. This process is of decisive importance, because in this occurrence, a letter of the executive language, an amino acid, is assigned to a word of the legislative language of the genetic code, to an anti-codon. In this way, the anti-codon or the amino acid now contains a "meaning". It is therefore possible to state that the semantic aspect of genetic information becomes recognisable here. The basic key to this allocation is the subject of current research.

3. In the ribosome, the amino acids are then brought into a sequence by means of the ability of codons and anti-codons to recognise each other. A polypeptide chain is formed, a text of the executive language of the genetic code. We can talk here of the syntactic aspect of genetic information. We can form a set function A_1 (C) for the recognition process, in which A_1 stands for the anti-codons and C for the codons. It is antisymmetric.

4. Complex laws apply to the further "behaviour" of the cells during the formation of a living being, which are not yet fully understood. What we can say, however, is that the pragmatic aspect of genetic information can be recognised here, but that the relevant findings are still incomplete.

Using Fig. 3, we will now attempt to discover for atoms conditions similar to those that exist for cells.

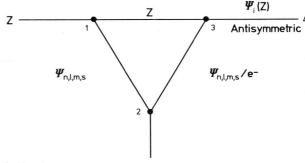

Fig. 3. Information processing in the atom

1. We imagine first of all that, similar to what happens in cells, atomic information is stored in the nucleus of the atom. On the one hand, the atomic charge Z, which we refer to as a messenger, implies a field. On the other hand, the Pauli principle ensures that there are discrete quantum conditions in this field ψ_i (Z), which can be occupied by electrons and represented in the form of Slater determinants. The individual quantum conditions $\psi_{n,l,m,s}$, now characterised by the quantum numbers n,l,m,s, are considered to be transfer elements, words of the atomic code legislative language, corresponding to the anti-codons.

2. Occupation of any of the quantum conditions $\psi_{n,l,m,s}$ by an electron then corresponds to the catalytic process of the genetic code. Even here, there is "meaning", and the electron in a certain quantum condition becomes the letter of the atomic code executive language. One deviation between the genetic and atomic codes is that for the genetic code there are 20 different amino acids as letters of the executive language, whilst for the atomic code indistinguishable electrons, by the occupation of a quantum condition $\psi_{n,l,m,s}$, first become distinguishable and a letter of the executive language at the same time. However, this does not appear to be a major difference.

3. The text of the atomic code executive language is formed by the occupation of the complete electron shell, in other words, of all conditions ψ_i (Z). The function ψ_i (Z) is anti-symmetric like the previously mentioned set function for the genetic code.

4. The text of the atomic code executive language, i.e., the structure of the electron shell, determines the chemical behaviour of the atom. But not just the so-called outer shell electrons (valency electrons) are responsible for this. If this were so, carbon and silicon, for example, would behave in an identical fashion and we would perhaps have some sort of organic chemistry for silicon. In order to obtain a really complete and full theory on chemical valency or bonds, it is probably necessary to achieve a better understanding of all the texts of the atomic code executive language. Thus, as in the case of genetic code, there is further work to be done – above all, with respect to the pragmatic aspect of information.

In the following paragraph, we will again compare what we have found with respect to the atomic and genetic codes.

	Atomic code	Genetic code
Letters of the legislative language	Possible vlaues of quantum numbers n,l,m,s	Bases A,G,C,U,T
Words of the legislative language	Combinations of the 4 quantum numbers	Triplets (codons, anti-codons)
Letters of the executive language	Quantum conditions ψ_i occupied by electrons	Amino acids

We can describe the atomic code as follows:

Legislative language	Executive language
Letters: possible values of 4 quantum numbers $$n = 1,2,3,4,\ldots.$$ $$l = 0,1,2,3,\ldots.,(n-1)$$ $$m = -1,\ldots,-3,-2,-1,0,+1,\ldots,+1$$ $$s = +1/2,-1/2$$	
Words: combinations of 4 quantum numbers occurring just once in each case	Letters: conditions ψ_i of the electron shell, occupied by electrons
Sentences: combinations having the same main quantum number n, the same secondary quantum number l and the same magnetic quantum number m	Words: conditions occupied by electrons having the same main quantum number n, the same secondary quantum number l and the same magnetic quantum number m
Paragraphs: combinations having the same main quantum number n and the same secondary number l	Sentences: conditions occupied by electrons having the same main quantum number n and the same secondary quantum number l
Chapters: combinations having the same main quantum number n	Paragraphs: conditions occupied by electrons having the same main quantum number n

Here, we quite consciously use the scale model for the electron shell, in the same form that even today this is still being used to explain our periodic system of elements. In so doing we do not overlook, for atoms with $Z > 1$, what consequences the interrelations amongst electrons have for the electron shell structure. That is, we do not overlook Hund's Rule.

We will do without the Russel-Saunders-description as this directly gives the total atom condition, i.e., a "text" in our terms, when what we are interested in is a structural description of this text consisting of letters, words, sentences and paragraphs. In this, the Pauli principle plays the crucial part; and in order to represent this principle, we use the one-particle-code with quantum numbers n,l,m,s as elements of structural build-up rules for atoms, realising that this is a simplification of what actually happens.

To complete the atomic code, for each Z the total conditions of all electrons must be indicated, taking into account, firstly, the Pauli principle, and, secondly, the additional rules mentioned above. Here, the electrons lose their individuality, and the quantum numbers determine total atom condition, as stated above.

Thus, for atomic code executive language "texts", as expressed by the Russel-Saunders-description, the structural rules of these "texts" cannot be recognised. This is a further analogy to gentic code, for even the polypeptide chains as texts of the genetic code executive language do not carry the legislative language triplets.

Figure 4 should clarify the tabulated summarisation, depicting the atom of phosphorus.

Above, the difference has already been mentioned between the chemical behaviour of carbon and that of silicon. In the atomic code executive language the "text" of carbon has two "paragraphs", whilst silicon has three. In the case of silicon, a paragraph has been inserted, as it were (see Fig. 5). The similarity of the respective (first and) final paragraphs is the reason for the chemical

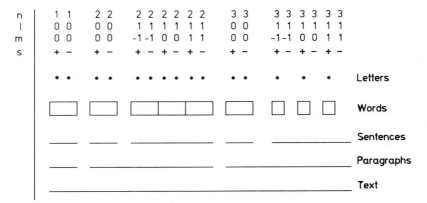

Fig. 4. Information at its various levels in the phosphorus atom

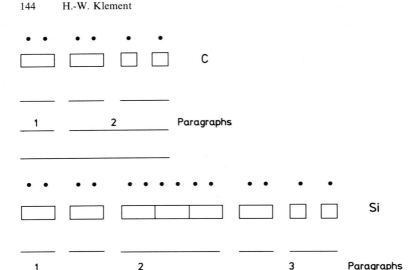

Fig. 5. Comparison of information in carbon and silicon atoms

likeness of both elements. The more comprehensive text of silicon is the reason for its different behaviour to that of carbon. Naturally, here we are not concerned with the discovery of these conditions, but with their somewhat unusual presentation. And we will quite expressly leave open the question as to whether this form of presentation can provide additional insights into chemical behaviour.

4 The Hydrogen Atom As an Information Processing System

If what we have considered in the previous section is correct, then the hydrogen atom was the first information processing system at the level of inorganic matter, and, if this was the case, formation of the hydrogen atom, the so-called re-combination of hydrogen, must have been associated with an entropic loss of electrons caught by the protons. In fact, cosmology teaches us that this process occurred at 3000 to 4000 K (the relevant literature is somewhat divided on this point), and that the ensemble of previously free electrons lost its entropy to cavity radiation [10].

The applicable formula is

$$\text{Temperature in K} \approx \frac{10^{10}}{\sqrt{\text{time in sec}}}$$

where time is the time elapsed since the big bang. At a cosmic temperature of 3251K around 300 000 years had elapsed since the big bang, and it is after this time that information processing systems first evolved on the level of inorganic matter.

During this process the ensemble of electrons emitted entropy (entropy is a term derived from statistical physics), whilst the individual atom gained information. Thus, the relation between entropy and information does not appear to be so straightforward that negative entropy can simply be equated with information, at least not here.

Generally, it seems to hold true that neither the reduction of entropy on its own, nor breaks in symmetry cause information to be created. The formation of symmetrically arranged frost crystals, for example, has nothing to do with the creation of information. The creation of information is obviously associated with the loss of symmetry and entropy at the same time [11].

In this connection I should like to repeat what has already been said in the first section: we are not using the term "matter" for elementary particles. Although these are building blocks for matter, matter as such only comes into being with the atom. There are three roots to its creation: the formation of rest mass and electrical charge, loss of symmetry and loss of entropy. With the atom, the first level of systems holding a form of information that may be defined as endogenous information is created. Figure 6 will illustrate what has just been said.

Within the context of our model, this implies more than just a linguistic convention. Although matter is not synonymous with information in this model, in principle, it holds information, which cannot be said of elementary particles.

In the following, we shall now attempt to make a statement on the quantitative information of the hydrogen atom. In doing this, we regard the electron

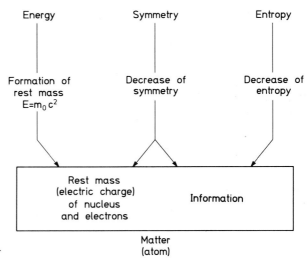

Fig. 6. Characteristics of matter

as a "symbol" and the possible conditions $\psi_{E,\eta}$ of an energy level E as a total of its configuration possibilities at this level. The information content of such a total entity can be expressed by the distribution of its *a priori* probabilities. Here, the *a priori* probability is the probability which is due to its condition, irrespective of the energy conditions in question; in our case irrespective of kT, k being the Boltzmann constant and T the absolute temperature. Although the *a priori* probability applies to a certain energy level, it expresses a principle of order, which exists independently of the actual energy conditions *a priori*.

The probability of measurement for the energy E of the electron inside the hydrogen atom is

$$W_E = \frac{g(E) \cdot e^{-E/kT}}{\sum_E g(E) \cdot e^{-E/kT}} \tag{1}$$

where k is the Boltzmann constant and T the absolute temperature and g(E) denotes the degeneration factor. It means that there are g(E) conditions of the same energy E. The probability that an electron is in a condition $\psi_{E,n}$ at energy E is

$$P_{E,\eta} = \frac{e^{-E/kT}}{\sum_E g(E) \cdot e^{-E/kT}} \tag{2}$$

One can interpret this probability to mean that it describes an ensemble of identical systems, indicating the proportion of systems which are found at a given point in time in the microscopic condition $\psi_{e,\eta}$. The same probability can also be interpreted to mean that it describes an individual system. In this case, it indicates the proportion of instances in which the system is found in the respective condition on frequent examination. We shall use this latter interpretation, since we are interested in the individual electron or hydrogen atom.

From equations (1) and (2) we can derive

$$P_{E,\eta} = \frac{1}{g(E)} \cdot W_E \tag{3}$$

The factor $1/g(E)$ in this equation represents a statistical weight, which is to be interpreted as *a priori* probability. We now express the mean information content H_E of a condition using the energy level E

$$H_E = - \sum_{\eta=1}^{g(E)} \frac{1}{g(E)} ld \frac{1}{g(E)}, \quad \sum_{\eta=1}^{g(E)} \frac{1}{g(E)} = 1 \tag{4}$$

whereby, in mathematical terms, we add up through marginal distribution. Assuming equal probability for all conditions using the energy level E, we can express the information content $I_{E,\eta}$ of a condition $\psi_{E,\eta}$ by

$$I_{E,\eta} = ld\, g(E) \tag{5}$$

The basic condition of the hydrogen atom (n = 1) in expressed by

$$g(E) = 2n^2 = 2 \qquad (6)$$

and therefore it is

$$I_{E, \eta} = ld\ g(E) = ld2 = 1 \qquad (7)$$

For atoms where Z > 1, the relevant sums have to be done.

5 Structures and Processes, Codes, Reduction and Emergence

Following our detailed study of the atom, the hydrogen atom in particular, as an information processing system, we will now return to our interdisciplinary examination of the four levels of evolution. Let us remind ourselves of Fig. 1. There, we defined the systems of the four levels as being atoms, cells, living beings and societies of living beings. Each of these systems has a special material structure, a specific code working at that level within each one of the systems, with numerous regular processes occurring in association.

For the cell, e.g., the synthesis of the m-RNA and the t-RNA (and incidentally for the r-RNA also), the catalysis of aminoacyl t-RNA and the synthesis of protein chains occurring in the ribosome should be mentioned.

One quickly realises that the structures of one level can be reduced to the structures of the next level down. Thus, in principle, it is possible to describe a cell by indicating the atoms and their structural arrangement, from which the cell is built, even if this would be a very complicated process.

In relation to processes, on the other hand, our modern knowledge of the emergence of laws applies. In particular, we are interested in the roles of the individual codes in this interplay of reducible structures and emerging laws. Fundamentally, of course, what has already been stated applies: the letters and words of both the legislative and the executive languages are reducible. It is possible, for example, to describe a triplet of RNA, a base or an amino acid by atoms arranged in a certain manner, and these, for example, by the orbits of their electrons. On the other hand, emergence plays a part in the relevant processes to do with a code, e.g., in the synthesis and catalysis processes mentioned previously when discussing the genetic code. It is obvious that codes and emerging laws jointly lead to the formation of reducible structures. Figure 7 is an attempt to depict this process, the full lines symbolising the evolution of reducible structures and the dotted lines the emerging laws.

The codes are an artifice which nature employs to transfer the order of a less complex structure such as perhaps the structure of the RNA to a more complex structure such as the protein chains, and thus to create not just a higher level but quantitatively greater order or information. This applies equally to the order of electrons within an atom $\psi_i(Z)$. The order thus created determines

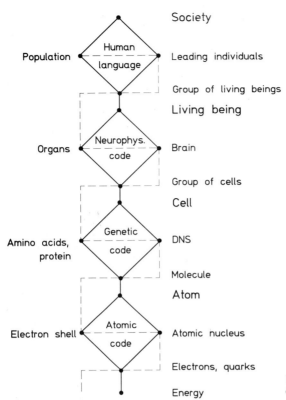

Fig. 7. "Coder" for the four levels of evolution of information processing

the behaviour of the respective system when forming the applicable next higher level of evolution. The electron shell, in general, determines the behaviour of atoms in the formation of molecules and, in particular, of molecules contributing to cell formation. The protein chain of a cell generally determines the behaviour within groups of cells and, in particular, within cell groups of a living being. In a further illustration, we now want to look at the systematic character of this raising of orders/arrangements with its semantic, its syntactic and its pragmatic aspects (see Fig. 8).

The numbered arrows of this illustration indicate:

1 The semantic aspect of atomic information (a possible condition $\psi_{n,l,m,s}$ "signifies" the "position" for an electron $e_{n,l,m,s}$ within the electron shell of the atom).
2 The syntactic aspect of atomic information (function $\psi_i(Z)$ indicates the sequence in which the electrons are arranged within the electron shell of the atom).
3 The pragmatic aspect of atomic information (the structure of the electronic shell determines the chemical behaviour of the atom).

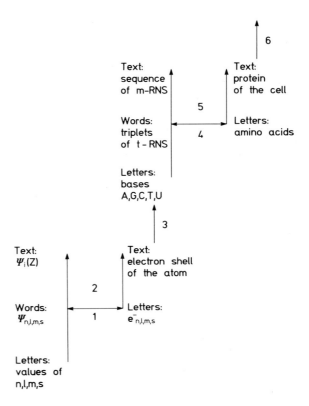

Fig. 8. Syntax, semantics, and pragmatics at the atomic and the genetic level

4 The semantic aspect of genetic information (a triplet of the t-RNA "signifies" an amino acid).

5 The syntactic aspect of genetic information (the sequence of the m-RNA indicates the sequence in which the amino acids are arranged for the protein chain).

6 The pragmatic aspect of genetic information (the cells "behave" according to their protein chains in the formation of a living being).

How do we now get from cell level to the level of a living being? Our knowledge about the physiology of living beings, especially mankind, is comprehensive. That applies equally to the organisation and function of the nervous system and the brain, including the electro-chemical processes which can be observed here. Yet we know virtually nothing about the code operating on this level, which effects the translation of received and stored information into the behaviour of living beings, in the course of evolution leading finally to the formation of a society. This deficiency results from the very great degree of complexity already existing at this level. Nevertheless, we should assume, in line with everything so far stated, that "letters" of the neurophysiological code are not produced by individual nerve cells, but by groups of cells. It is known in fact that nerve cells combine into groups having the simplest of functions. "Letters" are obviously spatially distributed, potentials occurring in time. Indeed, some individual relations between signal input and reaction are known, yet, as we have already seen, the operative code here has not yet been found.

6 Information Processing and Information Processing Systems

These days when we speak of information processing we primarily mean electronic systems (computers). We are aware of the fact that such a system comprises equipment for the input of data, storage systems, a central processing unit and output devices. Accordingly, we therefore refer to input, storage, processing and output of data or information. On the one hand, we feed data into these systems for processing, and, on the other hand, programmes, i.e., specifications for processing the data. In so doing, we use various computer languages. We talk of information systems in connection with substantial technically oriented applications. Thus, for example, there are sales information systems, material information systems, personnel information systems, etc. When we talk of systems within the context of our model, such systems, of course, are not what we mean. Rather, we refer to the term "system" more with respect to the processing and storage unit for a given level.

A certain structural similarity exists between technical systems and living beings, especially humankind. Human beings accept signals, which can be linguistically coded or not coded. They can store and process these signals, and the "output" is made in the form of linguistic information or other behaviour. Storage and processing is carried out according to rules, about which we still know very little.

When considering other levels of evolution, especially the level of atoms and cells, at which information processes occur too, we cannot expect to find such extensive similarities as exist between living beings and electronic computer systems. In particular, we should not just talk in terms of information whenever strong analogies happen to exist. Following the deciphering of the genetic code, nobody will seriously dispute that there is genetic information. Yet, ultimately, the processes which occur in a cell have only one thing in common with those that an electronic computer performs, and that is, that we are dealing with information to be processed according to certain rules and procedures, thereby arriving at certain results. It is a bit far fetched, if attempts are made to draw parallels between the processes associated with the fertilisation of an egg cell and the input of data.

We should not be surprised therefore that the atom has even less similarity with an electronic data processing system. In its basic state, it is primarily an information store. If it combines with other atoms under certain energy conditions, which, if desired, one could again denote as "input", its chemical behaviour alters according to the information stored in its electron shell, based on observance of certain physical rules. Any analogy with an electronic computer, therefore, would be only very rudimentary. In particular, there are no separate units for input, storage, processing and output. Nevertheless, even here information is processed following certain rules and procedures, and processing leads to certain results.

At the next higher level of evolution to that of living beings, the level of societies of beings, information is processed in many ways, yet one cannot compare the organisation of a society of living beings with that of a technical system. Input and output of information occurs, to a large extent, within the society. Storage and processing of information is carried out by the individuals from which the society is formed or by means of various media and technical equipment.

Thus, we finally arrive at the conclusion that similarity with the structure of a technical system does not provide a criterion for deciding whether the system characteristic for a certain level of evolution contains or processes information. Quite obviously, the basic criterion is: a code must be present, and additional order and additional asymmetry must be produced in relation to the relative prearranged system or condition.

7 Conclusion

In this paper, we have essentially stated that there are information processing systems on four levels of evolution, namely on the levels of inorganic matter – of atoms –, of the cell, of living beings and of societies of living beings, and that the codes of the lower relative systems are of fundamental significance for the evolution of higher relative systems. Accordingly, information does not just occur on the four levels of evolution, but it is the interface between the levels. It is an essential prerequisite for evolution. Our detailed considerations go further than the global statement that evolution means an increase in order and thus in information.

There is a distinct correlation of matter and information. On the surface, the impression could be gained that this paper argues for new-materialism by reducing information, i.e., an entity usually classified as belonging to the mental sphere, to material conditions. Although it is correct that, in atomic information, we perceive the root of genetic information as of higher information forms or information processing by living beings – if there were no atomic information which fundamentally influenced the chemical behaviour of atoms, neither cells nor brains nor communicating human societies containing information could have developed – it is a substantial difference whether something mental is being reduced to material processes or conditions or whether – as in this case – one states, that matter already contains information, that information is a component part of matter. This is not to reduce or even devalue the mental aspect, rather, it is an evaluation of creation as an entity comprised of matter and information, of matter and spirit.

It is not yet possible to assess how such an interdisciplinary view, going beyond its philosophical meaning, will affect scientific disciplines engaged in the study of the individual levels of evolution and the transitions between them. We hope that papers like the present one contribute towards establishing the study of information as an interdisciplinary subject cutting across all scientific disciplines. Possibly, this will cross-fertilize into individual disciplines where information plays a part, e.g., into chemistry or neurophysiology.

This paper is not really concerned with the evolution of information processing, but with evolution by information processing. At first, the behaviour of inorganic and organic matter as well as of humans was examined and studied in individual disciplines like physics and chemistry, biology and physiology, and the social sciences. There have been major breakthroughs and successes. But now we have begun to examine and study the information intrinsic already in inorganic matter, and then affecting all levels of evolution. We have only just embarked upon this study, and have only just opened the door to new insights into the nature of creation.

After the EIP workshop in Bremen from October 8 to October 11, 1990, and its discussions, there seems to be no reason for a fundamental modification of our model, dealing with four of the six levels of information processing which were the topics of the workshop.

The key points of our paper seem to be the description of the Atomic Code (pages 141–144), and in connection with it the statements concerning the role of codes for the evolution of information and evolution by information (pages 147–149 and Fig. 8).

References

1. Klement H-W (1974) "Halbordnung und Boolescher Verband zur Beschreibung der Evolution", Angewandte Informatik 2
   Klement H-W (1975) "Evolution und Bewußtsein" in Bewußtsein – ein Zentralproblem der Wissenschaften, AGIS-Verlag, Baden–Baden
   Klement H-W (1986) "Der Informationsgehalt des Atoms", Philosophia Naturlis 2, 23
2. Goser K (1989) "Das Gravitationsgesetz und das Coulomb-Gesetz aus der Sicht der Informationstheorie", Frequenz 6, 43
3. Hartmann N (1964) Der Aufbau der realen Welt, de Gruyter, Berlin
4. Lorenz K (1973) Die Rückseite des Spiegels, Piper Verlag, München, Zürich
5. Riedl R (1982) Evolution und Erkenntnis, Piper Verlag, München, Zürich
6. Krueger FR (1984) Physik und Evolution, Verlag Paul Parey, Berlin, Hamburg
7. Haken H (1979) Synergetics – An Introduction, Springer-Verlag, Berlin, Heidelberg, New York
8. Prigogine I (1979) Vom Sein zum Werden, Piper Verlag, München, Zürich
9. Laszlo E (1987) Evolution – Die neue Synthese, Europa Verlag, Wien
10. for instance, Neugebauer G (1980) Relativistische Thermodynamik, Vieweg u. Sohn, Braunschweig, Wiesbaden
11. see also Nicolis G, Prigogine I (1987) Die Erforschung des Komplexen, Piper, München, pp 113, 194 and 253

The Concept of Information Seen from the Point of View of Physics and Synergetics

Hermann Haken

Basic Concept 153

1 Some General Considerations 153

2 Self-Creation of Meaning 156

3 Effects of Information 161

4 Concluding Remarks 168

References 168

Basic Concept. The word information is used with quite different meanings, for instance in ordinary language in the sense of message, etc., or in the scientific sense of Shannon. In the present contribution information is understood as a property emergent in complex systems and several possibilities for its interpretation are discussed. In order to attribute semantics to information, the way a message acts on a receiver is studied in the frame of dynamical systems. Another approach is suggested by synergetics where the concepts of order parameters and enslaving can be used to study the emergence of an information field which acts on the individual parts of the same system. In a way, here the receiver is identical with the emitter of information. A comparison between the role of this sort of information in the physical system "laser" and the biological system "slime mold" exhibits interesting aspects with respect to the interpretation of the meaning of information. As it appears, the meaning of information or its semantics is not fixed in an objective manner but depends on our human interpretation.

1 Some General Considerations

The concept of information is used more and more in a variety of disciplines. For instance this concept is used in biology, or in sociology where one speaks of the information society. The question put by K. Haefner in how far the concept of information can be found in the various fields of science and which role it plays there is of great importance and very timely. In this respect we must be aware of the fact that this concept is used in a variety of ways, e.g., in common language or in the sense of Shannon information as used in communication theory. But even if we agree on a specific interpretation of this concept, we still have to study whether it plays the same or a different role in different sciences. Information and information processes can be found in the human brain as well as in computers and one may speak of information transfer and information processes also in societies.

The question put by Haefner as to how far this concept can be used down to physics is of great interest also from an epistemological point of view. Can one imagine that elementary particles are like small computers having an internal structure, which measures their position and speed and then determines their further trajectory in a given potential field? While this is a highly challenging question, I personally have a different point of view. My point of view is based on the principle of simplicity or on the principle of Occam's Razor.

For me it is tempting to start from elementary building blocks. But what are the elementary building blocks in physics? At first sight one may think of elementary particles, but this is not the point of view I share as I am going to explain. For me the elementary building blocks are protons and electrons and neutrons as far as they are constituents of atoms. If we set aside speculations in the frame of the grand unified theories with respect to the finite life-time of protons, electrons and protons are the stable particles. In a way they can be considered as given. In my opinion, a hydrogen atom cannot store information in its ground-state. When we think of a gas of hydrogen atoms in their ground-states, no information is stored there, leaving aside statistical considerations with respect to entropy. On the other hand, once some hydrogen atoms are excited, they become different from the rest of the hydrogen atoms in their ground-states and thus information can be stored. However, these hydrogen atoms have only a finite life-time because of this coupling to the radiation field. This life-time can be considerably enhanced if the hydrogen atoms are enclosed in a cavity in which only few electromagnetic modes can exist. In this example, information is stored by parts of a system that is brought out of thermal equilibrium.

When electrons, protons, or atomic nuclei collide at high energies, new particles can be produced. How do these particles relate to the concept of information in any sense? One may adopt two different attitudes: either one states that the newly generated particles, or at least some of them (for instance quarks), are elementary building blocks of, say, protons; in this way the data or the information on these elementary particles enables us to deduce the properties of, say, protons; or one may, on the other hand, think that the elementary particles of the kind described above are in reality (whatever that means) excited states of a quantum field. Under this aspect, the description of these new particles requires information to be deduced from the quantum field. These remarks may be related to the fact that the structure of the vacuum seems to become more and more complicated the more experiment and theory proceed.

These comments are also related to the problem of information compression. In the sense of algebraic complexity, we may attempt a description of nature in the sense of the Turing machine, in which the initial data and the program are as short as possible. As we know, there is no general algorithm in principle which will allow us to determine such a shortest string of data. However, in specific cases we may find a shorter string of data and program by a special algorithm or approach. To me it is at least an open question whether the concept of elementary particles on the one hand, or of excited states of a field on the other hand, will lead to a more concise description, i.e., to the most compressed information on the basic structure of matter.

Let me return now to the study of information under the idea that atoms are elementary building blocks which by their internal excitation or their cooperation produce more and more complex phenomena. In this sense I then consider it a main task of physics to explain the emergence of new properties including that of information. I must mention that I do not share the naive reductionist point of view which believes that we can deduce complex phenomena simply from underlying fundamental laws. Concurrent with the emergence of new properties, we need new concepts, for instance the concept of the order parameters of synergetics to which I will come below. In my opinion, nevertheless, one may start in a certain sense from a lowest hierarchical level of physical particles, from where we may consider the concept of information as an emergent property in chemistry and biology up to society or to, say, theoretical constructs, etc. It is important to mention that even within physics new properties may emerge. This is quite obvious when fluids or crystals are formed where we describe the system by new kinds of concepts which are alien to an atom. For instance, an individual atom cannot transport sound, whereas sound can occur when the concentration of molecules is sufficiently high. Correspondingly, solids show properties such as elasticity which is alien to an atom in the strict sense of the word.

In the following we will understand by information a property which, in the proper sense of the word, can only be defined indirectly. Information will have the following properties: it can be stored, it can be transferred, it can be multiplied, it can be read off reliably, it can be processed and – this seems to be important in the sense of semantics or pragmatics – it shall be able to cause specific actions in a receiver. This is a concept of information which is quite different from that of Shannon. Information can be stored in a static way, for instance by the position of impurity atoms or molecules in crystals. Note that a perfect crystal cannot store information. It is the very deviation from the regularity which may store information. This has another interesting consequence, namely that because of thermodynamics, specific configurations of impurity atoms which are necessarily connected with a low entropy will give way to a whole variety of configurations being connected with a higher entropy whereby the originally stored information is destroyed. Thus even in a crystal, information cannot be stored forever but only for a limited amount of time.

Interestingly enough, information can be stored dynamically and it is here where, at least in principle, we may account for an infinitely long life-time provided the system is kept under a continuous input of energy. For instance, when an electric current is on or off, one bit of information can be stored. As we shall discuss below, physical systems far from thermal equilibrium, i.e., open systems, may share some properties with biological systems.

The emergence of new properties in open systems has been systematically studied in the interdisciplinary field of synergetics. Lack of space does not allow me to develop this theory here in detail so I refer the reader to the literature. However, the example of the light source laser below may serve to introduce some of the basic concepts of this field.

2 Self-Creation of Meaning

As was mentioned before, synergetics may be considered as a theory of the emergence of new qualities at a macroscopic level. By an adequate interpretation of the results of synergetics, we may thus study the *emergence of meaning* as the emergence of a new quality of a system, or in other words the *self-creation of meaning*. In order to study how this happens we want to compare a physical system, namely the laser, with several model systems of biology. Let us start with some general remarks on the role of information in biological systems.

One of the most striking features of any biological system is the enormous degree of coordination among its individual parts. In a cell, thousands of metabolic processes may go on at the same time in a well regulated fashion. In animals millions or billions of neurons and muscle cells cooperate to bring about well ordered locomotion, heartbeat, breathing or bloodflow. Recognition is a highly cooperative process, and so is speech and thinking in humans. Quite clearly, all these well coordinated, coherent processes become possible only through the exchange of information, which must be produced, transmitted, received, processed, transformed into new qualities of information, and communicated between different parts of the system and at the same time between different hierarchical levels, as we shall see. We are thus led to the conclusion that information is a crucial element of the very existence of life.

The concept of information is a rather subtle one and it will be the goal of this section to elucidate further some of its aspects. As we shall see, information is not only linked with channel capacity and with orders given from a central controller to individual parts of a system – it can also acquire the role of a "medium" to whose existence the individual parts of a system contribute and from which they obtain specific information on how to behave in a coherent, cooperative fashion. At this level, semantics may come in.

Let us first take a look at physics. In closed systems, the second law of thermodynamics holds, which tells us that structures decay and systems become more and more homogeneous, at least on a macroscopic level. At the microscopic level complete chaos occurs. For these reasons, information cannot be generated by systems in thermal equilibrium, which is established in closed systems. But a system in thermal equilibrium cannot even *store* information. Let us consider a typical example, namely a book. At first sight, it seems to be in thermal equilibrium, and indeed we can measure its temperature. But inspite of that, it has not reached its final state of complete thermal equilibrium. In the course of time, the printer's ink of the individual letters will diffuse away until a homogeneous state is reached.

Another example is provided by the carrier of genetic information, DNA. Because it contains a specific sequence of nucleotides, its entropy can be expected to be higher than that of an arbitrary sequence. Thus this special DNA will deteriorate. To maintain its structure, the organism, the phenotype is required.

These simple examples teaches us that any memory consisting of a closed system is out of thermal equilibrium, and at any rate we have to ask *how long* information can be stored in each specific case. Let us therefore consider open systems which are kept far from thermal equilibrium by an influx of energy and/or matter into the system. Note that any biological system is an open system. In open systems, even in the inanimate world, specific spatial or temporal structures can be generated in a self-organized fashion. Examples are provided by the light source laser which produces coherent light, by fluids which can form specific spatial or temporal patterns, or by chemical reactions which can show continuous oscillations, spatial spirals or concentric waves. As we shall see, even at this level to some extent we can speak of creation or storage of information. On the other hand, we can hardly attribute words like relevance, purpose or meaning to these processes.

Let us discuss the laser in some more detail because it allows us to introduce terminology which is also most useful for biological and other systems. In the laser a number of atoms are embedded, for instance, in a crystal like ruby. After excitation of these atoms from the outside, the atoms my emit individual light wave tracks. Thus, each atom emits a signal, i.e., it creates information which is carried by the light field. In the laser the emitted wave tracks may hit another excited atom and cause it to amplify the original wave tracks. In this way, the signal is enhanced and the information serves the purpose of enhancing the signal. Because the individual excited atoms may emit light waves independently of each other which may be amplified by other excited atoms, a superposition of uncorrelated, though amplified wave tracks results and a quite irregular pattern is observed.

But when the signal reaches a sufficiently high amplitude, an entirely new process starts. The atoms begin to oscillate coherently and the field itself has become coherent, i.e., it is no longer composed of individual uncorrelated wave tracks, but has become a practically infinitely long sinusoidal wave.

We have here the typical example of self-organization where the temporal structure of the coherent wave is emerging without interference from the outside. Order is established. The detailed mathematical theory shows that the emerging coherent light wave serves as an order parameter which forces the atoms to oscillate coherently, or in other words, it enslaves the atoms. Note that we are dealing here with circular causality: on the one hand, the order parameter enslaves the atoms, but on the other hand, it is generated by the joint action of the atoms.

From the viewpoint of information, the order parameter serves a double role: it informs the atoms how to behave, and in addition, it informs the observer about the macroscopic ordered state of the system. While an enormous amount of information is needed to describe the states of the individual atoms, once the ordered state is established, only a single quantity, namely the phase of the total light field is necessary, i.e., we have an enormous compression of information. We may call the order parameter an "informator". Over the past years, it was shown that these concepts apply to a large number of quite different physical, chemical and biological systems.

To elucidate the role of information exchange at the level we are presently considering, let us consider an example from biology, namely the slime mold (dictiostelium discoideum). Usually the cells live individually on a substrate but when the food becomes scarcer they assemble at a specific point. The specific mechanism of this kind of self-assembly is as follows: the individual cells start to emit a substance, cyclic Adenosinemonophosphate (cAMP), i.e., they send out a signal or a message – information. Once cAMP molecules hit other cells, they are caused to increase their production in very much the same way as the laser atoms amplify the incoming signal. Quite clearly, the elements themselves are not aware of the meaning of the information but by the interplay between emission, amplification and diffusion of the cAMP molecules, a spiral pattern of concentration of cAMP is formed, i.e., information at a higher level is generated. Because this information is produced by the cooperativity of the system, we may call it *synergetic information*. The spiral waves form some kind of gradient field (the informator) which can be measured by the individual cells which then move to the highest point within the gradient field. Clearly we can distinguish here between the production of information, the information carrier and the information receiver, which in our case would be cell-cAMP-cell. However, at the next level, we observe that a new meaning has arisen, namely that the established pattern of a molecular concentration serves the purpose of guiding the cells to the center of their assembly. This established pattern may be called an information field.

Basically the same idea holds for the concept of positional information used in theories of morphogenesis. These theories attempt to explain cell differentiation whereby organs are formed. These processes were studied by special experiments on ontogenesis. For instance, when in a mouse embryo at an early enough stage cells from the brain region are transplanted to the stomach region, they become stomach cells, while at a later stage they will develop into, say, cells of an eye. Thus cell differentiation depends on position within the tissue. To account for these findings, it is assumed that the individual cell within a tissue receives its information from a pre-pattern chemical field which has been established by diffusion and production of chemicals. In general, two kinds of molecules are assumed, namely activator and inhibitor molecules. Where activator molecules have a high concentration, it is assumed that specific genes can be switched on which cause the differentiation of a cell. In this way the pre-pattern plays the role of the informator.

It is useful to recall what we have established so far. Quite evidently, there is a hierarchy of informational levels. At the lowest level, the individual parts can emit information which hits other parts of the system. Such information can be transferred by a general carrier. An example for the first case are nerve fibres each connecting two neurons, and examples for the second case are provided by hormones released into the blood, or by pheromones released into the air.

Although in all of these cases the beginning of the exchange of information may occur at random, a competition or cooperation between different kinds of signals sets in, and eventually a new collective state is reached which differs qualitatively from the disordered or uncorrelated state present before. Thus, a new state is described by an order parameter or a set of order parameters or equivalently by one or several informators. The states of the individual parts are determined by means of the slaving principle. However, one may express this occurrence in another way, namely that a specific consensus was reached among the individual parts of the system, or that self-organization has happened. At the same time, information compression takes place. The information appears manifest at a macroscopic level and, in many cases, increases the reliability or efficiency or both of the system, or serves other specific purposes as mentioned above.

This new collective level becomes observable to the outer world, and by establishing this context a new semantic level is reached. By the way, the context may be established with the outer world or equally well within the same system. Here then, words like useful, useless, or relevant can be applied. This is quite evident from the example of the laser where the cooperative state reaches a high efficiency. In the analogous case of a biological system, such behaviour is then useful for the whole system. Beyond instability points, the system can acquire different possible states and it needs additional information on which state to choose. A simple example is provided in physics by the spontaneous formation of structures in fluids. When a fluid in a vessel is heated from below, it may form a macroscopic structure in the form of rolls. But the orientation of the rolls can be arbitrary and must be fixed by initial conditions or, in their absence, by fluctuations. One may speculate that biology uses a similar trick. A good deal of morphogenetic processes occur by self-organization using internal mechanisms and only some cues are needed to select among different structures. One possibility is that this information is provided genetically, or by constraints established by other parts of the system. But often in such a case of degeneracy, the surroundings play an important role, or in other words, the context judges the value of the kind of state to be established. More recently we were able to substantiate these ideas by a brain model, which we realised with the synergetic computer. Here, specific connections between cells (neurons) are formed when the system learns, i.e., when it is in connection with its surroundings. It is, in my opinion, there where information in the biological sense starts. Through instability a collective state is formed, but it acquires its meaning only with respect to the surroundings and, in a way, to its value for the survival of the whole system.

These remarks also apply to the genetic code, though its very origin is not yet too well clarified. One may speculate that first, fluctuations occur which create some biological macromolecule with specific properties, the most important of which is that it can be multiplied in an autocatalytic fashion. The value of information conveyed by this molecule to its phenotype is then judged by the surroundings to which also possibly other molecules with their phenotypes belong. By the interplay of mutation and selection new types of molecules and their phenotypes are then generated, and in this way we observe the creation of new information. But whether this information is useful or not can be tested only by the interaction of this species with its surroundings.

In our above considerations, we described the first steps of the formation of ordered or structured collective states. But in contrast to the physical systems we quoted above, such as lasers, fluid dynamics, or chemical reactions, a new feature occurs in biology, namely a solidification. For instance, when genes of a cell are switched on by activator molecules, the cell differentiates into a specific cell which now is no longer modifiable, or can no longer be transformed back into the original cell. In a way, dynamical processes may lead to solid structures like bones or organs. In a similar way, information is laid down in a rigid manner in DNA, i.e., in the genetic code. We must bear in mind, however, that the "solid structures" do not persist forever, but can maintain their structures only by means of the support of open systems with which they are in contact.

On the other hand, in higher animals, in addition to rigid wiring of the nervous system a good deal of self-organization appears. The interaction of the system with its surroundings, jointly with the genetic information laid down in the system, leads to the formation of new information. Through the continuous testing of the new information stored and created in the brain by the surroundings, new contexts are established and thus a new kind of semantics occur. But we may also expect that "solidification" occurs at various hierarchical levels of semantic information and serves to make the system more reliable, and to store information (memory). While the concept of Hebb's synapse, which is strengthened by its use, may be a correct concept, the building up of semantics requires a high degree of cooperativity within the system and a repeated interaction with the outer world. In this respect, semantic information is not a static property, but rather a process in which contexts and relevance are checked, reinforced or dismissed again and again. By the way, I believe that consciousness is not a static state, but a process in which information is continuously transferred between various parts of the brain and processed there again and again.

At this time, a word on pattern recognition may be in order. Lower animals immediately react to inputs such as light flashes and only few characteristics are needed, such as threshold of intensity and so on in order to respond to a signal. In higher animals, however, the incoming information will certainly be compared with stored information. However, our picture is slightly changing as to how this comparison is done.

Quite often it is assumed that the incoming pattern is compared with templates. However, the storage of a template would require quite a large amount of information. Therefore, one may consider whether in the sense of synergetics only specific characteristic features in the form of order parameters are stored which, when called upon, can generate a detailed picture. This point of view is supported by the synergetic computer mentioned above. In this sense pattern recognition becomes an active process in which patterns are formed in a self-organized fashion by the brain using certain hypotheses and checking them repeatedly against the incoming patterns. For instance, it is well known that when people look at faces, they focus their attention on specific parts like eyes, or nose, or mouth and look at them again and again.

Let us finally discuss a point which is particular to humans. In contrast to animals, they can transfer information not only by the genetic code, but by teaching, which in the world of animals takes place only in a very limited way. So a good deal of our culture is based on this new way of transferring information from one generation to the next. But here an enormous difficulty arises because of the tremendous amount of knowledge which has been accumulated by humanity. Therefore, quite in the spirit of synergetics, it will be important to find unifying ideas and principles to cope with the world.

In addition, our approach provides us with a picture rather different from pictures conventionally drawn from biological systems. There, it is assumed that there is one single steering center, say in the brain, which then organizes all of the behaviour. The model we are strongly supporting calls for processes of self-organization instead, and more recently we were able to prove this hypothesis by our quantitative theory of specific experiments on the correlation of hand movements and their changes. In these experiments, performed by S. Kelso, test persons were asked to oscillate their fingers in parallel. At an increased oscillation frequency an involuntary change to an anti-parallel oscillation occurred. The way this transition occurs can be represented in every detail by the assumption of self-organization of the behaviour of neurons and muscles.

This is certainly an extreme case, but presumably the information production and transfer in biological systems must be considered in two ways: the first is the conventional one in which specific motor programmes serve for specific actions; other phenomena occur in an entirely self-organized fashion. We may hypothesize that self-organization in information processing in biological systems occurs to a large extent. This is for instance borne out by the great flexibility of biological systems and their adaptability and plasticity.

In my opinion, the study of information in biological systems is of interest also to modern society whose proper functioning rests on the adequate production, transfer, and processing of information. Perhaps the most important aspect which emerged was that of circular causality, which results in a collective state, which may in turn, represent in sociology a social climate, a general public opinion, a democracy or a dictatorship.

3 Effects of Information

In this section we wish to introduce a new approach which is a step towards a concept of information that includes semantics. We are led to the basic idea by the observation that we can only attribute a meaning to a message if the response of the receiver of that message is taken into account. In this way we are led to the concept of "relative importance" of messages as we wish to show in the following.

Let us consider a set of messages each of which is specified by a string of numbers. The central problem consists in modelling the receiver. We do this by invoking modern concepts of dynamic system theory or, more generally, by concepts of synergetics. We model the receiver as a dynamic system. For our present purpose a few general remarks will suffice. We consider a system, e.g., a gas, a biological cell or an economy, whose states can be characterised at the microscopic, mesoscopic or macroscopic level by a set of quantities, q, which we shall label by an index j, i.e., q_j. Over the course of time, the q_j's may change. We may lump the q_j's together into a state vector $\mathbf{q}(t) = (q_1(t), q_2(t), \ldots, q_N(t))$. The time evolution of $\mathbf{q}$, i.e., the dynamics of the system, is then determined by differential equations of the form

$$\frac{d\mathbf{q}}{dt} = \mathbf{N}(\mathbf{q}, \alpha) + \mathbf{F}(t) \tag{1}$$

where $\mathbf{N}$ is the deterministic part and $\mathbf{F}$ represents fluctuating forces. All we need to know, for the time being, is the following: if there are no fluctuating forces, once the value of $\mathbf{q}$ at an initial time is set, and the so-called control parameters α are fixed, the future course of $\mathbf{q}$ is determined uniquely. Over the course of time, $\mathbf{q}$ will approach an attractor. To visualize a simple example of such an attractor consider a miniature landscape with hills and valleys modelled by paper. Fixing α means a specific choice of the landscape, in which a ball may roll due to the force of gravitation (and subject to a friction force). Setting $\mathbf{q}$ at an initial time means putting the ball initially at a specific position, for instance on the slope of a hill. From there it will roll down until it arrives at the bottom of the valley: this is then an attractor. As the experts know, dynamic systems may also possess other kinds of attractors, e.g., limit cycles, where the systems perform an oscillation forever, or more complicated attractors, for instance "chaotic attractors". In the following, it will be sufficient to visualize our concepts by considering the attractor as the bottom of a valley (a so-called fixed point). When fluctuations $\mathbf{F}$ are present, the ball may jump from one attractor to another one.

After these preparations let us return to our original problem, namely to attribute a meaning to a message. We assume that the receipt of a message by the system means that the parameters α and the initial value of $\mathbf{q}_0$ are set by the message. For the time being, we shall assume that these parameters are then uniquely fixed. An extension of the theory to an incomplete message is straightforward. We first ignore the role of fluctuations. We assume that before the message arrives the system has been in an attractor which we shall call the neutral state. The attractor may be a resting state, i.e., a fixed point, but it could equally well be a limit cycle, a torus or a strange attractor, or a type of attractor still to be discovered by dynamic system theory. We shall call this attractor $\mathbf{q}_0$. After the message has been received and the parameters α and the initial value $\mathbf{q}$ are newly set, in principle two things may happen. Let us assume that we are allowed

Fig. 1.

to wait for a certain measuring period so that the dynamic system can be assumed to be in one of its possible attractors. Then either the message has left the system in the q_0 state, in which case the message is evidently useless or meaningless, or the system goes into a new attractor. We first assume that this attractor is uniquely determined by the incident message. Clearly, different messages can give rise to the same attractor. In this case we will speak of redundancy of the messages.

Finally, especially in the realm of biology, it has been a puzzle so far how information can be generated. This can be easily visualized, however, if we assume that the incident message provides the situation as depicted in Fig. 1, which is clearly ambiguous. Two new stable points (or attractors) can be realised depending upon a fluctuation within the system itself. Here the incident message contains information in the ordinary sense of the word, which is ambiguous, and the ambiguity is resolved by a fluctuation of the system. Loosely speaking, the original information is doubled because now two attractors have become available.

In the case of biology these fluctuations are realised by mutations. On the other hand, in the realm of physics we should instead speak of symmetry breaking effects. Taking all these different processes together we may find the elementary schemes of Fig. 2. Of course, when we consider the effect of different messages, more complicated schemes such as those of Fig. 3 may evolve.

We shall now treat the question of how we can attribute values to the incident messages or, more precisely speaking, we wish to define a "relative importance of the messages". To this end we first have to introduce a "relative importance" for the individual attractors. In reality, the individual attractors will be the origin

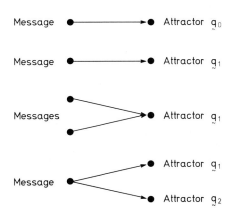

Fig. 2.

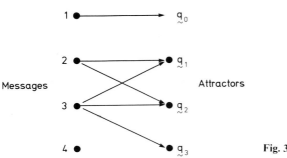

Messages   Attractors

Fig. 3.

of new messages which are then put into a new dynamical system, and we can continue this process *ad infinitum*. However, for practical purposes, we have to cut this hierarchical sequence at a certain level and at this level we have to attribute values of relative importance to the individual attractors. Since our procedure can already be clearly demonstrated if we have a one-step process, let us consider this process in detail.

Let us attribute a "relative importance" to the individual attractors, where the attractor 0 with $\mathbf{q}_0$ has the value 0, while the other attractors may have values $0 \leq p'_j \leq 1$, which we normalize to

$$\sum_j p'_j = 1. \tag{2}$$

The setting of p'_j depends on the task the dynamic system has to fulfill. We may think of a specific task which can be fulfilled just by a specific attractor or we may think of an ensemble of tasks whose fulfillment is of a given relative importance. Clearly the relative importance of the messages p_j does not only depend on the dynamic system but on the tasks it has to fulfill. The problem now is: what are the values p_j of the incident messages? To this end we consider the links between a message and the attractor into which the dynamical system is driven after receipt of this message. If an attractor k (including the 0 attractor) is reached after receipt of the message j, we attribute to this process the matrix element $M_{jk} = 1$ (or $= 0$). If we allow for internal fluctuations of a system, a single message can drive the system via fluctuations into several different attractors which may occur with branching rates M_{jk} with $\sum_k M_{jk} = 1$. We define the "relative importance p''_j" by

$$p_j = \sum_k L_{jk} p'_k = \sum_k \frac{M_{jk}}{\sum_{j'} M_{j'k} + \varepsilon} p'_k \tag{3}$$

where we let $\varepsilon \to 0$. (This trick serves the purpose that the ratio remains determined in the case where denominator and numerator vanish simultaneously.) We first assume that for any $p'_k \neq 0$ at least one $M_{jk} \neq 0$. One readily convinces oneself that p_j is normalized, which can be shown by the steps

$$\sum_j p_j = \sum_{kj} \frac{M_{jk}}{\sum_{j'} M_{j'k} + \varepsilon} p'_k \qquad (4)$$

$$= \sum_k \left(\sum_j \frac{M_{jk}}{\sum_{j'} M_{j'k} + \varepsilon} \right) p'_k \qquad (5)$$

$$= \sum_k p'_k \qquad (6)$$

where the bracket in (5) is equal to 1.

Now consider the case that some $M'_{jk} = 0$ for $p_{k'} \neq 0$ and all j. In this case in the sums over k in (4) and (5) some coefficients of $p'_k \neq 0$ vanish and, since $\sum p'_k = 1$, we obtain $\sum_j p_j < 1$. If this inequality holds, we shall speak of an **information deficiency**.

In a more abstract way we may adopt the left hand side of (3) as a basic definition where we assume

$$\sum_j L_{jk} \leqq 1 \qquad (7)$$

where the equality sign holds in the case of absence of an information deficiency.

We note that instead of the requirement $M_{jk} = 1$ in case of a single final attractor for an incident message, M_{jk} can be generalized to

$$0 < M_{jk} \leqq 1 \qquad (8)$$

The form of (3), left hand side, allows us to immediately write down the formulas when several systems are coupled one after the other. For instance in the two step process we immediately obtain

$$p_j = \sum_j L^{(1)}_{jk} p'_k = \sum_{kk'} L^{(1)}_{jk} L^{(2)}_{kk'} p''_{k'} \qquad (9)$$

where one can convince oneself very easily that $\sum p_j = 1$ provided $\sum p'_k = 1$ $\sum_j L_{jk} = 1$. The individual steps read

$$\sum p_j = \sum_{j\,kk'} L_{jk} L_{kk'} p''_k = \sum_{kk'} \underbrace{\left(\sum_j L_{jk} \right)}_{=1} L_{kk'} p''_{k'} \qquad (10)$$

$$= \underbrace{\sum_{k'\,k} L_{kk'} p''_k}_{=1} = 1. \qquad (11)$$

We may define

$$L'_{jk'} = \sum_k L^{(1)}_{jk} L^{(2)}_{kk'} \qquad (12)$$

Because the L's are positive we find

$$L'_{jk} \geqq 0 \tag{13}$$

and because of the normalization properties (in case of absence of information deficiency)

$$\sum_j L'_{jk'} = \underbrace{\sum_k \sum_j L^{(1)}_{jk} L^{(2)}_{kk'}}_{} \left.\begin{array}{c} \\ \end{array}\right\}$$
$$= \sum_k L^{(2)}_{kk'} = 1 \left.\begin{array}{c} \\ \end{array}\right\} \tag{14}$$

we readily obtain

$$L'_{jk} \leqq 1 \tag{15}$$

so that L'_{jk} obeys the inequality

$$0 \leqq L'_{jk} \leqq 1 \tag{16}$$

We mention that the recursion from p'' or still higher order $p^{(n)}$ to p may depend on the paths.

Our above approach does not only introduce the new concept of relative importance of a message but it also provides us with an algorithm to determine p_j which has some conceptual and practical consequences. With a certain task or ensemble of tasks given, this algorithm allows us to select the message to be sent, namely the one with the biggest p_j. If there are several p_j of the same size it does not matter which message is sent. From the conceptual point of view we may decide whether information is annihilated, conserved or generated by dynamical systems. To this end we make use of the concept of information in the sense of conventional information theory. But instead of the information content caused by the relative frequency of symbols we use the relative importance within a set of messages, i.e., we introduce the quantities

$$S^{(0)} = -\sum p_j \ln p_j \tag{17}$$
$$S^{(1)} = -\sum p'_k \ln p'_k \tag{18}$$

where p_j, p'_k have been introduced in (18). If $\sum_k p'_k = 1$, as is always assumed here, and $\sum p_j < 1$, an information deficiency is present. In the case $\sum p_j = 1$ we shall speak of annihilation of information if

$$S^{(1)} < S^{(0)} \tag{19}$$

holds, of conservation of information if

$$S^{(1)} = S^{(0)} \tag{20}$$

holds and of generation of information if

$$S^{(1)} > S^{(0)} \tag{21}$$

holds. The meaning of this definition quickly becomes clear when we treat special cases. If, for instance, two messages lead to the same attractor there is a redundancy in the system and the information content (in the traditional technical sense of the word) becomes smaller. It is reduced from

$$S^{(0)} = -K(\tfrac{1}{2}\ln(1/2) + \tfrac{1}{2}\ln(1/2))$$
$$= K\ln 2 \tag{22}$$

to

$$S^{(1)} = -K \cdot 1 \cdot \ln 1 = 0 \tag{23}$$

In the case of a one-to-one mapping of p_j onto p'_k we find the transfer of $\{p_j\}$ into the same set $\{p'_k\}$, except maybe for the permutation of indices, i.e., for different enumeration of states. In such a case clearly (20) holds. In case (21), finally, for instance one $p_j = 1$ and all others $= 0$, are transferred into, e.g., $p' = p'' = 1/2$, all others equal 0. Then $S^{(0)} = -K \cdot 1 \cdot \ln 1 = 0$ is enlarged to

$$S^{(1)} = -(\tfrac{1}{2}\ln(1/2) + \tfrac{1}{2}\ln(1/2))$$
$$= K\ln 2. \tag{24}$$

Of course these examples are not meant to prove the definitions (19) to (21) but rather to illustrate their meaning.

Our approach based on synergetics has some further pleasant features. Semantics has become a problem of the study of the response (attractors) of the dynamic system. The system may be error-correcting (or supplement lacking information). If the incident message does not set the initial state **q** *on the attractor* (i.e., *not correctly*), it may set the initial state **q** within the *basin of the attractor*, i.e., on the slope of the hill surrounding the bottom of the specific valley which represents the attractor (fixed point). In this way the system pulls the state vector into the attractor corresponding to that basin, i.e., into the *correct* state. An interesting problem will be to determine the minimum number of bits required to realise a given attractor (or to realise a given value of "relative importance").

Within our present scheme, the learning of a system can also be modelled. A system can be "sensitized" or "desensitized" with respect to messages j, e.g., by letting more or fewer parameters react on specific messages.

In our above treatment we have assumed that the value of the messages is measured with respect to the *same* initial state of the receiver. In the next step of our considerations we may assume that messages apply to a receiver in *another* initial state which has been set for instance by a previous message. In such a way we obtain an interference of messages and the relative importance of a message depends on messages the receiver has received before. In particular, in the general case the relative importance of the message will depend in a non-commuting way on subsequent messages. In this way the receiver is transformed by messages again and again and clearly the relative importance of messages will become a function of time.

Another remark might be useful particularly with respect to synergetic processes. A synergetic system need not only be a dynamical system showing, e.g., limit cycle or chaotic behaviour but it might also be one in which irreversible processes leading for instance from an unorganized liquid state into a structured solid state may happen.

4 Concluding Remarks

Our considerations may be summarized as follows: when we consider information and information processing, quite obviously we have to ask the question on which level we consider these processes. It seems important to me that when we proceed from simple to more complex systems, we have to consider different steps of information and information processing in the sense of emergent properties. It appears to me that synergetics offers an adequate starting point because it shows how new qualities may emerge by means of collective behaviour, for instance the coherent light field in the laser, or the highly ordered concentration field of molecules when slime mold is formed or specific patterns of excitation in the brain in animals and man. It appears that the order parameters act as a specific kind of informator which determines the behaviour of parts of the systems, and in turn is determined by the individual parts of the system. These internal states are changed by signals or messages coming from the outside where in most cases the external signals cause changes of the parameters of the system rather than act as direct forces. As we know from synergetics, even small changes of so-called control parameters may cause macroscopic changes of the structure of the system.

In conclusion we may state that the concept of information occurs on the different levels of complex systems where information is an emergent property caused by the collective behaviour of the individual parts of a system. In biological systems, by means of mutation and selection, more and more complex living beings can be formed where information plays a central role, though it is becoming increasingly difficult to define the concept of information universally. This is because we must not consider the concept of information independently of the systems, especially receiving systems, when we proceed from "pure" (Shannon) information to semantics and pragmatics.

References

Haken H (1983) Synergetics, An Introduction, 3rd ed, Springer, Berlin
Haken H (1988) Information and Self-organization, Springer, Berlin
Haken H Haken-Krell M (1989) Entstehung von biologischer Ordnung und Information, Wissenschaftl. Buchges

Dynamical Systems, Instability of Motion and Information Processing

Gregoire Nicolis

1   Introduction   169

2   Bifurcation, Entropy and Information   170

3   Finite Coarse-graining and Chapman-Kolmogorov Equation
    in Deterministic Chaos   173

4   Generation of Information-Rich Strings of Symbols   177
4.1 Logistic Map in the Fully Chaotic Region   177
4.2 Dissipative Flows   178
4.3 Creation of a Language   180

5 Discussion   182

References   183

1 Introduction

In this essay we develop some connections between information processing (IP) on the one hand, and self-organization and nonlinear dynamics on the other hand. The principal idea we want to convey is that IP is to be viewed as an emerging property associated with a self-organization phenomenon. In other words, IP is not only a prerequisite allowing a system to undergo self-organization, but is also the end point of a complex evolutionary process leading from the state of "no information" to the state of "information".

The starting point of our development is to realise that non-linear dynamical systems driven by nonequilibrium constraints are capable of generating a variety of complex behaviours on the basis of two universal mechanisms:

(i) the *bifurcation* of new branches of solutions following the loss of stability of a certain reference state, and

(ii) *deterministic chaos*, reflected by an aperiodic, irregular evolution of the system's variables in space and time as a result of the ongoing instability of motion and the associated exponential divergence of nearby trajectories.

Both of these mechanisms are ultimately related to the phenomenon of *sensitivity*: sensitivity with respect to the parameters, as far as bifurcation is concerned; and sensitivity to initial conditions in the case of deterministic chaos. This immediately entails as a corollary, that *random elements* should intervene in the description in an essential manner. Our principal objective will be, then, to develop methods for casting, in a systematic manner, the evolution of unstable dynamical systems into a well-defined stochastic process. We shall see that the probabilistic description adds a completely new dimension, since it opens the possibility of viewing the evolution of a dynamical system as a succession of *information-rich strings of symbols*. We shall identify the salient features of such sequences and relate their probabilistic structure to the parameters of the underlying deterministic dynamics, thereby establishing an intimate connection between instability of motion and information.

In Sect. 2 the connection between bifurcations and information is considered. By working out an extended version of thermodynamics incorporating the fluctuations we show that there are universal relations governing the behaviour of Shannon's entropy across a bifurcation point. In Sect. 3 we consider a class of dynamical systems giving rise to deterministic chaos and show how one can map, through finite coarse-graining, the dynamics into a stochastic process described by a Chapman–Kolmogorov equation. The generation of information-rich strings of symbols from the unstable dynamics on the basis of this formalism is discussed in Sect. 4 and the main conclusions are drawn in Sect. 5.

2 Bifurcation, Entropy and Information

The standard form of the laws governing the evolution of the macrovariables of a physico-chemical system is:

$$\frac{d\mathbf{X}}{dt} = \mathbf{F}(\mathbf{X}, \lambda) \tag{1}$$

where $\mathbf{X}$ is the state vector $\mathbf{X} = (X_1, \ldots X_n)$, $\mathbf{F}$ a dissipative operator (generally a nonlinear functional of $\mathbf{X}$) and λ a set of control parameters built into the system. There is a vast literature [2] devoted to the bifurcation phenomena generated by (1). Suffice it to recall here that two of the simplest forms of bifurcation are:

(i) the pitchfork bifurcation, whereby two stationary state solutions are generated beyond the instability of a unique "reference" state (Fig. 1), and
(ii) the Hopf bifurcation, whereby beyond the instability of a stationary state the system exhibits periodic behavior in time (Fig. 2).

A most remarkable feature is that the nature of the eigenvalues of the linearized operator of (1), $\mathscr{L} = (\delta\mathbf{F}/\delta\mathbf{X})_{\mathbf{X}_s}$ around a reference state $\mathbf{X}_s$ conditions, to a large extent, the behaviour of the original system even in the nonlinear range. In particular, in the vicinity of a *simple criticality* in which $\mathscr{L}(\lambda)$ has a simple real eigenvalue or a pair of simple complex conjugate eigenvalues whose real part is negative for $\lambda < \lambda_c$, zero at $\lambda = \lambda_c$ and positive for $\lambda > \lambda_c$, it turns out that the dynamics can be cast in a *universal normal form*. The latter displays

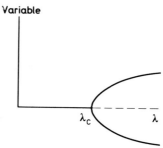

Fig. 1. Bifurcation diagram of a supercritical pitchfork bifurcation

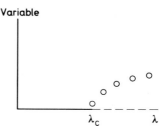

Fig. 2. Bifurcation diagram of a supercritical Hopf bifurcation

a limited set of privileged variables, the *order parameters*, to which the system's original variables are connected through time-independent relationships. In other words, any dynamical system operating near the simple criticality defined above is described by the same normal form, whatever its detailed structure might be. The subspace of the initial phase space spanned by the order parameters is referred to in the mathematical literature as the *center manifold*.

For the two "classical" bifurcations considered above the normal form equations become [2]:

$$\frac{dz}{dt} = (\lambda - \lambda_c)z - uz^3 \tag{2}$$

(pitchfork bifurcation)

$$\frac{dz}{dt} = [(\lambda - \lambda_c) + i\Omega]z - (u_1 + iu_2)|z|^2 z \tag{3}$$

(Hopf bifurcation),
where u, Ω, u_1, u_2 are real numbers. Notice that the center manifold is one-dimensional in the case of pitchfork bifurcation (order parameter z is real) and two-dimensional in the case of Hopf bifurcation (order parameter z is complex).

In view of the high sensitivity associated with the loss of stability of the reference state, and in agreement with the comments made in the introduction, we can now set up an augmented description of dynamical systems incorporating random elements. One universal source of randomness is provided by the thermodynamic fluctuations, which are present in all physicochemical systems.

An alternative mechanism is that of the stochastic perturbations that the environment continuously inflicts on a real-world system. Without going into detail we recall here that there are two methods that allow one to analyse the role of these phenomena [3–5]: the *Master equation description* in which detailed information on the individual processes involved in the stochastic dynamics is incorporated; and the *generalized Langevin equation* approach, in which stochastic effects are accounted for globally by adding a random white noise force term in the equations of evolution. It can be shown that below and up to a small neighbourhood of the bifurcation point λ_c both formalisms are equivalent and describe the augmented dynamics as a first order Markov process. Let us summarize briefly the main results to which they lead.

Setting

$$\mathbf{X} = \mathbf{X}_s + \mathbf{x} \tag{4}$$

where $\mathbf{x}$ represents the fluctuation around the reference state $\mathbf{X}_s$, one obtains, in the limit of long time, solutions of the Master equation or of the Fokker–Planck equation associated to the Langevin description in the form

$$P(\mathbf{x}) \sim \exp[-U(\mathbf{x})] \tag{5}$$

In the vicinity of a bifurcation point the quantity $U(\mathbf{x})$, referred to as *stochastic potential*, can be split into a sum

$$U = U_{cr} + U' \tag{6}$$

where U_{cr} is a function of the critical variables $\mathbf{z}$, which in the deterministic limit lie on the center manifold, whereas the part U' contains the deviation of the non-critical variables $\mathbf{x}'$ around a reference value which in general depends on the critical variables. The probability distribution (5) can thus be written in the form

$$P = P_{cr}(\mathbf{z})P(\mathbf{x}'|\mathbf{z}) \tag{7}$$

in which the second factor is the conditional probability of $\mathbf{x}'$ given that $\mathbf{z}$ attains a specific value. A bifurcation shows up by a non-Gaussian dependence of P_{cr} on $\mathbf{z}$, while typically $P(\mathbf{x}'|\mathbf{z})$ is Gaussian. Equation (7) provides us, therefore, with the stochastic counterpart of the center manifold theorem.

For the pitchfork bifurcation near criticality, Figs. 1 and 2, P_{cr} becomes

$$P_{cr} = Z^{-1} \exp \frac{1}{\varepsilon} \left[(\lambda - \lambda_c) \frac{z^2}{2} - u \frac{z^4}{4} \right] \tag{8}$$

where Z is the normalization factor and ε a small parameter related to the variance of the random force (Langevin formalism) or to the inverse of the system size (Master equation description of thermodynamic fluctuations). A similar expression holds for the Hopf bifurcation.

Having cast the augmented dynamics in the vicinity of the bifurcation into a Markovian stochastic process we may now introduce the entropy of the process

$$S_1 = -\sum_{\mathbf{x}} P(\mathbf{X}, t) \ln P(\mathbf{X}, t) \tag{9}$$

which, as is well known from Shannon's theory, describes the amount of information (in bits) needed to specify a particular state among the states accessible to the system [6]. Let us study the behaviour of S_1 across a pitchfork bifurcation. using (8) one has

$$S_1(\lambda, \varepsilon) = \frac{\partial}{\partial \varepsilon}(\varepsilon \ln Z(\lambda, \varepsilon)) \tag{10}$$

where $-\varepsilon \ln Z$ plays formally a role analogous to one of the thermodynamic potentials familiar from equilibrium statistical mechanics.

One can easily evaluate (10) perturbatively for $u > 0$ (supercritical bifurcation), in a range of values of λ close to λ_c given by

$$\lambda - \lambda_c \cong \varepsilon^{1/2 + \delta} \quad (\delta > 0) \tag{11}$$

The result is

$$S_1(\lambda, \varepsilon) = S_1(\lambda_c, \varepsilon) + \frac{1}{\sqrt{2}} \frac{\Gamma(\frac{3}{4}) 1}{\Gamma(\frac{1}{4})\sqrt{u}} \varepsilon^{-1/2}(\lambda - \lambda_c) \tag{12}$$

It shows that the information entropy increases as the system enters into the region of multiple steady states [7, 8]. This can be understood by realising that in the region of multiple states the system is more delocalized in phase space; a greater amount of information is thus needed if one is to specify its configuration at a given time. A similar result can be established for the Hopf bifurcation. Since the normal forms of pitchfork and Hopf bifurcation on the basis of which S_1 has been evaluated are universal, we can summarize by asserting that *bifurcation increases information*. Notice that the thermodynamic entropy and the other familiar thermodynamic state functions show no universal trends across bifurcation.

The above result establishes a first explicit connection between dynamics and information. It also highlights two important points: firstly, fluctuations are the natural carriers of information; and secondly, it is only in the regime of multiple solutions that fluctuations can realise fully their potentialities and allow the system to behave as an information processor.

3 Finite Coarse-Graining and Chapman–Kolmogorov Equation in Deterministic Chaos

We now turn to nonlinear dynamical systems giving rise to deterministic chaos. Owing to the exponential divergence of nearby trajectories, beyond a time of the order of the inverse of the largest Lyapunov exponent predictions or individual trajectories lose their operational significance in such systems. We shall try to circumvent this fundamental limitation by setting up a probabilistic description of chaos which, while being an *exact image* of the dynamics allows one to explore the large scale features of the system and to capture properly its intrinsic complexity.

The approach will be illustrated on a class of recurrent dynamical systems which we write in the generic form

$$\mathbf{x}_{n+1} = f(\mathbf{x}_n, \lambda) \tag{13}$$

It will be assumed that in the region of chaotic solutions of (13) the system possesses a non-singular invariant measure $\mu(\mathbf{x})$.

Let $\rho_n(\mathbf{x})$ be the probability density for finding the system, at time n, at point $\mathbf{x}$ of phase space. It follows as an immediate consequence of (13) that ρ_n evolves in time according to the Liouville or Perron-Frobenius equation [9]

$$\rho_n(\mathbf{x}) \equiv U\rho_{n-1}(\mathbf{x}) = \int_F d\mathbf{y}\,\delta(\mathbf{x} - f(\mathbf{y}))\rho_{n-1}(\mathbf{y}) = \rho_{n-1}(f^{-1}(\mathbf{x})) \tag{14}$$

where Γ is the region of phase space available and f^{-1} the (generally multivalued) inverse of the mapping f. It is well-known that, despite its probabilistic appearance, (14) is equivalent to the deterministic description that amounts to following the trajectories of the system along their natural motion in phase space. Technically, this equivalence is reflected in the fact that the "transition probability" featured in (14) is singular, as it is given by the Dirac delta function. Following the generally accepted terminology, we may also refer to the description afforded by (14) as the "fine-grained" description.

It is a matter of observation that, whatever the conditions under which we communicate with a complex system might be, this interaction is limited by a finite (although possibly very large) precision of the measurement process. Operationally, this is manifested by the fact that the instantaneous value of a macroscopic observable $A(\mathbf{x})$ can only be known with an error margin Δa, where

$$a < A(\mathbf{x}) < a + \Delta a \tag{15}$$

If the dynamics of the system, Eq. (13), is stable, the above limitation has no further consequences. However, in the presence of unstable motions giving rise to deterministic chaos, small errors like Δa are amplified exponentially; as a result, the deterministic "fine-grained" description (14) based on the monitoring of individual trajectories loses its operational significance.

In order to come to terms with this fundamental limitation, we set up a "coarse-grained" description incorporating from the very beginning the idea that a physically accessible state corresponds to a finite region rather than to a single point of phase space. Let $\{C_i, i = 1, \ldots, K\}$ be the set of cells in phase space constituted by these regions, assumed to be connected and non-overlapping. Obviously, $\{C_i\}$ defines a partition, P, of phase space, on which the dynamics, Eq. (13), will induce transitions between the cells (see Fig. 3), thereby generating sequences of K symbols, which may be regarded as the letters of an "alphabet".

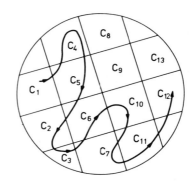

Fig. 3. Time evolution of a phase-space trajectory viewed as a sequence of transitions between the cells of a "coarse-graining" partition $C_1 \rightarrow C_5 \rightarrow C_4 \rightarrow C_5 \rightarrow C_6 \rightarrow C_2 \rightarrow C_3 \rightarrow C_7 \rightarrow \ldots$

The question to which we are naturally led is, therefore, whether this *shift process* can be characterised in a sharp manner.

The first step toward this goal is to introduce a "physical" probability distribution function as a projection of the full phase-space distribution $\rho_n(\mathbf{x})$ onto the partition

$$\mathbf{P}_n = \mathbf{E}\rho_n(\mathbf{x}) = \sum_i \frac{1}{\mu(C_i)} \int_{C_i} \rho_n(\mathbf{x}) d\mu(\mathbf{x}) 1_{C_i} \tag{16}$$

where 1_{C_i} is the characteristic function of cell i, $\mu(C_i)$ the measure of the set C_i, $\mathbf{P}_n$ the row vector $\{P_n(i)\}$.

The question to which we now have to turn our attention is: what is the kind of evolution that (14) induces for the discrete probability vector $\mathbf{P}_n$? Let the initial probability distribution $\rho_0(\mathbf{x})$ be coarse grained in the sense

$$\rho_0(\mathbf{x}: \mathbf{x} \in C_{i'}) = P_0(i) \tag{17a}$$

In principle, ρ_1, where

$$\rho_1 = U_1 \rho_0 \tag{17b}$$

need not be coarse-grained. One may, of course, consider its projection onto the partition,

$$\mathbf{P}_1 = \mathbf{E}\rho_1 = \mathbf{E}U\rho_0 \tag{18a}$$

but, contrary to (17a), one will typically have

$$\rho_1(\mathbf{x}: \mathbf{x} \in C_i) = P_1(i) \tag{18b}$$

The above argument can now be carried out consecutively n times. Two different views naturally emerge:

(i) follow the evolution of ρ_0 for n time units and consider the projection of the final distribution function onto the partition

$$\mathbf{P}_n = \mathbf{E}U^n \rho_0 = \mathbf{E}U^n \mathbf{E}P_0 \tag{19a}$$

(ii) after each time unit, consider the projection of the fine-grained distribution onto the partition and follow this sort of evolution for n consecutive steps. This leads to the probability vector

$$\tilde{P}_t = \underbrace{EU_rE\cdots EU_rEP_0}_{n \text{ times}} = EU_rE)^n P_0 \tag{19b}$$

In general the construct $\tilde{P}_n$ will be different from P_n unless the following condition is satisfied:

$$EU_r^nE = (EU_rE)^n \tag{20}$$

Using the explicit form of the projector **E**, (16), one can show that this condition is equivalent to either of the following statements:

(i) the conditional measure $\mu(C_j^{-n}|-C_i)$ satisfies

$$\mu(C_j^{-n}|C_i) = \sum_{k_1\ldots k_{n-1}} \mu(C_j^{-1}|C_{k_{n-1}})\mu(C_{k_{n-1}}^{-1}|C_{k_{n-2}})\cdots\mu(C_{k_1}^{-1}|C_i) \tag{21}$$

where C_j^{-n} is the *n*th pre-image of cell C_j and

$$\mu(B_1|B_2) = \frac{\mu(B_1\cap B_2)}{\mu(B_2)} \tag{22}$$

(ii) the matrices $\mathbf{Q}^{(n)}$, whose elements $q_{ij}^{(n)}$ are given by

$$q_{ij}^{(n)} = \mu(C_j^{-n}|C_i) \tag{23}$$

constitute a sequence of stochastic matrices satisfying the Chapman–Kolmogorov condition

$$\mathbf{Q}^{(n)} = (\mathbf{Q})^n \tag{24}$$

Using (20), (21) or (24) one obtains for the evolution of the probability vector $P_n(i)$ a Master equation of the form [10, 11]

$$P_n(i) = \sum_j W_{ji} P_{n-1}(j) \tag{25a}$$

with a transition probability matrix W given by

$$W_{ij} = \frac{1}{\mu(C_j)}\mu(C_i\cap C_j^{-1}) = \mu(C_i|C_j^{-1}) \tag{25b}$$

Furthermore, if the matrix W is aperiodic and irreducible, (25) gives rise to a strictly monotonic H-theorem ensuring the approach of P_n to a unique stationary solution $(P_\infty(i))$.

An important class of processes compatible with the above requirements are Markov processes. It should be realised, however, that the converse is not necessarily true: the Chapman–Kolmogorov condition does not necessarily imply the Markov property.

Let us summarize. We have mapped a deterministic description mediated by a set of operators $U^{(n)} = (U)^n$ onto a probabilistic description mediated by a Chapman–Kolmogorov type equation. The feasibility of this mapping rests on the validity of conditions such as (20), (21) or (24). This in turn imposes constraints on the partition and on the dynamics itself. In particular, if the transition probability matrix W (25b) is to be aperiodic and irreducible, it will be necessary that the underlying dynamics (13) exhibit strongly chaotic properties. So far it has not been possible to identify the most general class of systems compatible with the above requirements, although a number of nontrivial examples have been worked out successfully. Still, we believe that the developments outlined in the present section show that there is an intrinsic way to introduce the concept of information in physics, through the mapping of the dynamics into a stochastic process.

4 Generation of Information-Rich Strings of Symbols

4.1 Logistic Map in the Fully Chaotic Region

A concrete example in which the coarse-graining procedure developed in the present article can be carried out in detail is the logistic map in the fully chaotic region:

$$x_{n+1} = 4x_n(1 - x_n), \quad 0 \leq x \leq 1 \tag{26}$$

This dynamical system possesses an interesting class of partitions satisfying (20) to (24) whose cells are separated by point of the unit interval belonging to

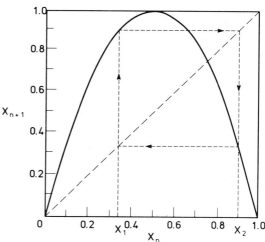

Fig. 4. Three-cell partition of the logistic map generated by the (unstable) period two orbit

an unstable cycle. For instance, the points of the period two orbit $x_1 = 0.345$, $x_2 = 0.905$ define a three-cell partition (see Fig. 4). The resulting three "states" σ, β, γ, which can also be viewed as "letters" of an alphabet, are then continuously transformed into each other by the dynamics according to a first-order Markov chain obeying a master equation (25a) whose conditional probability matrix (25b) turns out to be [10]

$$\mathbf{W} = \begin{pmatrix} \frac{1}{2} & \frac{1}{2} & 0 \\ 0 & \frac{1}{2} & \frac{1}{2} \\ 1 & 0 & 0 \end{pmatrix}. \tag{27}$$

4.2 Dissipative Flows

Consider a dissipative flow whose state variables x, y, z,... perform sustained (aperiodic) oscillations. We assume that when a variable crosses a certain predetermined level with, say, a positive slope a symbol forms and is subsequently "typed" (Fig. 5). One can envisage in this way a sequence of level crossing variables (symbols) standing as a one-dimensional trace of the underlying multidimensional flow [12]. The sequence is by necessity asymmetric as a result of the dissipative character of the flow in phase space. A numerical example is provided by the Rössler attractor (Fig. 6):

$$\begin{aligned} \dot{x} &= -y - z \\ \dot{y} &= x + ay \\ \dot{z} &= bx - cz + xz \end{aligned} \tag{28}$$

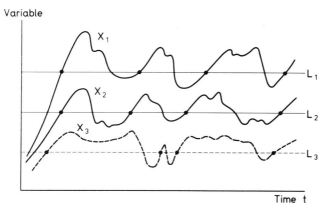

Fig. 5. Generation of strings of symbols by a dynamical system. Whenever the state variables $X_1,...$ cross levels $L_1,...$ with positive slope, a symbol representative of the variable is released. The resulting (one-dimensional) sequence provides one with a measure of the complexity of the (multidimensional) initial dynamics

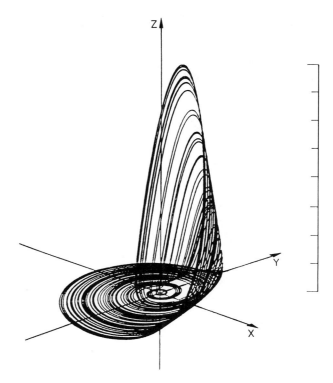

Fig. 6. Chaotic attractor generated by (28)

with $a = 0.38$, $b = 0.3$, $c = 4.5$, and thresholds $L_x = L_y = L_z = 3$. A typical
sequence generated by this mechanism is

$$zyx \; zxyx \; zxyx \; zyx \; zxyx \; zyx \; zyx \; zx \; zyx \; zyx \; zxyx \; zyx \dots \qquad (29a)$$

Remarkably, one can verify that the above sequence can be formulated more
succinctly by introducing the hypersymbols

$$\alpha = zyx \quad \beta = zxyx \quad \gamma = zx \qquad (29b)$$

giving rise to

$$\alpha\beta\beta\alpha\beta\alpha\alpha\gamma\alpha\alpha\beta\alpha\dots \qquad (29c)$$

A statistical analysis reveals strong correlations in the sequence (29a) which
to a very good approximation can be fitted by a fifth-order Markov process
[12]. On the other hand, the hypersymbol sequence is definitely a more random
first order chain, indicating that the "compression of information" achieved by
the "minimal" coding through the hypersymbols has indeed removed much of
the structure of the original sequence.

4.3 Creation of a Language

We can now show, using again for concreteness the examples of the logistic map and of the Rössler attractor, that starting from the "alphabet" induced in the shift space, one can generate sequences of "words" having some well-defined statistical properties [13].

The starting point is to choose one of the symbols of our alphabet to be the pause (blank space). As the dynamics unfolds in the shift space, the remaining $N - 1$ symbols are then organized in words C_L, of varying lengths L, interrupted by the pause. We want to find the probability $P(C_L)$ of formation of such words. Notice that the sequence C_L is in general a *non-Markovian* process.

We first carry out the analysis on the simple example of the three-cell partition of the logistic map. Choosing β to be the pause limits the "language" to words involving a single nontrivial letter α, since the role of γ is trivial ($W_{\gamma\alpha} = 1$). We obtain, using the explicit form of the transition probability matrix (27)

$$P(C_L) = W_{\alpha\alpha}^{L-1} = (1/2)^{L-1}, \quad L \geqq 2. \tag{30a}$$

If, on the other hand, γ is used as the pause, a richer language involving two non-trivial symbols α and β is created. Arguing as above, we obtain

$$P(C_L) = \sum_{m=1}^{L-1} W_{\alpha\alpha}^m W_{\alpha\beta} W_{\beta\beta}^{L-1-m}$$

$$= (1/2)^L (L - 1), \quad L \geqq 2. \tag{30b}$$

Although (30a) and (30b) differ significantly for small integer values of C_L, they tend to the same asymptotic form for long words, $L \to \infty$. Stated differently, in this limit the word processor is universally penalized in an exponential fashion with the length of the word generated. The argument can clearly be extended to partitions involving more cells, the difference being merely the occurrence of higher powers of L multiplying an exponential of the form a^L, a being a suitable combination of elements of the conditional probability matrix.

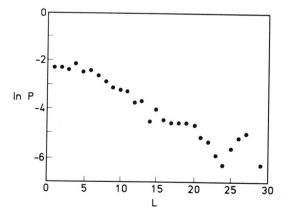

Fig. 7. Probability of words of length L plotted against L. The words are generated by Rössler's model (28) in the hypersymbol space, γ being used as the pause

Figure 7 depicts the dependence of the logarithm of $P(C_L)$ versus L obtained by iterating the dynamical system of (26) and subsequently monitoring the frequency of appearance of strings of symbols of given length between two pauses. The agreement with (30a) and (30b) is very satisfactory.

Following Mandelbrot's discussion of the concept of "lexicographical tree", we now define the rank r_L of a word of length L as the sum of all words of length equal to or less than L

$$r_L = 1 + \sum_{i=1}^{L} K^i \tag{31}$$

K being the number of symbols in the alphabet other than the pause. For long words, such that $r_L \gg 1/(K-1)$, one can easily deduce from (31) that

$$r_L \sim K^{L-L_0} \tag{32a}$$

where L_0 is defined through

$$K/(K-1) = K^{-L_0} \tag{32b}$$

Let us apply this to our previous examples. We have $K = 2$, i.e., $L_0 = -1$ and $r_L \sim 2^{L+2}$. Consequently, from (30a) – (30b) we obtain for large values of L

$$P(r_L) \sim 1/r_L \tag{33}$$

This is nothing but a particular case of Zipf's law of experimental linguistics [14] relating the probability $P(r)$ of appearance of words in a natural language to their rank r

$$P(r) \sim 1/r^{\mu} \tag{34}$$

Essential in the above reasoning was the fact that our dynamical system releases the successive symbols at regular time intervals. In a dissipative flow such as Rössler's system (28) this will not be the case. Figure 7 depicts the numerically computed $\ln P(C_L)$ versus L for the hypersymbol sequence of this model using γ as the pause. The dependence is now more complicated than in Fig. 6, despite certain similarities in the general trend.

One can verify that the mean time of formation of a word increases with its length. On the other hand, there exists a large dispersion around the mean, leading to crossovers in the times of observed deviation in a given realisation of particular words of different lengths. We conjecture that this phenomenon might be at the origin of the observed deviation from Zipf's law.

An interesting question is whether one can identify a function measuring the "quality" of the three processors corresponding to (30a), (30b) and the Rössler model. In the theory of dynamical systems it is customary to characterise the qualitative properties of the trajectories by the topological entropy [9]. However, being an invariant, this quantity cannot differentiate between strings of symbols involving alphabets with different numbers of letters. For instance, the processors corresponding to (30a) and (30b) both have the same topological entropy $h = \ln 2$, which is nothing but the topological entropy of the logistic map. A similar remark holds for the Rössler system. It has been shown that in the region in which the Rössler chaotic attractor exists, the successive iterates of variable x arising from the intersection of the flow by the plane ($y = 0$, $x < 0$, $z < 1$) give rise to a cube-like map involving three monotonous segments. This gives a topological entropy $h = \ln 3$, which is again an invariant.

It is well known that topological entropy is an upper bound of the measure-theoretic entropy, the latter being the maximum of the rate of change of the entropy of an initial partition of state space in the limit of infinite refinement. In the present paper we are not concerned with such a limit, since we are able to obtain, through coarse graining, a well-defined stochastic process involving a finite number of states. We therefore introduce the information entropy of the stochastic process $\{C_L\}$

$$S_1 = - \sum_L P(C_L) \ln P(C_L)$$

Contrary to h, this quantity is not an invariant, but depends on the partition, that is, on the algorithm generating the stochastic process $\{C_L\}$ from the original dynamical system. Notice, however, that S_1 should not be identified with the entropy of the partition: the latter involves the probability $P(J)$ of being in cell j (25) rather than $P(C_L)$.

Computing S_1 using the analytical expressions (30a) and (30b) and the numerical values corresponding to Fig. 7 yields $S_1 = 2 \ln 2 \sim 1.39$ for (30a), $S_1 \sim 1.88$ for (30b) and $S_1 \sim 2.87$ for Rössler's model. This trend, which must be contrasted with the invariance of h, reflects, in a sense, the increasing richness of the "repertoires" of the corresponding languages.

5 Discussion

One of the first questions arising in connection with IP concerns the kinds of messages that can be perceived as information. Now, an important aspect of the comprehension of a message is the presence of a particular set of symbols unfolding in time or, more generally, along the direction of reading. Although perfectly well defined and reproducible once known – think, for instance, of the genetic code embodied in the DNA structure – such a sequence is basically unpredictable in the sense that its global structure cannot be inferred from the knowledge of a part of it, no matter how large. In this respect it can be regarded as a stochastic process *and it is this recognition that allows one to speak of information.* For instance, what we perceive as information arising from the reading of Euclid's *Elements* or Newton's Principia would be very different if we were able to infer by a simple algorithm the second half of these treatises from reading the first half! In short, by revealing an object or a message that the "reader" could not infer to begin with, information is intimately associated with randomness.

On the other hand, the very existence of randomness raises in turn a new fundamental issue, since we know from algorithmic information theory [15, 16] that the number of random sequences of length $\rho \gg 1$ involving K symbols is exceedingly large, of the order of $N_\rho \sim K^\rho$. This makes it extremely improbable to encounter, on *a priori* grounds, the particular class of sequences that is likely to play the major role in a given phenomenon. As an example, to sort out a particular amino-acid sequence in a protein of $\rho = 100$ aminoacids in length, it would be necessary to scan the tremendous number of $N_\rho \sim 20^{100} \sim e^{300}$ sequences. Put differently, the spontaneous appearance of this particular sequence would happen with a probability as low as e^{-300}, which simply precludes this event on a time scale of the order of the age of the universe. In short, an efficient *selection mechanism* is needed in order that the information contained in a random sequence actually be revealed.

Now, the states of matter generated by nonlinear dynamics of systems far from equilibrium provide us with precisely the sort of balance between randomness and selection that is needed, on the grounds of the above arguments, in any information processing system. Most important among these states, for our present purposes, is chaotic dynamics. Indeed, the instability of motion associated with chaos allows the system to explore its state space continuously, thereby creating information and complexity in the sense discussed in Sects. 3 and 4. On the other hand, being the result of a physical mechanism, these states are produced with high probability and exhibit, as a rule, long range correlations; the problem of selection of a particular sequence out of a very large number of *a priori* equally probable sequences is thus accounted for automatically. In a way, the dynamical system generating chaos acts as an efficient selector that rejects the vast majority of random sequences and keeps only those compatible with the underlying rate laws. For instance one can verify that out of the 3^7 possible sequences long of 7 symbols in an alphabet of 3 letters, only 21 are actually generated by the Rössler attractor ((28)). In concluding this essay it appears therefore legitimate to assert that we have identified a physical origin and some plausible mathematical models of information processing systems.

References

1. See, for instance, Nicolis G, Prigogine I (1989) Exploring Complexity, Freeman, New York
2. Guckenheimer J, Holmes Ph (1983) Nonlinear oscillations, dynamical systems and bifurcations of vector fields, Springer, Berlin
3. Nicolis G, Prigogine I (1977) Self-organization in nonequilibrium systems, Wiley, New York
4. Gardiner C (1983) Handbook of stochastic methods, Springer, Berlin
5. Haken H (1977) Synergetics, Springer, Berlin
6. See, e.g., A. Khinchin: Mathematical foundations of information theory, Dover, New York

7. Nicolis G, Altarès V (1988) Physics of Nonequilibrium systems, in Synergetics and dynamical instabilities, North Holland, Amsterdam
8. Haken H (1986) Information and information gain close to nonequilibrium phase transitions; numerical results. Z Physik **B62**, p. 255
9. Collet P, Eckmann JP (1980) Iterated maps of the interval as dynamical systems. Birkhäuser, Basel
10. Nicolis G, Nicolis C (1988) Master equation approach to deterministic chaos. Phys Rev **A38**, p. 427
11. Nicolis G, Nicolis C (1990) Chaotic dynamics, Markovian coarse-graining and Information. Physica **A163**, p. 215
12. Nicolis G, Rao G, Rao S, Nicolis C (1989) Generation of asymmetric, information-rich structures in far-from-equilibrium systems, in Structure, Coherence and Chaos in Dynamical Systems, Manchester Univ Press, Manchester
13. Nicolis G, Nicolis C, Nicolis JS (1989) Chaotic dynamics, Markov partitions and Zipf's law. J Stat Phys **54**, p. 915
14. Zipf G (1949) Human behavior and the principle of least effort, Addison-Wesley, Reading, MA
15. Chaitin G (1987) Algorithmic information theory, Cambridge Univ. Press, Cambridge
16. Nicolis JS (1986) Dynamics of hierarchical systems, Springer, Berlin

Pragmatic Information in Nonlinear Dynamo Theory for Solar Activity

Jürgen Kurths, Ulvike Feudel and Wolfgang Jansen

Basic Concept 185

1   Introduction 185

2   A Nonlinear Dynamo Model for Solar Activity 186
2.1 The Concept of Solar Dynamo Theory 187
2.2 Observations of Solar Activity 188
2.3 A Nonlinear Dynamo Model 189

3   Qualitative Analysis of Solar Activity 190
3.1 Analysis of the Model 190
3.2 Analysis of the Wolf Numbers 192

4   Pragmatic Information (PI) 194
4.1 Measures of PI 194
4.2 Application to Solar Activity 197

5   Conclusions 197

References 198

Basic Concept. Active magnetic stars act as complex Information Processing Systems (IPS) because their dynamics results from a complicated feedback between the magnetic fields and the motion of charged particles. With the aid of a truncated nonlinear model obtained from dynamo theory we can reproduce some rather peculiar features which have been observed particularly for the global activity of our sun.

To interpret such complex systems as IPS, a new quantity of pragmatic information (PI) based on a thermodynamical description of fractals has been introduced. This approach provides a distinction between different kinds of deterministic chaos. We have pointed out that from this point of view multi-fractal dynamics is the only source which can continously produce PI.

The corresponding study of our model shows that this system internally produces PI as a result of self-organization.

1 Introduction

Cosmic objects of astrophysical interest are usually far away from thermo-dynamical equilibrium. It became apparant that in such systems a hierarchy of various symmetry breaking transitions can take place when the distance from equilibrium is increased by external or internal parameters (cf. Haken, 1982). Thus, these objects behave as complicated information processing systems (IPS).

In order to study astrophysical processes in the framework of IPS, the first task is to find the dominant forces. Gravity seems to be the most important physical force. As is generally accepted in cosmology, it drove the early phases of evolution in our universe. The principle of gravitation, stating that everything is informed about everything else and everything refers to everything else, is rather simple and acts without a memory (Haefner, 1988).

In later stages, where highly complex macroscopic structures such as galaxies, stars or molecular clouds have been formed, electromagnetic processes increasingly determine the behaviour of these subsystems. The corresponding IPS are characterised by the interaction between the motion of charged particles and magnetic fields. The structures occurring in such systems as the result of self-organization range from regular patterns to chaotic turbulence. The dynamical regime depends on the distance from equilibrium which is measured by dimensionless parameters such as Reynolds or Rayleigh numbers. It is widely accepted that many of these cosmic objects can be described as self-excited dynamos (Krause and Rädler, 1980). The dynamics of solar activity is such an example which is additionally important for our life. Some effort has been made to explain the variability of activity of our sun as an object of our particular interest with the aid of dynamo theory (Ruzmaikin, 1987).

The purpose of the present paper is to investigate the IPS of a nonlinear dynamo model describing the dynamics of solar activity. The layout is as follows: Sect. 2 describes the model. Section 3 contains a qualitative analysis of the model and a comparison with observations of the Wolf numbers. In Sect. 4 we discuss some measures of pragmatic information (PI) as a key parameter of IPS. Taking a thermodynamical formalism of fractals into account we introduce a new quantity for PI and apply this concept to our model.

2 A Nonlinear Dynamo Model for Solar Activity

The energy output of the sun as the main basis of life on the earth is nearly constant. However, the visible solar surface, the photosphere, is far from being uniform. It always consists of a granulation pattern which is produced by buoyancy-driven convection. Furthermore, we observe sunspots. These are dark regions, in which the luminosity is diminished compared to the general solar surface. In these regions strong magnetic fields (1 to 10 kilogauss) appear, whereas the global magnetic field of the sun is on the order of only 1 gauss. As was discovered by the druggist H. Schwabe in 1843, the spatial distribution of sunspots reflects on average an 11-year cycle of solar activity (Priest, 1982).

2.1 The Concept of Solar Dynamo Theory

This activity of the sun can be explained in terms of dynamo theory (Krause and Rädler, 1980). The basic idea of solar dynamo theory is that the magnetic field of the sun is amplified and maintained by its rotation. The rotation of a shell, the solar convection zone, with angular velocity in the magnetic field B leads to an electric field in a radial direction. Due to this field the charged particles move outward and in the conducting matter a current begins to flow. This current causes a new magnetic field itself, whose direction is equal to that of the original magnetic field B. The result is an amplification of the original field B. This principle of a self-excited dynamo is assumed to be the basic process for understanding the magnetic cycle of the sun.

It is obvious that feedback between the motion of charged particles and magnetic fields is essential to describe solar activity, i.e., this process acts as a complicated IPS. The only physical mechanism known to produce such magnetic variations are hydrodynamic plasma motions, that means rotation and convection. Therefore the equations of magnetohydrodynamics are the basis of the theoretical treatment of the processes within the convective zone. The changes of the field B are described by the induction equation

$$\frac{\partial \vec{B}}{\partial t} = \nabla \times (\vec{u} \times \vec{B}) + \eta \nabla^2 \vec{B} \tag{1a}$$

(where η is ohmic diffusivity). The changes of the particle velocity u depending on gravity g, the Lorentz force, and the pressure gradient ∇_ρ are described by the Navier–Stokes equation

$$\rho \left(\frac{\partial \vec{u}}{\partial t} + (\vec{u}\nabla)\vec{u} \right) = \rho g - \nabla_P + \vec{j} \times \vec{B} + \rho v \nabla^2 \vec{u} \tag{1b}$$

(where j is electric current density, ρ is mass density and v is kinematic viscosity). The flow of the plasma in the convective zone is a combination of large-scale motions like differential rotation with rather random turbulent convection. Thus, the main achievements of the solar dynamo theory are due to the deduction and the study of Eq. (1) for mean magnetic fields. The mean magnetic field transport is governed by mean flow values. The main characteristics which determine the mean field transport are differential rotation, turbulent convection and mean helicity. Differential rotation expresses the fact that the material near the equator rotates more quickly than near the poles of the sun. In the convective zone heat is transported by convection in special cells where the hot material rises at one point and cooler material sinks down at another point. The third process, the mean helicity, is due to the Coriolis force. Rising matter will expand and rotate because of the action of the Coriolis force yielding left-handed helical motion. On the other hand, sinking matter is compressed and forced by the Coriolis force to rotate in the opposite direction.

All these processes influence the changes of the magnetic field. The differential rotation generates a toroidal magnetic field from a poloidal one. The magnetic lines are stretched by the differential rotation leading to a toroidal field in the convective zone. On the other hand, the mean helicity generates a poloidal magnetic field from a toroidal one. Imagine a magnetic line of a toroidal field going from left to right. Due to the convective flow an omega-shaped loop is built up. Afterwards a poloidal component is created since this loop is twisted by the Coriolis force. The interaction of these processes gives rise to oscillatory magnetic fields coinciding at first glance with the solar magnetic cycle of about 22 years. The 11 year cycle results from it because absolute values only are observed.

2.2 Observations of Solar Activity

Despite its apparent successes, there are many aspects of the solar cycle that this dynamo model cannot really predict. To measure the dynamics of the solar activity Wolf introduced the sunspot relative numbers W by simply counting the number S of sunspots and the number G of groups of sunspots, and arrived at

$$W = w(S + 10G)$$

where the weighting factor w takes the influence of individual observers and instruments into account. These numbers can be regarded as a rough measure of the evolution of the sun's magnetic field (Priest, 1982).

Note that the dynamics of the Wolf numbers is far from being periodic (Fig. 1). We observe a complicated amplitude-frequency modulation. Furthermore, intervals of extremely low solar activity occur. The Maunder minimum (1645–1715) was the latest sharp weakening of solar activity reported in literature (Eddy, 1976). To explain these peculiar features of the solar cycle, nonlinear dynamo theory is necessary.

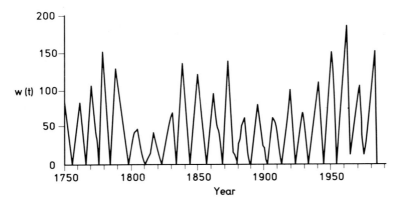

Fig. 1. Yearly sunspot relative numbers W(t) (Wolf numbers)

2.3 A Nonlinear Dynamo Model

A rather simplified nonlinear model has been obtained by truncating the partial differential Eq. (1) for the mean magnetic field (Malinetsky et al., 1986). It can be rewritten if the mean magnetic field is split into two parts, the poloidal component B_P and the toroidal component B_t. The poloidal component is expressed as the rotation of a purely toroidal vector potential A_t.

$$\vec{B} = \vec{B}_t + \nabla \times \vec{A}_t \qquad (2)$$

This way we get equations for the dynamics of A_t and B_t

$$\frac{\partial \vec{A}_t}{\partial t} = \alpha \vec{B}_t + \Delta \vec{A}_t$$

$$\frac{\partial \vec{B}_t}{\partial t} = D\nabla \vec{A}_t + \Delta \vec{B}_t \qquad (3a)$$

including the influence of the mean helicity α. The dynamo number D depends directly on the differential rotation and the mean helicity and is inversely proportional to the square of the coefficient of turbulent diffusivity. This linear system yields regular dynamo waves and enables the explanation of a periodic solar cycle. Additionally, we include the back-action of the magnetic field upon the helicity. Therefore α has to be replaced by $(\alpha + C)$ in Eq. (3a). An additional equation describes the changes of the helicity:

$$\frac{\partial C}{\partial t} = -rC + p\vec{A}_t\vec{B}_t - q(\alpha + C)\vec{B}_t^2. \qquad (3b)$$

The parameters r, p and q control the relaxation of C and the influence of the nonlinear terms. C is the deviation of the helicity from its value in the absence of the magnetic field. Thus we get a nonlinear dynamo model. To make it tractable the equations are truncated using a first order mode starting point

$$A_t = a_1 \cos k\theta - a_2 \sin k\theta$$

$$B_t = b_1 \cos k\theta - b_2 \sin k\theta$$

$$C = c_0 + c_1 \cos 2k\theta - c_2 \sin 2k\theta \qquad (4)$$

yielding a system of 7 autonomous differential equations depending on seven coefficients

$$\dot{a}_1 = -\sigma a_1 + (\alpha + c_0)b_1 + 0.5(c_1 b_1 + c_2 b_2)$$

$$\dot{a}_2 = -\sigma a_2 + (\alpha + c_0)b_2 - 0.5(c_1 b_1 - c_2 b_2)$$

$$\dot{b}_1 = -b_1 - Da_2$$

$$\dot{b}_2 = -b_2 + Da_2$$

$$\dot{c}_0 = -\gamma_0 c_0 + p(a_1 b_1 + a_2 b_2) - q((\alpha + c_0)(b_1^2 + b_2^2) + 0.5 c_1 (b_1^2 - b_2^2) + c_2 b_1 b_2)$$

$$\dot{c}_1 = -\gamma c_1 + p(a_1 b_1 - a_2 b_2) - q((\alpha + c_0)(b_1^2 - b_2^2) + c_1 (b_1^2 + b_2^2))$$

$$\dot{c}_2 = -\gamma c_2 + p(a_1 b_2 + a_2 b_1) - q((\alpha + c_0) 2 b_1 b_2 + c_2 (b_1^2 + b_2^2)) \tag{5}$$

3 Qualitative Analysis of Solar Activity

3.1 Analysis of the Model

In order to investigate the qualitative behaviour of the dynamo model (5) dependent on characteristic parameters, this 7-dimensional system of ordinary differential equations has been analysed by means of numerical methods. For this purpose the software system CANDYS/QA designed for the qualitative analysis of nonlinear dynamical systems was used (Jansen and Feudel, 1988). CANDYS/QA offers the calculation of invariant sets, which are mapped onto themselves for all times. Two different types of invariant sets are taken into account, namely stationary points and periodic solutions of a given nonlinear dynamical system, and these are then computed, including their stability. The stationary points are computed as the solutions of the nonlinear algebraic system which is obtained by reducing the derivatives with respect to time. Using Poincaré's method of first return map, the search for periodic solutions is also transformed into an algebraic problem. Therefore, the calculation of steady-states and one point of a cycle are based in principle upon the same numerical methods. Linearizing the system around the stationary point or periodic solution found, the stability of this solution is determined by evaluation of the eigenvalues of the corresponding Jacobian.

To study the bifurcation phenomena of the nonlinear system the stationary points as well as the cycles are continued by variation of a chosen model parameter. During the continuation critical points at which their number and/or their stability and/or their kind (turning point versus cycle) changes are detected and computed. Among the critical points possible we look for turning and bifurcation points, period doublings, Hopf bifurcations and torus bifurcation points. Moreover, all new branches of invariant sets are determined. This leads to the construction of a bifurcation diagram reflecting the qualitative behaviour of the nonlinear system dependent on the chosen characteristic model parameter.

For the model (5) the dynamo number D is the crucial bifurcation parameter. Moreover, the special values D at which qualitative changes occur mainly depend on the parameter q quantifying the influence of the higher order terms. Hence, our results are shown in a corresponding D-q-bifurcation diagram (Fig. 2). It exhibits 6 regions with different kinds of invariant sets. It is obvious that the system (5) possesses a trivial stationary point where all variables are

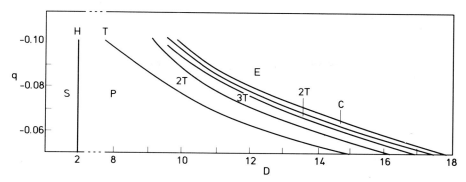

Fig. 2. Bifurcation diagram for the nonlinear dynamo model (5) dependent on the dynamo number D and parameter q, where S, is stationary solution; P, periodic solution; H, Hopf bifurcation; 2T, stable two-torus; 3T, stable three-torus; C, chaos; E, exploding

equal to zero, which corresponds to a vanishing magnetic field. At D_0 the system undergoes a Hopf bifurcation: the trivial stationary point loses its stability and a stable periodic solution occurs. This stable cycle is continued with increasing dynamo number. At a critical value D_1 this periodic solution loses its stability also, and we find a stable quasi-periodic motion on a torus. This torus is stable within a certain parameter region but it becomes more and more geometrically complicated. For even higher dynamo numbers there is a further torus bifurcation leading to a stable three frequency torus. Stable three frequency tori have been reported in only a few cases for maps (Kaneko, 1986) and periodically driven systems such as two-coupled Van der Pol oscillators (Battelino, 1988). As is known from special maps (Kaneko, 1986), this torus breaks down to a two-frequency torus if D increases further. For higher D a new instability sets in, leading to a chaotic region.

Since the attractor is somewhat complicated, we cannot decide from Poincaré plots whether the motion is quasiperiodic or chaotic. A conclusive way to distinguish between torus and chaos is to estimate the Lyapunov exponents. They express this by the exponential convergence or divergence of initially close trajectories, i.e., for the distance of two trajectories, after an evolution time T,

$$d(T) \sim e^{\lambda T}$$

holds, where $\lambda < 0$ means stability and $\lambda > 0$ points to an unstable behaviour. If the maximum Lyapunov exponent λ_{max} is positive, this stage of the system is defined to be chaotic. In the case of instability the initial distance has to be infinitely small, otherwise this distance would grow beyond all boundaries. The evolution of infinitely small deviations of a trajectory is computed by $\frac{\partial}{\partial t} x(t) = H(t)$ where the matrix H(t) fulfils the differential equation

$$\frac{d}{dt} H(t) = J(x(t)) H(t)$$

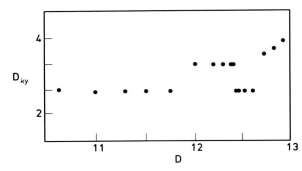

Fig. 3. Kaplan-Yorke dimension D_{KY} of (5) dependent on D for $q = -0.075$

with $H(0) = I$ (unity matrix) and J the Jacobian of the system. This leads to the following definition of

$$\lambda = \lim_{t \to \infty} \frac{1}{t} \log \frac{\|H(t)v\|}{\|v\|} \tag{6}$$

where v is a vector in the direction of initial deviation of the trajectory. Oseledec (1968) has shown that dependent on v in an n-dimensional system, at most n different limits λ_i can occur.

The numerical procedure to compute the λ_i, as proposed by Shimada and Nagashima (1979), is based directly on this definition.

Between the λ_i of a trajectory of a system and the dimension of the corresponding attractor there exists a relation found by Kaplan and Yorke (1979). If the λ_i are ordered descendingly,

$$D_{KY} = k + \frac{\sum_{i=1}^{k} \lambda_i}{-\lambda_{k+1}} \tag{7}$$

provides a good approximation of the Hausdorff dimension. k is the greatest number with $\sum_{i=1}^{k} \lambda_i > 0$.

Calculating the λ_i, we indeed find a region for D and q with chaotic motion. In the case of stable tori, D_{KY} is equal to the number of incommensurable frequencies, but it takes fractal values in the chaotic regime (Fig. 3). The transition from quasiperiodicity to chaos found for the system (5) is the typical Ruelle-Takens-Newhouse route to chaos (Newhouse et al. 1978).

3.2 Analysis of the Wolf Numbers

We have shown that the nonlinear dynamo model (5) generates a large variety of different types of behaviour from regular to chaotic motion. If we compare the trajectory of one component in the chaotic regime (Fig. 4) with the Wolf

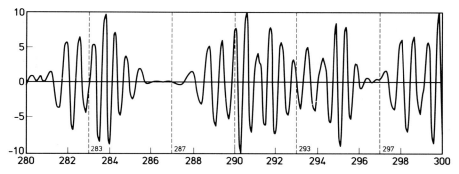

Fig. 4. Component a_1 of a trajectory of (5) in the chaotic stage

numbers observed (Fig. 1), some qualitative similarities are conspicuous at first glance. In both cases there is a complicated amplitude-frequency modulation. Furthermore, the model (5) generates periods of low activity as observed during the Maunder minimum. To get a more objective comparison between the model and the data we have to calculate dynamical invariants, such as attractor dimensions or Lyapunov exponents, from the Wolf numbers, too. Applying some techniques of time series analysis recently developed in the theory of nonlinear systems (cf. Ruelle, 1989) it comes out that in fact a low-dimensional attractor exists which seems to be chaotic (Kurths, 1987). We find a correlation dimension $\bar{D}_2 = 2.1 \pm 0.3$ from the data. Other analyses confirm this result (Ajmanova and Makarenko, 1988). The difference in the attractor dimensions obtained from the Wolf numbers to those from model (5) results from the fact that the rather short observational interval of the Wolf numbers does not cover grand minima so far, whereas model (5) also describes these long-term variations (cf. Fig. 4).

A study on forecasting the sunspot numbers by means of learning nonlinear dynamics, as proposed by Farmer and Sidorovich (1987), gives further evidence

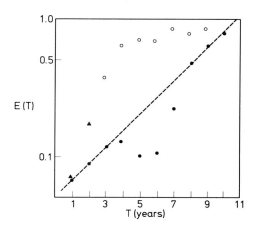

Fig. 5. Average prediction error E(T) of the yearly Wolf numbers as obtained from fitting ● a model learning nonlinear dynamics, × a global linear model. T is the forecasting time

for the chaotic nature of solar activity (Kurths and Ruzmaikin, 1990). We have shown that this rather simple approach yields relatively good results for short-term forecasts (less than 11 years). Moreover, this procedure outperforms the linear prediction models. On the other hand, the average prediction error grows considerably for longer forecasts (Fig. 5). This is a typical property of chaotic systems which preclude long-term predictions. These findings suggest that the dynamics of global solar activity is low-dimensional chaos.

4 Pragmatic Information (PI)

Interpreting this complex dynamics of solar activity as an IPS, we have to extend the classical viewpoint of information, entropy and syntax to the semantic and pragmatic aspects of information. The crucial parameter of this concept is the PI (Haefner, 1988) measuring the urge for action which results from the meaning of a message received by the system. As has been shown for some nonlinear physical systems (cf. Atmanspacher and Scheingraber, 1990, for a multimode cw dye laser), PI is generated at the critical values where instabilities occur. It is important to note that meaning is self-created by such systems because the structural changes are due to self-organization. These authors used efficiency, defined as the ratio of a special output and input power, as a key parameter, since it can be related to PI. This way we can conclude that the different bifurcations found for the dynamo model (5) increase the PI.

However, this approach fails to characterise chaotic regimes of an IPS. Comparing some measures discussed in the literature, we introduce another measure of efficiency which can in general be applied to chaotic dynamics.

4.1 Measures of PI

Nonlinear systems, even as simple as the logistic map

$$x(n + 1) = rx(n)(1 - x(n)) \tag{8}$$

can generate several kinds of chaotic behaviour. Chaos sets in firstly after an accumulation of period doublings at $r_\infty = 3.57\ldots$. In the interval $(r_\infty, 4)$ there is an uncountable set of r leading to a chaotic regime, i.e., $\lambda > 0$. For $r = 4$, fully-developed chaos with $\lambda = \log 2$ occurs whose dynamics is in some sense similar to white noise. Unfortunately, the dynamical invariant λ reaches its maximum at this stage of randomness. This is in contrast to the general property of PI to be zero for such a message containing wholly novelty, as well as for pure confirmation, such as periodic behaviour (Weizsäcker, 1974). A measure of PI is required to be maximum if novelty and confirmation are optimally mixed. Thus, we have to look for other parameters.

(i) Measures of complexity

- The *algorithmic complexity* is defined as the number of bits of the shortest algorithm which generates a message or a pattern (Kolmogorov, 1965). Again, this parameter assigns a higher complexity to a pure random sequence than to any ordered structure. It is really a definition of randomness.
- The *logical depth* of a number is the number of machine cycles that it takes a computer to calculate the number from the shortest possible program (Bennett, 1986). This measure has been generalized to physical systems as thermodynamical depth (Lloyd and Pagels, 1988). It satisfies the requirements mentioned above that wholly ordered and wholly random systems are not deep, i.e., they have a low complexity. On the other hand, this parameter is difficult to calculate.
- Crutchfield and Young (1989) proposed a procedure to construct an automaton, referred to as the ε-machine, that recognises the language of messages to within some approximation. Additionally, this leads to a new invariant, the graph complexity, for dynamical systems. This measure satisfies the requirements of PI. Furthermore, they gave an effective technique to calculate it.

(ii) Thermodynamics of fractals

A fractal set is an object having structures on all length scales described by power laws with noninteger exponents. In the simplest examples of the mathematical monster gallery, such as the Cantor set or the Koch curve, these scaling exponents are identical at all points of the object (Mandelbrot, 1982). But it happens very often that dissipative chaotic systems are characterised by different scaling indices which are complicatedly interwoven on their attractor. We observe such a nonuniform or multifractal behaviour even in the dynamics generated by the logistic map, for example for $r = 3.8$, whereas $r = 4$ produces a single scaling. Recently there has been much work on developing ways of quantitatively describing such multifractals.

Starting from the distribution of pointwise dimensions $D_\mu(x)$, i.e., the dimension of the set contained in a small ball centered at x, over all points x on the attractor, Halsey et al. (1986) introduced the $f(\alpha)$ spectrum of fractal dimensions

$$f(\alpha) = D_H(x/D_\mu(x) = \alpha) \tag{9}$$

where D_H denotes the Hausdorff dimension of a set. For practical purposes it is important to note that $f(\alpha)$ is associated with the hierarchy of Renyi dimensions $D(q)$ via a Legendre transformation

$$\tau(q) = (q - 1)D(q)$$
$$f(\alpha) = \alpha q - \tau(q) \tag{10}$$

with $\alpha = d\tau/dq$.
Note that this q has nothing to do with that in (5).

Another approach toward multifractal systems is based on the local divergence rate of initially close trajectories of the system, i.e., the growth of the initial distance $d(t_i)$ after an evolution time T to $d(t_i + T)$, which is obtained by monitoring a reference trajectory over a long time. Instead of only taking the average of these local divergences leading to λ_{max} via (6), we calculate a whole set of moments which yields the generalized entropies K(q) (Grassberger, 1988)

$$\log\left(\frac{1}{N}\sum_i\left(\log\frac{d(t_i + T)}{d(t_i)}\right)^{-q}\right)^{1/1-q} \sim TK(q) \tag{11}$$

Setting

$$\tau_0(q) = (q - 1)K(q) \tag{12}$$

Grassberger introduced analogous to (9) a $f_0(\alpha)$ spectrum of local growth rates which is closely connected with the $f(\alpha)$ spectrum for systems with one positive λ_i.

Bohr and Tel (1988) applied thermodynamical formalism to describe multifractals. They proved that the generalized dimensions and the spectra of scaling indices can be expressed in terms of known thermodynamical functions. Using the relationship between the internal energy E and the dimension function $\tau(q)$ we find a new way to calculate the efficiency η of chaotic systems:

$$\eta \sim - d^2\tau/dq^2. \tag{13}$$

Because the multifractal nature is especially indicated in the decrease of τ or τ_0 near $q = 0$, we take η from this range only.

This measure of efficiency satisfies the main properties required for PI. As is immediately seen from the logistic map (8), it is zero for both whole randomness or novelty $(r = 4)$ and regular behaviour or pure confirmation $(r \leq r_\infty)$. Moreover, we calculate $\eta > 0$ for intermediate values, such as $r = 3.8$, which generates a multifractal (Kurths et al., 1990). Thus, we interpret η as a measure of PI. From this viewpoint multifractal dynamics refers to a source producing new pragmatic information continuously. Homogeneous fractals, on the other hand, exhibit either a rather simple order, called self-similarity, which is equivalent to confirmation, or produces pure randomness without any pragmatic usefulness.

Additionally, (13) allows the comparison of different stages of nonuniformity. According to Gernert (1985) we define *dynamical pragmatic information* (DPI) as the distance of one efficiency η_1 and another efficiency η_2 obtained from the dynamics at different system parameters:

$$DPI = (\eta_2 - \eta_1)/\eta_2 \tag{14}$$

Thus, (13) determines the internal information processing of an open self-organizing system, whereas the DPI includes a change of system parameters.

Compared to the other measures of PI discussed above, the quantities (13) and (14) have some advantages. They give an insight into chaotic systems

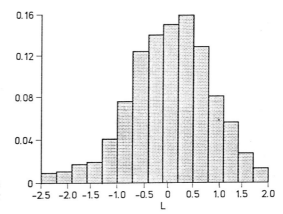

Fig. 6. A rough histogram of the local expansion rates calculated for (5) with $D = 10$ and $q = -0.1$

regarded as IPS and they are easy to calculate. Therefore, we apply this approach to solar activity which is in a chaotic regime.

4.2 Application to Solar Activity

From the analysis of the nonlinear dynamo model (5) we have found some sets of parameters which reproduce the dynamics of global solar activity in quite a good manner. Next, the local expanding rates of trajectories produced by (5) for $q = -0.1$ and $D = 10$ were calculated. The resulting histogram is shown in Fig. 6. The moments of this estimated distribution are far from being constant. Thus, the different values of the generalized entropies via Eq. (11) indicate the multifractal nature of this system. Finally, we get an efficiency significantly greater than zero, meaning that system (5) acts as a source of pragmatic information for these parameters.

5 Conclusions

The nonlinear dynamo model for solar activity studied here exhibits a rich dynamical behaviour from steady state via some bifurcations to a chaotic regime. It is interesting to note that there is a non-classical route to chaos. A comparison with the Wolf numbers suggests that the dynamics of global solar activity is low-dimensional chaos.

To interpret this system as an IPS, several measurements of PI have been discussed. Furthermore, we have introduced a new quantity of PI based on a thermodynamical description of fractals. This is a suitable tool for studying different kinds of chaos in the framework of IPS. Applying this technique to our model we find that this system internally produces pragmatic information as a result of self-organization.

Finally, we should like to emphasize that the suitability of this model for describing global solar activity is not a well-established fact despite the above indications. The investigation of such dynamics in the framework of IPS is still in its infancy and should be continued by including other stars with activity cycles and other nonlinear dynamo models.

References

Ajmanova GK, Makarenko NG (1988) The correlation dimension of the solar cycle, AZ No 1527, 27

Atmanspacher H, Scheingraber H (1990) Pragmatic information and dynamical instabilities in a multimode cw dye laser, Can J Phys **68**, 728

Battelino PM (1988) Chaotic attractors on a 3-torus, and torus break-up..., Phys Rev A **38**, 1495

Bennett CH (1986) On the nature and origin of complexity in discrete homogeneous, locally – interacting systems, Found Phys **16**, 585

Benettin G, Galgani L, Strelcyn J (1976) Kolmogorov entropy and numerical experiments, Phys Rev A **14**, 2338

Bohr T, Tel T (1988) Thermodynamics of fractals, In: Direction in Chaos, World Scientific, Singapore 194

Crutchfield JP, Young K (1989) Inferring statistical complexity, Phys Rev Lett **63**, 105

Eddy JA (1976) On the Maunder minimum, Science **286**, 1198

Farmer JD, Sidorovich JJ (1987) Predicting chaotic time series, Phys Rev Lett **59**, 845

Jansen W, Feudel U (1989) CANDYSIQA – A software system for qualitative analysis of the behaviour of the solutions of dynamical systems, In: Systems Analysis and Simulation, Springer-Verlag, New York

Gernert D (1985) Measurement of pragmatic information, Cognitive Systems **1**, 169

Grassberger P (1991) Information and complexity measures in dynamical systems, In: Information Dynamics, Atmanspacher H, Scheingraber H, eds: Plenum Press, New York

Haefner K (1988) Evolution of information processing, Preprint

Haken H (1982) Synergetics – An Introduction, Springer, Berlin

Halsey TC, Jensen MJ, Kadanoff LP, Procaccia I, Shraiman BI (1986) Fractal measures and their singularities: the characterization of strange sets, Phys Rev A **33**, 1141

Kaneko K (1986) Collapse of Tori and Genesis of Chaos in Dissipative Systems, World Scientific, Singapore

Kaplan JL, Yorke JA (1979) Chaotic behaviour of multi-dimensional difference equations, In: Functional Difference Equations and Approximations of Fixed Points, Peitgen HO, Walter HO eds: Springer, Berlin

Kolmogorov AN (1965) Three approaches to the quantitative definition of information, Inf Trans **1**, 3.

Krause F, Rädler K-H (1980) Mean-Field Magnetohydrodynamics and Dynamo Theory, Akademie-Verlag, Berlin

Kurths J (1987) An attractor dimension of the sunspot relative numbers, Preprint

Kurths J (1990) Zeitreihenanalyse in Geo- und Astrophysik, Doctoral Dissertation

Kurths J, Ruzmaikin AA (1990) On forecasting the sunspot numbers, Solar Phys **126**, 403

Kurths J, Schwarz U, Witt A (1991) The analysis of a spatio-temporal pattern during a solar flare, Solar Phys (subm.)

Lloyd S, Pagels H (1988) Complexity as information depth, Ann Phys **188**, 186

Malinetsky GG, Ruzmaikin AA, Samarsky AA (1986) A model for long variations of the solar activity, Preprint

Mandelbrot BB (1982) The Fractal Geometry of Nature, Freeman, San Francisco

Newhouse S, Ruelle D, Takens F (1978) Occurrence of strange axiom A attractors near quasi periodic flows on T^m, $m \geq 0$, Comm Math Phys **64**, 35

Oseledec VI (1968) A multiplicative ergodic theorem – Lyapunov characteristic numbers for dynamical systems, Trans Math Soc **19**, 197

Priest E (1982) Solar Magnetohydrodynamics, Reidel, Dordrecht

Ruelle D (1989) Chaotic Evolution and Strange Attrctors, University Press, Cambridge

Ruzmaikin AA (1981) The solar cycle as a strange attractor, Commun Astrophys **9**, 85

Shimada J, Nagashima T (1979) A numerical approach to ergodic problem of dissipative dynamical systems, Progr Theor Phys **61**, 1605

Weizsäcker EU (1974) Pragmatische Information, In: Weizsäcker EU Offene Systeme I, (ed). Klett Stuttgart

IV Information Processing in Biological Systems

Thermal Proteins in the First Life and in the "Mind-Body" Problem

Sidney W. Fox

1 Introduction 203

2 Essential Molecular Advances 205

3 Premises 207

4 The Self-Sequencing of Amino Acids As the Initial Bioinformational Process 208

5 The Mechanism of Self-Sequencing of Amino Acids 212

6 Future Studies of Sequence 214

7 Self-Organization of Thermal Proteins to Cells 215

8 The "Mind-Body" Product 217

9 Artificial Neurons 217

10 The Evolutionary Direction 220

11 Biofunctions 221

12 Bidirectional Transfer of Information 223

13 Informational Electrical Responses 224

References 226

1 Introduction

Our general understanding is that biological information first arose in pre-biological molecules. The two kinds of macromolecule that are suitable candidates for initial bioinformation are nucleic acids and proteins, since these are the informational molecules of present-day biological systems. A logical case with some qualifications can be made *a priori* for each of these types of macromolecule as the original source of information (Schmitt, 1962; Lehninger, 1975). It is nevertheless true that special difficulties arise when one seeks to understand how either DNA or RNA could originally have served not only as a source of information but also as processing agents for that information, e.g., to synthesize nucleic acids and proteins, and cellular systems with or without template. Also, no one has explained satisfactorily by experiment how DNA or RNA could have arisen without prior protein (Fox, 1988, Waldrop, 1989).

Proteins are easily understood as processing agents because they serve such functions so richly in modern systems, and those abilities are found in isolated systems in the laboratory. A main question has been, rather, how proteins could function initially also as a source of information conceptually as well as in processing agents, e.g., enzymes. When looked at in a strictly evolutionary way, one could test the origin of information in proteins from precursors of proteins, amino acids. The origin of information from nucleic acids, which are indeed storage molecules for information in modern systems, has not had comprehensive testing. Proteins have been thoroughly tested in this role (Fox, 1988) although not as the result of the above line of reasoning. The total results indicated that one kind of protein (thermal protein; protoprotein) is able to function as various processing agents: enzymes, inhibitors, direct precursors of cells, etc. and also as an adequate initial source of information.

A pragmatic way of looking at the alternative possibilities is to recognise that in the period 1960 to 1990 the route of protein-first has yielded a complete theory of the transition from inanimate matter to polybiofunctional cells, whereas in the same period nothing comparable has emerged from experiments with nucleic acids of any kind, especially not with primordial types. In order to explain an initial DNA Crick (1981), for example, has invoked a catalytic mineral on an

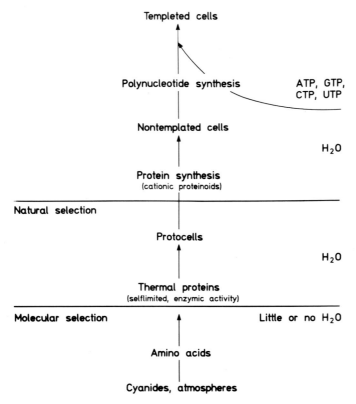

Fig. 1. Flowsheet of the proteinoid (thermal protein) emergence of life leading into its evolution. The self-ordering of amino acids (molecular selection) has made it possible

unnamed planet. While a modern RNA has been shown to have a kind of catalytic self-repair activity (Cech, 1985), there has been no support for the origin of metabolism from RNA broadly speaking, no satisfactory explanation for how the original RNA arose, no explanation for how cells appeared and no broad explanation for how various RNAs could have served as processing agents for information other than storage (to which, however, nucleic acids are well adapted because of their stability). The flowsheet from proteins-first is in Fig. 1.

The reason for attempting to make a protein before producing a cell was initially simply the attempt to understand how protein *per se* might have come into existence, cell or no cell. The answer to how the cell could have arisen thrust itself upon the experimental scene. When water entered the flasks containing thermal protein, as it almost inevitably did, the thermal protein *organized itself* into cells. The self-organization of thermal protein was one of two essential advances that had to precede a laboratory protoorganism with qualities of life as modern biologists list them.

2 Essential Molecular Advances

The two chemical advances were:

1. *The self-sequencing of amino acids to thermal proteins.* Some chemists recognised that the composition of mixtures of amino acids heated to yield thermal proteins would necessarily differ from the compositions of the resultant proteins. They were however not prepared for the idea that the sequences in the polymers might be ordered by the compositions in the mixture. Three reasons for delay in thinking can be given:
   (a) there is no sequence in a mixture of free amino acids to compare with,
   (b) the thinking is preconditioned by knowledge of modern systems, which rely on nucleic acids, and
   (c) the sequences in the polymers formed had first to be analysed.
2. *The self-organization of thermal proteins to cellular structures.* In his lecture at the Sorbonne in 1864, Pasteur had asked, "Can matter organize itself? In other words are there beings that can come into the world without parents, without ancestors? That is the question to be resolved."

In 1956, Schmitt reported the self-organization of precipitating collagen to form microfibrils. In 1959, quite independently, our laboratory (Fox et al., 1959) reported the ready conversion, by contact with water, of thermal protein to cellular structures. This tendency is more characteristic of laboratory protoprotein than of modern protein, probably because of (limited) branching in the thermal protein molecules (Harada and Fox, 1975).

The production of cellular structures from thermal proteins by steps that were themselves understood allowed determination of the properties of such units through the next 20 or more years (Fox, 1981, 1989). Since the cell-like structure possesses properties of metabolism, growth, reproduction, evolvability, and many other biofunctions it deserves to be regarded as a synthetic *proto*organism. Demanding special attention was the finding that growth of a cell from protein is quite strictly programmed in size, and that this limitation is not mediated by nucleic acids. Once that is recognised, one can understand that the formation of buds, a necessary step in primitive reproduction, would be a continuation from parental growth.

The significance of self-organization for the construction of flowsheets beginning with the origin of cellular life was recognised early (Fox, 1960a, b) and even explained by request to scientists associated with the Vatican (Fox, 1985). That a greater significance of self-organization in wider vistas of biology and computerology would arise was not foreseen.

The synthetic production of a protoorganism from noncellular precursors validates the conceptual flowsheet representing the route from amino acids to the first organism (Young, 1984). The amino acids are much like those identified as precursors in the Apollo program, especially the dicarboxylic amino acids that enable polycondensation (Fox, 1960b). Because of the experimentally established properties of this organism, as determined in the laboratory between 1960 and 1980 and beyond, no other experimental or theoretical flowsheet has been proposed. Nor has any other sequence of studies recited the need for self-organization, although Eigen (1971) suggested it for RNA. No self-organization of RNA has been demonstrated.

The initial information base for the reaction sequence in the flowsheet is provided by the method of self-sequencing of amino acids. The self-sequencing mechanism is the one that Ivanov and Förtsch (1986) have inferred as being "universal" for modern organisms. They have reached this conclusion from study of 2898 protein sequences in the databank at the Max Planck Institute for Biochemistry in Martinsried, in association with the late Professor Gerhard Braunitzer. In this connection, it should be noted that the starting point for the entire study that developed into an investigation on the emergence of life was out of the laboratory efforts that are credited with stimulating the present era of amino acid sequence determination (Fox, 1945) before Sanger's work (history in Rosmus and Deyl, 1972; Oparin and Rohlfing, 1972, and others). That line of study was aimed at elucidating the informational, i.e., bioinformational, beginning of evolution based on proteins. Ivanov and Förtsch used data *derived* by statistical analysis from amino acid sequence data. The study has thus come full circle methodologically and this fact can introduce a unity of thinking into the larger view. It supports the existence of biospecificity derived from varied amino acid sequences, by an integrated approach of both chemical analysis and synthesis, leavened with evolution.

3 Premises

The premises of information processing in the evolutionary context include the assumption that all molecules contain information. Molecules possess this information due to their individual stereoelectronic configurations. Each molecular type has its own shape, including those in some arrays of tautomers.

The informed system is capable of making choices because it possesses information. The conversion of one type of molecular system to another is information processing. We recognise the result of interaction of informed macromolecules with other molecules, other macromolecules, or with other systems as manifestations of making choices.

Although premises of randomness or determinism could not at the outset of this research in the 1940s and 1950s be analysed as clearly as they can now be, they provided contexts to include randomism, near-randomism, a state approximately equidistant between randomism and determinism, near-determinism, and determinism. The premise of randomism in evolutionary beginnings, for example, was assumed by Eigen (1971), although he has more recently departed somewhat from that position (Eigen, 1986) on the basis of theoretical analysis.

More emphatically, Tyagi and Ponnamperuma (1990b) have studied the formation of polymers from amino acyl adenylates which Krampitz and Fox (1969) showed reacted selectively (Tyagi and Ponnamperuma, 1990a). Ponnamperuma's study of the condensation compared the selectivity in polymerization of amino acid residues with that of polynucleotides from the same starting compounds, various amino acyl nucleotide anhydrides. All of the specificity was found to be in the amino acids, without influence from the nucleotide residues in the same molecules. Thus, amino acids select their own sequences whether polymerized by heat, by reaction through the activated adenylates, guanylates etc., or via the more complex modern protein synthesis mechanism (Ivanov and Förtsch, 1986).

Since his graduate student days (1938 to 1940), the author of this paper has believed in genetic determinism (Wilson, 1978), due especially to discussions with the chairman of his Ph.D. committee, Thomas Hunt Morgan. The premise of genetic determinism has now been extended forward from molecular determinism, a view somewhat easily arrived at by a chemist who is concerned with transferring the high degree of repeatability of chemical reactions into a more philosophical, evolutionary context. The most recent synthesis of these views has led to the suggestion of a "daisy-chain" of (a high degree of) determinism throughout the hierarchy and throughout the evolutionary sequence (Fox and Nakashima, 1984).

As a result of this deterministic context, one may not speak strictly of the "origin of the genetic code", but rather of the "origin of the genetic coding mechanism". The genetic code is an expression of the selective interactions of nitrogen bases in triplets for amino acid residues; this attraction is inherent in the molecules. The evolutionary sequences developed in this study are expressed, in fact, more fully in the sequence of molecular structures than in the processes; this is believed to be due to the fact that the processes are derived from the molecular structures.

Evolutionary Continuity. The concepts of evolution underlying conversion of information gain validity as they develop into a lengthening sequence of processes having connections that have been disciplined by testing, especially experimental demonstration. The basic assumption is that evolution of systems or processes must be stepwise as in a staircase.

4 The Self-Sequencing of Amino Acids As the Initial Bioinformational Process

Biological molecules sufficiently large to contain the amount of information needed for processes in organisms appear to be nucleic acids (DNA and RNA), proteins, and polysaccharides. Polysaccharides are readily eliminated from consideration because they tend each to be composed of a single type of monomer, e.g., glucose (in starch or cellulose) or mannose (in mannans). The essential conditions for informational nucleic acids and informational proteins is the variety of information derived from variety of monomer types in polymer chains.

Since proteins and nucleic acids are polymer types that contain monomers in variety and do not resemble monotonous chains, they are richly informational. Lehninger (1975) has treated both nucleic acids and proteins as informational molecules although it is possible to hear or read treatments by others in which nucleic acids seem to be the only informational biomacromolecules, by a kind of default of consideration. In some of the thinking in this field, in fact, the focus of attention on DNA or RNA is so intense that proteins seem to have been tacitly disqualified by being sloughed off.

A demonstrable model for sequence formation of nucleotide monomers in polynucleotides without prior processing by already informed protein has not been provided. Such fundamental questions as how DNA and RNA came into existence without prior informed protein, how such nucleic acids were informed, how they produced cells, how they were translated into protein, and many other questions have not been answered experimentally. Nor have they been answered theoretically. On the other hand, the emergence of informed protein has been explained experimentally, as has the emergence of informed nucleic acids as a result of the action of informed protein. The answer stemming from informed protein has developed many interdigitated spinoffs, and meets the requirement of evolutionary continuity – in an extended series.

Two kinds of experiment support the rigour of the protein-first perspective for information. One is the structural nonrandomness. The other is the finding of biofunctions in many of the thermal proteins: antiaging of neurons when treated in culture (Hefti et al., 1991), memory enhancement when injected into mice (Fox and Flood, 1991), etc. Random polymers could not be expected to yield biofunctions and to yield them repeatably (Eigen, 1971b).

Among the biofunctional results from the proteins-first investigations are: an array of enzymatic activities, specificities of interactions between thermal proteins and substrates, linking of enzymatic events to provide metabolic chains, self-assembled cells, experiments indicating how an initial genetic coding mechanism could have arisen, the onset of bioelectricity, comprehensive membrane properties, reproduction, assemblies of cells, and others (Fox and Dose, 1977; Fox, 1980; Fox, 1988).

Related to the primary question of which kind of informational biomacro-molecule, protein or nucleic acid, arose first, is the question of whence came that information? The one answer, which has several significances, is that informational content for the first protein molecules was derived from the amino acids that reacted to form thermal proteins. This follows from the finding that amino acids can be reacted to form proteins directly (Fox and Harada, 1958) (Fig. 2).

It is of interest that Fritz Lipmann (Kleinkauf et al., 1988) found he could produce a presumably primitive polypeptide without direct involvement of DNA or RNA; only protein and amino acids were used. Incidentally, Lipmann several times (1972, 1984) acknowledged motivation from the present program for his active interest in his studies, which became a principal interest in his latest years.

Lipmann, however, attributed the directive effects on amino acids in large part to a protein template that functioned instead of the nucleic acid template usually

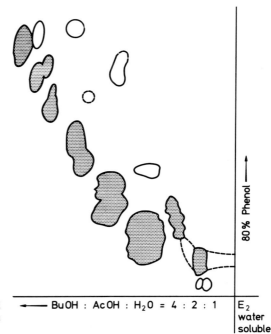

Fig. 2. Chromatogram of hydrolyzed sample of polymer of the common amino acids

BuOH : AcOH : H_2O = 4 : 2 : 1

E_2 water soluble

80 % Phenol

invoked. If he had looked further into the mechanism, Lipmann might have as well or better attributed the main "patternization" effects to the self-sequencing of the free amino acids, much as Ivanov and Förtsch (1986) later did.

If protein and nucleic acid template were each superfluous, as the experiments suggest, how did amino acids line up if there was no template for them to line up on? The answer, as has already been mentioned, is that as many as twenty kinds of amino acid order, or sequence, themselves into a line. This mechanism is an extremely economical one, and seems most appropriate to prebiotic times. It can be best understood on the basis that the family of amino acids consists of a remarkable group of siblings, to which there is nothing comparable in the biochemical world, and the further inference that there was nothing comparable in the protobiochemical world. The amino acids are all related in having both amino and carboxyl groups (which makes polycondensation possible) but in being varied by having different side chains and different configurations of electrons. Their condensation results in predominantly linear molecules, while the stereoelectronic effects of the side chains builds in specificities of interaction in the growing polymer chain.

For the self-sequencing of amino acids there are several kinds of evidence (Fox and Nakashima, 1984); the best is undoubtedly that obtained by sequence analysis of peptides formed (work from which this entire project began; Fox, 1945, 1956). They include the difference in composition between the reacting amino acid mixture and the composition of the polymers obtained, sharply limited heterogeneity in the components of the polymers obtained by fractionation, large differences between compositions of the polymers and the compositions of the N-terminal and the C-terminal portions, and nonrandom sequences in the components as determined by standard methods (Fox, 1945; Fox and Dose, 1977).

The results of fractionation of such polymers on DEAE-cellulose as compared to fractionation on DEAE-cellulose of turtle serum proteins are seen in Fig. 3.

In Fig. 4 a two-dimensional chromatogram of the product of heating of glu, gly, and tyr is presented. This result was surprising in that the individual spots (stained by α-nitrosonaphthol) were found to be as discrete as they are. More surprising was that each one of them showed upon hydrolysis and analysis stoichiometric ratios between the contained amino acids.

Spot 3–3 contains by far the largest amount of any of the peptides formed. It contains all of the tripeptides (Nakashima et al., 1977) except for a very small amount of <glu-tyr-tyr subsequently found in Martinsried (1988). Otherwise 3–3 is only two *tri*peptides: <glu-tyr-gly and <glu-gly-tyr (<signifies pyro or cyclo-).

In Table 1 the peptides obtained are listed, and also those expected if the polycondensation were random, as would usually be anticipated. The results obviously indicate a great nonrandomness. Dr. John Jungck calculated the ratio nonrandomness/randomness to be 19/1 (Nakashima et al., 1977). *This was pro-*

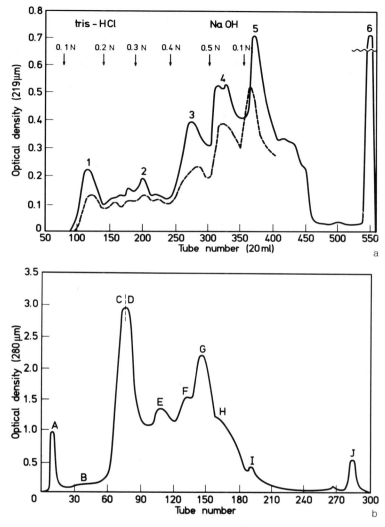

Fig. 3a. Fractionation of amidated 1:1:1-proteinoid on DEAE-cellulose column. A few major fractions are observed. Repetitions yield similar patterns. The dotted line represents one of these; **b** Distribution of turtle serum proteins on DEAE-SF-cellulose column in sodium phosphate buffer (Block and Keller, 1960, in Fox and Nakashima, 1984)

bably a first quantitative evaluation of nonrandomness. Since no template is present when the polycondensation begins, we infer that the reactant amino acids have constructed their own sequences. This is to be understood on the basis that each of the twenty kinds of amino acid has its own steric and electronic identity, as stated.

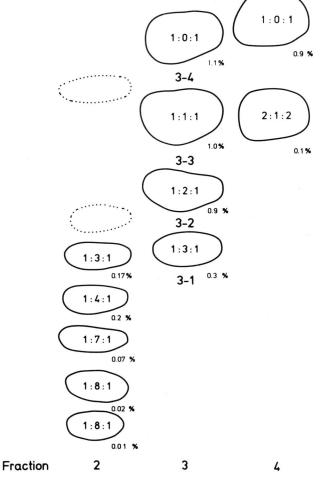

Fig. 4. Discrete peptides by paper chromatography of thermal product of glutamic acid, glycine, and tyrosine (Nakashima et al. 1977; Hartmann et al. 1981). Dominant fraction is 3–3; it represents all tripeptides formed

5 The Mechanism of Self-Sequencing of Amino Acids

The fundamental answer to the question of the outset of evolution of information and its processing is, as already noted, the self-sequencing of amino acids. The answer to the question of the necessary mechanism has been partially given, but it is not complete. What has been learned is that pyroglutamic acid, which results

Table 1. Tyrosine-containing tripeptides found vs. those expected on the basis of the random hypothesis

Expected from *Random* Polymerization		*Found* from *Nonrandom* Polymerization
αUαUY	YαUU	
αUγUY	YγUU	
γUαUY	YαUG	
γUγUY	YγUG	
αUGY	YαUY	
γUGY	YγUY	
αUYU	PαUY	
γUYU	PγUY	
αUYG	PGY	*PGY*
γUYG	PYU	
αUYY	PYG	*PYG*
γUYY	PYY	
GαUY	YGU	
GγUY	YGG	
GGY	YGY	
GYU	YYU	
GYG	YYG	
GYY	YYY	

Note: The dominant fraction obtained from the thermal copolymerization of glutamic acid, glycine, and tyrosine proved to be an equimolar complex of pyroglutamylglycyltyrosine and pyroglutamyl-tyrosylglycine (Nakashima et al., 1977; Hartmann et al., 1981) U = glutamic acid residue, Y = tyrosine residue, G = glycine residue, P = N-pyroglutamyl from unfractionated peptides, each had but a single amino acid in each terminal residue. In the N-terminal position, the only amino acid was pyro-glutamic acid. Three individual peptides could, of course, have no more than three C-termini in total. Three of the six conceptual possibilities were thus found. Each C-terminus was singular: glycine, alanine, or leucine. While multiple C-terminal types are found analytically in unfractionated pro-teinoids (Harada and Fox, 1975), the singularity suggests three single peptides. Proline and phenyl-alanine were totally absent from the C-terminal and N-terminal positions. Glutamic acid, which in the pyro form is known as an N → C polymerization initiator (Fox 1980), is totally absent from all three C-termini

easily by warming glutamic acid, is an excellent N → C polymerization initiator (Melius and Sheng, 1975; Fox, 1980). The fact that N-pyroglutamic acid starts one or more, usually more, amino acid sequences results in induced internal arrangements in those sequences.

The main aspects of what was been learned about mechanism are that N-pyroglutamic acid is a polymerization initiator, that aspartic acid and glutamic acid are elongators, and tyrosine is a terminator. Since it is increasingly clear that protein alone could have served as a bridge from prebiotic matter to the first animate systems, the question of how that bridging protein received, or developed, its own internal information becomes extremely fundamental to understanding the emergence of life and its initial information content.

In pursuit of the answer to this question, the P.I. spent April to September 1988 in the Protein Department of Director Dr. Gerhard Braunitzer at the Max Planck Institute for Biochemistry, Martinsried, Germany. Braunitzer (1984) (who died on 27 May, 1989) had determined the primary sequences in haemoglobins from over 200 species. Braunitzer's deep interest in the history of science led him to surround himself with those whom he identified with the "ursprung" (origin) of his studies. For example, he made labs available for Pehr Edman of the Edman Sequenator. These labs were used by Edman until his death, and then by Dr. Agnes Henschen, widow and scientific colleague of both Edman and Braunitzer until the death of the latter. In his invitation to the P.I.–author of this paper he was acknowledging, as he told visitors, pioneering work on sequence determination (Fox, 1945, et seq.). In the course of this association, Braunitzer became further interested in the evolutionary origins of protein, having already aided work on the inheritance of the self-ordering mechanism by Ivano and Förtsch (*Origins Life* 17: 35–49, 1986), in which his suggestions and support are acknowledged.

Braunitzer and the P.I., with Dr. Peter Rücknagel, established or confirmed that:

1. The thermal polycondensation of amino acids is highly nonrandom.
2. The initial steps of thermal protein synthesis reproducibly yield a few component macromolecules.
3. The synthesis of larger molecules evidently involves a kind of "springboard" mechanism in which small and quite large peptides are more favoured than intermediate-sized ones.
4. Pyroglutamic acid is a frequent feature of thermal proteins, and also of a number of modern proteins.
5. Use of Melius' techniques (Melius and Sheng, 1975) to convert pyroglutamic acid residues to glutamic acid residues has made sequential studies a possibility through at least six residues (incomplete experiments in Martinsried).
6. As a step in evolution, the patterns obtained suggest diversity within near-unity rather than unity within diversity.

6 Future Studies of Sequence

1. Survey of modern N-pyroglutamyl proteins and peptides. Approximately 125 were identified in August 1988 at MIPS (Martinsried Institute for Protein Sequences).
2. Compositional values of amino acid types in the 2nd, 3rd, 4th, and 5th positions should be calculated.
3. Glutamic acid to be heated with various combinations suggested in the preceding evaluation, and under varied conditions. While the laboratory

model suggested the computer work that has been done, the computer studies may now lead to ongoing laboratory models.

4. Relating HPLC analyses of polyamino acids to oscilloscope patterns of proteinoid microspheres assembled from such polymers. This advance converts molecular electronics from a primarily theoretical exercise to an experimental avenue.

5. Spinoffs are: potential applications to computers more versatile than the present generation, clarification of the protein-cell-body-mind relationships, and therapy that has already begun to appear.

7 Self-Organization of Thermal Proteins to Cells

The spontaneous formation of sequences of amino acids starting with the amino acids themselves is a kind of self-organization (plus bond formation). It is a unique process inasmuch as the properties, especially the reactions, of the amino acids are found in a family of molecules without parallel in nature. The twenty types of amino acid that we know best are sibling molecules. They all bear the family imprint of amino and acid groups; they are distinguished one from the other by their characteristic side chains. While the amino and acid groups participate in the formation of peptide bonds, the contributions of the side chains are responsible for the rate at which any one type of amino acid couples with another or with the growing peptide chain.

The terminology employed with amino acid sequences reflects the dominance of the analytic approach over the synthetic approach in science. Strictly speaking, the sequences determined analytically are amino acid residue sequences. The use of primary analysis on peptides and proteins has come to be spoken of as "sequencing" the peptides or proteins. The verb "sequencing" belongs more properly to synthesis from amino acids, however.

While the amino acids themselves function to yield a spontaneous ordering into peptide chains, the peptides formed manifest their attractions by organizing themselves into cells through intramacromolecular binding. This is a crucial step, discovered in this program (Fox et al., 1959). The attractions of one amino acid for another when they are largely fixed in molecules of larger size can then be expressed by the pulling inward of parts of the peptide molecule. The tendency of polyamino acid molecules to organize themselves thus is believed to be at the heart of the original cell formation (Fox, 1985). This kind of expression of information has not received detailed study; it offers much promise (Fig. 5).

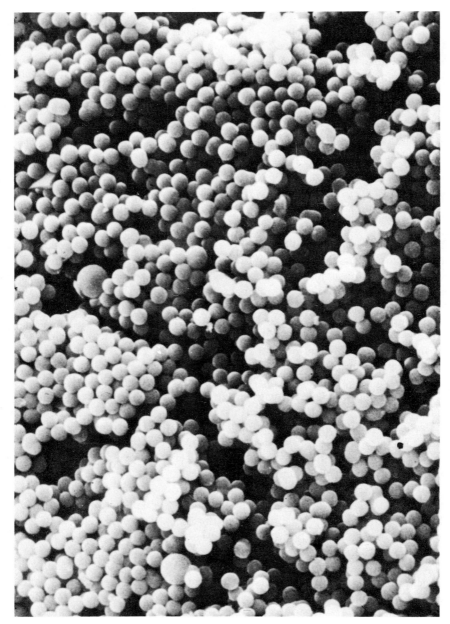

Proteinoid Microspheres

Fig. 5. Scanning electron micrograph of proteinoid microspheres. Note uniformity and huge number. These are nearly all of a narrow diameter size within a range of 1–3 microns. Illustration prepared by Mr. Steven Brooke of Steven Brooke Studios

8 The "Mind-Body" Product

In the beginning of this research, it was not visualized that the flowsheet would enlarge to the emergence of life and mind from, first, the study of the emergence of protein and, later, the study of the emergence of life from proteins. Once again, the *direction* of approach is found to be crucial. When one employs the standard approach of science, that of analysis, one cannot analyse life, and one cannot analyse mind, without dissecting them in a way that causes loss of their essential property of organization. The investigator can only experiment, somewhat as nature has done, to assemble putative components to yield, first the cell, and later, as suggested by that first exercise, the mind.

The "mind-body" problem (Popper and Eccles, 1977; Uttall, 1977) should, accordingly, be referred to as the *body-mind* problem. The "mind-brain" problem, likewise and for the same reason, should be referred to as the *brain-mind* problem. This understanding has been recognised since the early 1970s (Young, 1984); see Purves (1990).

Much as the dicovery that cells will assemble from wetted thermal protein, the properties of the units of the brain were found in the artificial neurons, which are simply in experimentation special properties of the artificial cells.

9 Artificial Neurons

Approximately thirty years of research have gone into cataloguing and understanding the emergence of cells. These are properly referred to, of course, as artificial cells. But they are not simply models. They represent retracement of the

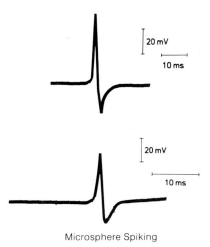

Fig. 6. Action potential resembling that of neuron. (Upper) Spiking in crayfish stretch receptor neuron. (Lower) Spiking in microsphere of 2:2:1-proteinoid. (By Dr. A. Przybylski.)

evolutionary pathway by which more and more modern cells arose. This is reassuring because we know in some respects quite thoroughly what it is that eventuated from evolution.

As indicated, the properties of the artificial cells have been rather thoroughly catalogued. The electrical properties, coupled with the connectional properties

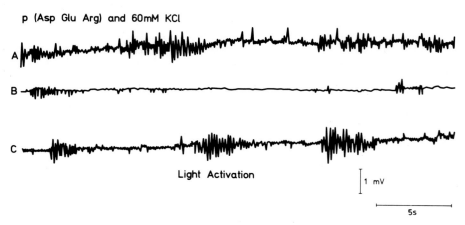

Fig. 7 A–C. Electrical activity begins with illumination (10 lux) at **A**. It dies off when light is extinguished (**B**), and it begins again with reillumination (100 lux) at **C**

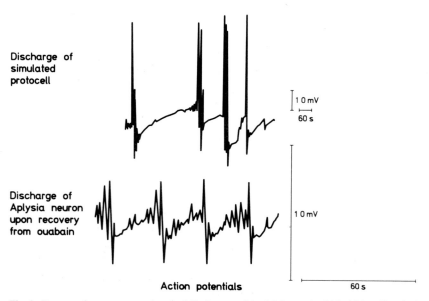

Fig. 8. Pattern of spontaneous electrical discharges of the 2:2:1 proteinoid-lecithin cell and of *Aplysia* neuron

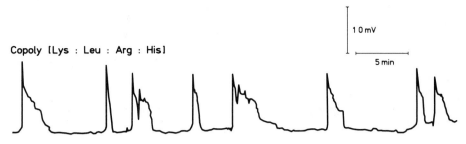

Fig. 9. Pattern of spontaneous electrical discharges of the proteinoid cell made of copoly (lys, leu, arg, his)

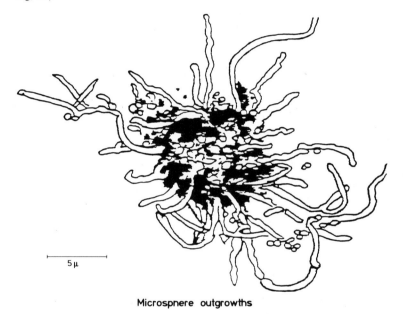

Microsphere outgrowths

Fig. 10. A microsphere made from leucine, proline-rich proteinoid projects outgrowths resembling those from neurons

of some of these cells (Figs. 6 to 10) has allowed them to be regarded by some experts in the sense of artificial neurons.

This perspective allows us to engage in the possibility of constructing, or engineering, an artificial brain through the artificial neuron and the self-organizing properties that were employed by evolutionary processes.

The starting point is informed thermal protein. Thermal protein has yielded cells in one step. As a result of studying the properties of these cells we realise that the details indicate that a large proportion of them, or all, are excitable. The physiologist Howland (1973) has characterised all cells as excitable. This agrees smoothly with what has been found in the work here. In fact, we can with much assurance state that the electrical behaviour is due to the protein portion of the cellular membrane and that such can be modified by an appropriate lipid.

The mind-body resolution is then one of thermal protein yielding cells, and the thermal protein also simultaneously yielding excitable cells. When the protein is of an appropriate compositional type the cells are neurons. *Starting thus with informed proteins, the appearance on the Earth of thermal proteins led quite directly into neurons, which then (experiments yet incomplete) evolved by self-association into brains.* This picture is of course greatly simplifying and greatly unifying, even as it omits innumerable details.

In the final section we will review some of the principal functions of proteinoid microspheres that permit them to be referred to as artificial neurons. It is especially simplifying that both cells in general and the excitable type that function as neurons in living systems are so closely derived from the same source. Crucial to that recognition was the finding that *informed* proteins could arise so easily and that *informed* cells could arise in turn by self-organization. Also crucial to this judgement is the finding of other cytofunctions.

10 The Evolutionary Direction

In many evolutionary studies the direction of the flow of events is largely overlooked. Science is predominantly analytic and approaches problems from the outside (e.g., into the cell). Evolution is however synthetic and in its flow of events moves from the inside out. For example, we know of protein by dismantling the cell and by charaterising the extracted materials. In order to learn about an evolution from protein to cell we had to synthesize protein from its junior precursors. That, however, is the evolutionary direction, opposite to the analytical direction.

From studies in the evolutionary direction, we could ascertain what we could not learn from analysis. Thereby it became possible to understand how first cells arose by self-organization of the thermal protein, whereas analytical studies of cells merely destroy the evolved organization.

Before this research began, the question could be asked: did the type of information processing influence later stages of evolution? When amino acid self-sequencing was suggested as the original source of bioinformation, it became possible to begin to recognise how that source influenced subsequent developments. In 1986, using data from the base in Martinsried, Ivanov and Förtsch explained that amino acid sequences in *modern* proteins are the consequence of self-sequencing. Also explained thereby were the results of Lipmann with polypeptide antibiotics, in which specific amino acids are added to the C-terminus one at a time.

The organization of thermal proteins into cellular structures is not limited to self-organization into cellular structures such has have served empirically as models for the protoorganism. The thermal proteins, especially those rich in basic amino acids, e.g., lysine, organize cooperatively with other compounds, such as lipids and polynucleotides. The kind of datum obtained has provided understanding of how the first and later ribosomes came into existence (Fox and Dose, 1977, pp 232–240; Waehneldt and Fox, 1968). This area may be especially valuable and ready for future study of information processing.

The parallelisms between natural protein-polynucleotide complexes and the artificial ones examined in this project also deserve further study. In one example, lysine-rich proteinoid has been treated with, on the one hand, DNA, and on the other hand with RNA. The one containing DNA is fibrous whereas the one containing RNA in this controlled experiment is globular (Fox and Dose, 1977, p. 234) (Fig. 5). In natural systems, too, DNA is known to yield fibrous complexes, whereas RNA yields globular ones.

11 Biofunctions

The term "evolution of information processing" connotes in this paper "evolution of biofunctional information processing". In preceding sections evidence has been reviewed from heuristic experiments, for which the results showing the emergence of increasingly complex molecular structures are regarded as demonstrating the origination and subsequent emergence of biomacromolecular information. That the information in these molecular structures is biofunctional has been assumed largely on the basis that the kinds of compound known to be the repository of modern biological information have been produced in the laboratory under widespread geological conditions that exist now, and therefore plausibly existed earlier.

The argument has focussed on the view that the necessary kind of compound to contain sufficient information must be a polymer. The two plausible kinds of polymer are the variegated ones – nucleic acids and proteins. The experiments, and classical biochemistry, have indicated biofunctional molecules to be proteins while the nucleic acids serve as repositories, coded repositories, of information. The new research on origins has been initially all chemical. However, evidence has accumulated that a limited variety of appropriate chemical structures, not just exact duplicates of modern protein, could have arisen from the evolution of such informational molecules. The common feature for information appears to be in each case the exact spectrum of constellations of amino acid side chains in polymers. What has been learned, then, is that the kind of molecule that represents biofunctionality has been shown to be capable of spontaneous origin and also, in more recent research, that the biofunctions are themselves present in such polymers. By 1989, a long list of specific and general biological functions has been located in fairly full measure or, in a few cases, in root amounts in thermal proteins and their aggregation products. The first kind of evidence, the structural, has been indirect whereas the second kind, that demonstrating biofunctionality, is direct. Both kinds of evidence are now abundant; the latter is reviewed in what follows.

Enzymatic. Shortly after the finding was reported that one could make polymers of α-amino acids in the laboratory by simple heating of α-amino acids (Fox and Middlebrook, 1954; Fox and Harada, 1958; Fox, 1960a), a number of colleagues asked for either samples or information so that they could test proteinoids for enzymatic activity. (The term thermal proteins was applied subsequently by Chemical Abstracts to replace the term proteinoid which they had used previously. The term thermal protein came into use in 1972.) What was found in a period of about ten years by various interested biochemists were representative activities of each of the major classes of enzymatic power as categorised by the International Union of Biochemistry. The table of enzymatic activities of thermal proteins has been reviewed a number of times (Rohlfing and Fox, 1969; Melius, 1982; Dose, 1984; Fox, 1980). The activities thus found are all weaker than those recorded for the modern counterpart enzymes, but this is explained on the basis that the power of an enzyme increased during evolution, probably with chiral purification of the monomers during evolution, an idea developed by W.A. Bonner, at Stanford University.

The enzymatic activities were found not only in the polymers but also in the microspheres assembled therefrom. Indeed, the activities are sometimes greater when the agents are in the microsphere form than when in aqueous solution. A few instances of coupled activities have been reported (Rohlfing and Fox, 1969; Fox, 1980) and have made their way into textbooks (Wessells and Hopson, 1988, pp. 443–444). These, then, are instances of the emergence of metabolic units, which are of more relevance than single enzymatic activities. Moreover, these examples include specificity in enzymatically active thermal proteins, e.g., lysine-rich proteinoid catalyzes the decarboxylation of oxaloacetic acid, whereas acidic proteinoid catalyzes the decarboxyalation of its reaction product, pyruvic acid. These two reactions already constitute a metabolic unit.

Classical cytofunctions. The activities that have been imputed for a long time by biologists to living systems are the properties of metabolism, growth and reproduction. These have been found in primitive form in proteinoid microspheres, as well as many other functions. (Fox, 1980, 1988). Except for the property of growth, each of these has been reproduced in textbooks by various authors. The property of growth is in some ways the most characteristic of living systems. In the experiments already published it can be seen that the growth of microspheres is quite exactly programmed. This then is a property of the protein molecules, without need to explain the control by invoking nucleic acids. The internal control is seen in the high uniformity of diameter of the microspheres.

Inasmuch as the self-sequencing of amino acids has been reported as a property of modern proteins by Ivanov and Förtsch (1986) the new data emphasizing internal control of cellular growth suggests looking for the controls of programmed growth and their metabolic destruction in pursuit of understanding cancer processes.

Among other major properties that have been found are bioelectricity, membrane selectivity, ability to synthesize peptide and polynucleotide bonds, etc.

Other Cytofunctions. Yet other functions have been found in primitive form and related to specific thermal proteins and the intermediate proteinoid microspheres. They include intercellular recognition, hormonal activity (MSH), photochemical response, electrotaxis, compartmentalization (much influenced by calcium), osmotic behaviour, motility, attraction and avoidance, Brownian motion, conjugation, protocommunication between microspheres (a) via endoparticles and (b) via electric signals; antiaging, and protosocial protection.

Roots of Cerebrofunctions. In addition to enzymatic functions and classical cytofunctions, a group of what belongs under other cytofunctions is here however singled out for special attention. Any behaviour that appears to relate to cerebral activity at its evolutionary roots is potentially in a special position in EIP. Most, but not all of these, have been studied in proteinoid microspheres impaled by microelectrodes (Przybylski and Fox, 1986).

Among the non-electrophysiological attributes are extension of outgrowths and prolongation of life of true neurons when appropriate "primordial polymers" are added to cultures (Fox et al., 1987; Hefti, 1991), induction of enhanced memory when similar polymers are injected into mice (Fox and Flood, unpublished), and formation of junctions between microspheres (Fox et al., 1988). Indications of oscillograms characteristic of the composition of the polymers is then more promising (Przybylski and Fox, 1986). In this connection it should be recalled that the evidence is that the composition determines the sequence.

12 Bidirectional Transfer of Information

A principal conceptual resistance to the inferences from the experimental findings that indicate that the original source of biological information was protein is undoubtedly the mindset induced by the Watson–Crick Central Dogma, as touched on earlier. Crick's (1958) exact words were "the transfer of information from nucleic acid to nucleic acid, or from nucleic acid to protein may be possible, but transfer from protein to protein, or from protein to nucleic acid is impossible."

In his esteemed book on Origins, Shapiro (1986, pp. 290–291) points out that Shapiro interviewed Crick to learn that Crick had meant Central Hypothesis rather than Central Dogma. Shapiro also quotes Crick (1970) who said that the Central Dogma "was intended to apply only to present-day organisms and not to events in the remote past, such as the origin of life or the origin of the code". This judgment is compatible with our inference that the information can flow in

Table 2. (turbidities)

Polyribonucleotide	Lys-Rich Proteinoid	Arg-Rich Proteinoid
Poly C	+ + + +	0
Poly U	+	+
Poly A	0	+
Poly G	0	+ + +
Poly I	0	+ + + +

either direction (experimental) but only in one direction when the mechanism arose in organisms. We see also that, in an analytical perspective, composition can be inferred from sequence, while in a synthetic perspective from experiments, *composition determines sequence.*

The experiments of Yuki and Fox (1969) are relevant for the prebiotic relationships. Those experiments with models of prebiotic nucleic acids and prebiotic proteins indicated that information could be transferred either from polynucleotides to polyamino acids or from polyamino acids (thermal proteins) to polynucleotides in noncellular systems. Since thermal proteins are models for prebiotic proteins (Crick's "remote past") the Central Dogma does not apply. Shapiro favours proteins as the first informational macromolecules; this may be especially noteworthy inasmuch as Shapiro is himself a nucleic acid chemist and is among those who have been unable to rationalize the unsupported assumption that nucleic acids arose first. The results of Yuki's experiments are in Table 2.

The possibility that information could have been transferred in a kind of "reverse translation" from proteins has been suggested by a number of authors apart from these experiments. Some evidence exists of the possibility of such a process (Fox and Nakashima, 1984) and it is supported by investigators such as de Duve (1988), Shapiro (1986) and Fox (Fox and Dose, 1972). This is however an area in which more comprehensive experimentation should be carried out.

13 Informational Electrical Responses

Communication between microspheres is recognised as occurring in a fast mechanism and a slow one, as in modern systems. The slow one in the laboratory involves the transfer of informational proteinoid endoparticles from one spherule to another. This observation was made in 1970 (Hsu et al., 1971) and it is believed to have preceded the finding of transfer from modern cell to modern cell, as described by the Western Reserve workers (Lasek et al., 1977). Its identification was preceded by the finding that macrospheres avidly form junctions with each other and the junctions are hollow (Hsu et al., 1971). The proteinoid "endoparticles" are informational, since proteinoids are known to react selectively with other proteinoids (Hsu and Fox, 1976), to react selectively with enzyme substrates (Fox and Dose, 1977), and to react selectively with polynucleotides (Yuki and Fox, 1969).

The fast communication in the laboratory is electrical. It was sought because of the earlier finding of the slow kind of communication (endoparticle transfer), because of the finding that the proteinoid membranes have electrical properties (Ishima et al., 1981; Przybylski and Fox, 1986) and because of the urging of the late physiologist, H. Burr Steinbach. To seek such properties first required awarenesses that had come along in the development of the research based on the evidence for proteins-first. One of the resultant inferences was the realization that the artificial protein was no more heterogeneous than unfractionated protein from most natural sources. The other change in mindset came from the finding that the electrical properties of the membrane are centered in the protein portion rather than the lipid, much as Nachmansohn had been insisting on in a rather lonely emphasis about 1970 (Nachmansohn, 1970).

Relatively specific readouts have been recorded for specific thermal proteins which were then assembled into microspheres and impaled with microelectrodes (Przybylski and Fox, 1986). This is the one known research route that at present offers promise of explaining how specific information contained in neuronal systems as involved in learning can be produced and monitored (Schmitt, 1962; Fox, 1988). What is needed here is correlation of amino acid composition with electrical readout and with a cellular property such as photoreactivity (perhaps with Dr. Gordon Tollin), or with supramicrosphere morphology correlated with amino acid composition (this work has begun with Dr. Aristotel Pappelis). This kind of correlative, collaborative research is believed by the writer to be an avenue of outstanding promise. It includes the possibility of computerizing creative thinking, and of retracing the construction of a truly artificial brain.

The ways in which neurons interact in networks can be studied in these retracement models due to the fact that the proteinoid microspheres readily, and sometimes avidly, interact to form clusters. They also form junctions readily (Hsu et al., 1971).

A further supporting spinoff is found in the observation that administration of selected members of the thermal protein group enhance memory in mice. The first experiments (with Professor Franz Hefti) were set up on the basis that the model protoorganism, when appropriately constituted chemically, extends outgrowths. In the light of the connectionist theory of Hebb (1949), the outgrowths were hypothesized as being also connecting units and they were therefore tested in mouse brain by Dr. James Flood for memory effects. Such polymers as exhibited the outgrowth effects in the model were found significantly to promote memory-enhancement when injected into mice intracerebroventricularly (Fox and Flood, 1991). This is to be viewed in conjunction with the finding of Hefti that similar polymers stimulate growth of true axons and furthermore promote extended life in the neurons (Fox et al., 1987; Hefti et al., 1991).

The results suggest the broadest kind of evidence for both the vitality and the informational capacity of the continuously evolving thermal protein system.

Acknowledgements. Gratitude is expressed to a number of agencies that have supported the research, especially the National Aeronautics and Space Administration, the National Foundation for Cancer Research, and others cited in individual papers. Guidance in thinking and assistance in formulating concepts of the evolution of biological information are due to Professor Dr. Klaus Haefner and the Bertelsmann Foundation. Special thanks are recorded to Julie K Rowell for major help in composing this paper.

References

Braunitzer G (1984) Sequencing of hemoglobins. In: Brussels Hemoglobin Symposium 1983, Shnek AG, Paul C (eds) Univ Bruselles, Belgium, pp 341–353

Brooke S, Fox SW (1977) Compartmentalization in proteinoid microspheres, BioSystems 9, 1 p 22

Cech TR (1983) RNA splicing. Three themes with variations, Cell 34, pp 713–716

Crick FHC (1985) Protein synthesis, Symp Soc Exptl Biol 12, pp 6–30

Crick F (1970) Central dogma of molecular biology, Nature 227, pp 561–563

Crick F (1981) Life Itself: Its Origin and Nature. Simon and Schuster, New York

de Duve C (1988) Prebiotic syntheses and the mechanism of early chemical evolution. In: The Roots of Modern Biochemistry, Kleinkauf H, von Döhren H, Jaenicke L (eds) de Gruyter Co, Berlin, pp 881–894

von Döhren H, Kleinkauf H (1988) Research on nonribosomal systems: biosynthesis of peptide antibiotics. In: The Roots of Modern Biochemistry, Kleinkauf H, von Döhren H, Jaenicke L (eds) de Gruyter Co, Berlin, pp 355–367

Dose K (1984) Self-instructed condensation of amino acids and the origin of biological information, Int'l Quantum Chem, Quant Biol Symp 11, pp 91–101

Eigen M (1971a) Self-organization of matter and the evolution of biological macromolecules, Naturwissenschaften 58, pp 465–523

Eigen M (1971b) Molecular self-organization and the early stages of evolution, Quart Rev Biophys 4, pp 149–212

Eigen M (1986) The physics of molecular evolution. In: The Molecular Evolution of Life, Baltscheffsky H, Jörnvall H, Rigler E (eds), Cambridge University Press, Cambridge, England, p 25

Fox SW (1945) Terminal amino acids in peptides and proteins. Advances Protein Chem 2, pp 155–177

Fox SW (1956) Evolution of protein molecules and thermal synthesis of biochemical substances, Amer Scientist 44, pp 347–359

Fox SW (1960a) How did life begin? Science 123, pp 200–208

Fox SW (1960b) Self-organizing phenomena and the first life, Yearbook Soc Genl Systems Res 5, pp 57–60

Fox SW (1977) Bioorganic chemistry and the emergence of the first cell. In: van Tamelen EE, (ed). Bioorganic Chemistry 3, Academic Press, New York, pp 21–32

Fox SW (1980a) The origins of behavior in macromolecules and protocells, Comp Biochem Physiol 67B, pp 423–436

Fox SW (1980b) Life from an orderly cosmos, Naturwissenschaften 67, pp 576–581

Fox SW (1981) Copolyamino acid fractionation and protobiochemistry, J Chromatogr 215, pp 115–120

Fox SW (1985) Protobiological self-organization. In: Structure and Motion: Membranes, Nucleic Acids & Proteins, Clementi E, Corongiu E, Sarma RH (eds) Adenine Press, Guilderland, New York, pp 101–114

Fox SW (1988) The Emergence of Life, Darwinian Evolution from the Inside, Basic Books, New York

Fox SW, Dose K (1972, 1977) Molecular Evolution and the Origin of Life, rev edition, Marcel Dekker, New York

Fox SW, Flood J (1990) Effect of thermal proteins on memory processing for T-maze footshock avoidance training, Submitted

Fox SW, Harada K (1958) Thermal copolymerization of amino acids to a product resembling protein, Science, **128**, p 1214

Fox SW, Harada K (1960) The thermal copolymerization of amino acids common to protein, J Amer Chem Soc **82**, pp 3745–3751

Fox SW, Middlebrook M (1954) Anhydrocopolymerization of amino acids under primitive terrestrial conditions, Federation Proc **13**, p 211

Fox SW, Nakashima T (1984) Endogenously determined variants as precursors for natural selection. In: Individuality and Determinism, Fox SW (ed) Plenum Press, New York, pp 185–201

Fox SW, Harada K, Kendrick J (1959) Production of spherules from synthetic proteinoids and hot water, Science **129**, pp 1221–1223

Fox SW, Hefti F, Hartikka J, Junard E, Przybylski AT, Vaughan G (1987) Pharmacological activities in thermal proteins: relationships in molecular evolution, Intl J Quantum Chem Quantum Biol Symp **14**, pp 347–349

Harada K, Fox SW (1975) Characterization of functional groups of acidic thermal polymers of α-amino acids, BioSystems **7**, pp 213–221

Hartmann J, Brand MC, Dose K (1981) Formation of specific amino acid sequences during thermal polymerization of amino acids, BioSystems **13**, pp 141–147

Hebb DO (1949) The Organization of Behavior. John Wiley & Sons, New York

Hefti F, Junard EO, Knüsel B, Strauss WL, Przybylski A, Vaughan G, Fox SW (1991) Promotion of neuronal survival in vitro by thermal proteins and poly(dicarboxylic) amino acids, Brain Research **541**: 273–283

Howland JG (1973) Cell Physiology, Macmillan, New York, p 309

Hsu LL, Fox SW (1976) Interactions between diverse proteinoids and microspheres in simulation of primordial evolution, BioSystems **8**, pp 89–101

Hsu LL, Brooke S, Fox SW (1971) Conjugation of proteinoid microspheres: a model of primordial communication, Curr. Mod. Biol (now BioSystems) **4**, pp 12–25

Ishima Y, Przybylski AT, Fox SW (1981) Electrical membrane phenomena in spherules from proteinoid and lecithin, BioSystems **13**, pp 243–251

Ivanov OC, Förtsch B (1986) Universal regularities in protein primary structure: preference in bonding and periodicity, Origins Life, pp 35–49

Kleinkauf G, von Döhren H, Jaenicke L (eds) (1988) The Roots of Modern Biochemistry, de Gruyter, Berlin

Krampitz G, Fox SW (1969) The condensation of the adenylates of the amino acids common to proteins, Proc Natl Acad Sci USA **62**, pp 399–406

Lasek R, Grainer H, Barker JL (1977) The glia-neuron protein transfer hypothesis, J Cell Biol **74**, pp 501–523

Lehninger AL (1975) Biochemistry, 2nd edn, Worth and Co, New York

Lipmann F (1972) A mechanism for polypeptide synthesis on a protein template. In: Molecular Evolution, Prebiological and Biological, Rohlfing DL, Oparin AI (eds), Plenum Press, New York, pp 261–269

Lipmann F (1984) Pyrophosphate as a possible precursor of ATP. In: Molecular Evolution and Protobiology, Matsuno K, Dose K, Harada K, Rohlfing DL (eds), Plenum Press, New York, pp 133–135

Melius P, Sheng (1975) Thermal condensation of a mixture of six amino acids, Bioorg. Chem **4**, pp 385–391

Nachmansohn D (1970) Proteins in excitable membranes, Science **168**, pp 1059–1966

Nakashima T, Jungck JR, Fox SW, Lederer E, Das BC (1977) A test for randomness in peptides isolated from a thermal polyamino acid, Intl J Quantum Chem Quantum Biol Symp **4**, pp 65–72

Oparin AI, Rohlfing DL (1972) Molecular evolution and Sidney W Fox. In: Molecular Evolution: Prebiological and Biological, Rohlfing DL, Oparin AI (eds), Plenum Press, New York, pp 1–5

Popper KR, Eccles J (1977) The Self and its Brain, Springer, Berlin

Przyblyski AT, Fox SW (1986) Electrical phenomena in proteinoid cells. In: Bioelectrochemistry, Gutmann F, Keyzer H (eds), Plenum Press, New York, pp 377–396

Purves D (1990) A constructionist view of postnatal brain development, Society for Neuroscience Program, St Louis MO, Oct 28–Nov 2, Abstract **255**, p 119

Rohlfing DL, Fox SW (1969) Catalytic activities of thermal polyanhydro-α-amino acids, Advances Catal **20**, pp 373–418

Rosmus J, Deyl Z (1972) Chromatographic methods in the analysis of protein structure. Methods for identification of N-terminal amino acids in peptides and proteins, BJ Chromatogr **70**, pp 221–339

Schmitt FO (1956) Macromolecular interaction patterns in biological systems, Proc Amer Philos Soc **100**, pp 476–486

Schmitt FO (1962) Psychophysics considered at the molecular and submolecular levels. In: Horizons in Biochemistry, Kasha M, Pullman B (eds), Academic Press, New York, pp 437–457

Shapiro R (1986) Origins, Summit Books, New York

Tyagi S, Ponnamperuma C (1990a) A study of peptide synthesis by amino acyl nucleotide anhydrides in presence of complementary homopolynucleotides. In: Prebiological Self-Organization of Matter, Ponnamperuma C, Eirich FR (eds), Deepak Publishing, Hampton, VA, pp 197–210

Tyagi S, Ponnamperuma C (1990b) Nonrandomness in prebiotic peptide synthesis, J Mol Evolution **30**, pp 391–399

Uttall WR (1978) The Psychobiology of Mind, Lawrence Erlbaum Associates, Hillsdale, NJ

Waehneldt TV, Fox SW (1968) The binding of basic proteinoid with organismic or thermally synthesized polynucleotides, Biochem Biophys Acta **160**, pp 239–245

Waldrop MM (1989) Did life really start out in an RNA world? Science **246**, pp 1248–1249

Wessells NK, Hopson JL (1988) Biology, Random House, New York, pp 443–444

Wilson EO (1978) On Human Nature, Bantam Books, New York

Young RS (1984) Prebiological evolution: the constructionist approach to the origin of life. In: Molecular Evolution and Protobiology, Matsuno K, Dose K, Harada K, Rohlfing DL (eds), Plenum Press, New York

Yuki A, Fox SW (1969) Selective formation of particles by binding of pyrimidine polyribonucleotides or purine polyribonucleotides with lysine-rich or arginine-rich proteinoids, Biochem Biophys Res Commun **36**, pp 657–663

Remembering and Planning: A Neuronal Network Model for the Selection of Behaviour and Its Development for Use in Human Language

Jenny Kien

Summary   229

1   Introduction   230

2   Planning in the Motor System   231
2.1 Temporal Organization in Behaviour Time Frames   231
2.2 A Model for Motor Organization Based on Time Frames   234
2.3 The Time Frame Model As a Neuronal Network   236
2.4 Simulation of the Proposed Neuronal Network   238

3   Memory and Motor Planning   239
3.1 Models of Memory   240
3.2 Memory in the Model for Motor Planning   241
3.3 Conclusion – the Animal Model for Planning and Remembering   244

4   Memory and Motor Planning in Humans   245
4.1 Re-Organization of Data Reduction May Lead to Evolution of Symbolization . . .   245
4.2 Use of Neural Symbols   246
4.3 Re-Organization of Human Memory   246
4.3.1 Elaboration of Context-Dependence in the Memory   247
4.3.2 Elaboration of the Memory to Hold Serial Order   247
4.4 Use of Symbols in the Motor System   248
4.5 Conclusion the Model for Human Memory and Motor Organization Systems   248

5   Neural Symbolization, Motor Organization and Language   249
5.1 Relation of Words to Neural Symbols   249
5.2 Attaching Meaning to Symbols and Words   250
5.3 Use of the Motor Organization System for Choosing and Planning
     of Strings of Words   251

6   Biological Evolution and Information Processing – the Limitations of Models   252

References   254

Summary. A description is given of the frame/content relationship between the various temporal orders of magnitude in behaviour, from long term contexts down to the duration of single synaptic potentials. The time frame model of motor organization, consisting of nested levels operating at different time scales, is based on this description and deals with how long term contexts influence instantaneous motor output. This model can be simulated by a neuronal network based on the Hopfield algorithm. The time frame approach and the same algorithm can also be used for memory, resulting in a general model for remembering and planning actions in animals. Simple elaborations of this model, consisting of replication and new interconnections of existing components, lead to symbolization and increased categorising power and so allow application of the model to human memory and motor organization. These elaborations are both evolutionarily plausible and supported by neurological findings. Further, it is shown that language can also be described in terms of time frames and that words may be organized in the same way as actions. Thus, the model can also be extended to remembering and planning choice of words in language.

1 Introduction

Fossils reveal the existence of invertebrates, such as molluscs, annelids and arthropods for 600–400 million years BP; vertebrates appeared circa 450 m.y. years ago (Meglitsch, 1972). In spite of their separation hundreds of millions of years ago these phyla show many similarities in behaviour, and in the components and organization of their nervous systems. Similarities of behaviour most probably derive from the fact that animals face similar physical conditions (e.g., gravitational field or the daily alternation of light and dark), must solve similar problems (e.g., control of movement, interaction with the world) and share similar needs (e.g., for food and reproduction).

The similarities in the organization of the nervous system may also have arisen from the similarity of problems to be solved. For example, the direction of visual movement is analysed by similar visual networks in insects and vertebrates (Kien, 1975) and the organization of their motor systems for preparing and co-ordinating movements such as walking (Altman and Kien, 1987a; Kien, 1990a, b; Kien and Altman, 1991a) are also similar. As well, the neuronal organization underlying choice of behaviour appears to be comparable in molluscs, arthropods and vertebrates (Kien, Winlow and McCrohan, 1991). It is these overall similarities amongst quite different phyla which allow us to derive both general descriptions of behaviour and models for the underlying neuronal organization.

One such general description of behaviour analyses the various temporal orders of magnitude or time frames in behaviour ranging from long term contexts down to the duration of single muscle contractions. A descriptive model deals with how these time frames influence momentary choice of motor output (Altman and Kien, 1991). Based on our finding that the neuronal systems involved in momentary motor planning can be described by a Hopfield (1982) algorithm (Altman and Kien, 1989–1991 in prep.; Kien and Altman, 1991a, b), we have extended this descriptive model to a neural network (Kien et al., 1991 in prep.). This is also derived from the Hopfield algorithm but now includes a temporal structure (Kien et al., 1991 in prep.).

I shall show here that this algorithm is useful not only for motor planning but also for memory. The animal model for remembering and planning can also be applied to human motor and memory systems but requires certain changes to accommodate unique human abilities such as symbolizing and serial ordering. I have already shown how the ability to symbolize can result from simple elaborations within the memory which are evolutionarily feasible (Kien, 1990c).

These elaborations lead to an enrichment in categorising power and are pre-requisites for storing serial order. The human motor and memory model which incorporates these abilities can also be used to model choice of words for it can be shown that language conforms to the same organizational principles as motor behaviour (see MacNeilage, 1987). Therefore, the general network model for remembering and planning actions in animals can be used, with simple modifi-cations, to model human language ability.

Due to the broad scope of the functions covered here, the various steps are presented as general arguments which are necessarily incomplete. The biological theory behind the neuronal network for the motor system, and its computer simulation and analytical testing are presented in detail elsewhere (Kien et al., 1991, in prep.; Altman & Kien, 1991 in prep.).

2 Planning in the Motor System

2.1 Temporal Organization in Behaviour – Time Frames

The world is full of changes, those arising unpredictably outside the animal and those produced by the animal itself. Animals interact with this changing world by producing coherent behaviours; e.g., individual movements are co-ordinated to approach food or to avoid noxious stimuli. The individual movements have no significance; it is only when they are embedded in a context, such as searching for food or fleeing from a noxious stimulus, that they acquire meaning for the animal. Thus, a basic problem which animals must solve is that of using a longer term context or time frame for selecting and organizing the individual movements which form the contents of this frame.

In our description of this frame/content organization (Kien et al., 1991 in prep.; Altman & Kien, 1991 in prep.), we have defined five time frames or levels of temporal organization (Fig. 1). All time frames operate simultaneously; frames in which processes of shorter duration are organized are embedded or nested in those of longer time scale.

The overall frame is given by the goal or global aim of behaviour. This global aim can be achieved by an appropriate strategy or group of behaviours. Each behaviour is made up of one or more routines, each of which is a pattern of move-ments produced by muscle contractions. Strategies may be maintained over hours or days, behaviours for minutes to hours and so on down to muscle contractions with durations of less than a second (Fig. 1).

These time frames can be illustrated by the example of an animal setting off in search of food – be it an ant on a food trail or a foraging vertebrate. The long-term frame in which all actions are selected is the global aim of satisfying hunger. This global aim leads to the selection of a strategy for finding food. This strategy can consist of a combination of behaviours such as searching for the food, getting

232   J. Kien

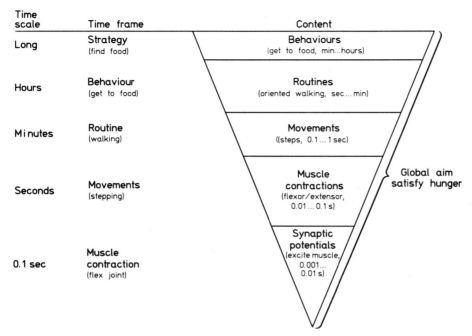

Fig. 1. Temporal levels or time frames in behaviour. Decreasing time scale from top to bottom is represented by a cone. The aim in each time frame sets the context for choice of its content at this level, the contents of each level forming the frame for the level below. All frames are simultaneous, so that the behaviours and routines of the longer time frames contain the motor outputs of the shorter frames. The duration of the frames is relative and may vary by several orders of magnitude

physical contact with the food and preparing it for eating. Each behaviour, such as locating food, consists of a number of routines, walking or climbing, and these routines consist of patterns of movements. Thus, the movement performed by the animal at each instant, e.g., raising a leg to step, takes place in the context of the routine of walking which is performed in the context of the behaviour of getting to food, and so on.

This frame/content organization can be shown schematically with the frames organized as concentric circles, strategy on the outside and muscle contraction in the middle (Fig. 2a). The point at the centre of the circles represents the instantaneous motor activity. The diminishing time scales and the interdependence of the levels can be emphasized by depicting them in the form of a cone (Fig. 2b). Because each level is the frame for the level below, it sets what has to be done in the lower level and so becomes its aim. The content of each level is how the aim should be accomplished, i.e., the plan. Thus, the circumference of the circles in Fig. 2a or the larger face of the cone segments in Fig. 2b represent aims and the areas of the circles or cone segments represent plans.

Each aim is maintained until it is achieved or until there is a sudden change of circumstances, e.g., a sudden change of weather which makes foraging

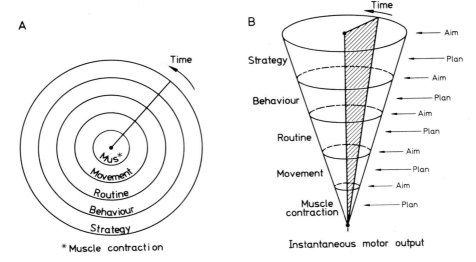

Fig. 2 A, B. Frame/content organization and nesting of the time frames in behaviour. **(A)** Representation as concentric circles to emphasize the nesting of the shorter in the longer frames. The circumference of each circle gives the aim of each time frame, the outer circumference is the overall aim. The plan is given by the area of each circle. The point at the centre of the circles represents the instantaneous motor output, the radius represents the context in which this output is embedded at any instant. The radius sweeps around the circles as time passes and the activity progresses. **(B)** Representation as a cone to emphasize the interdependence of the levels and the diminishing time scales

impossible. In this case new aims and plans must be selected. Once the aim in each time frame has been achieved, the next operation can begin and this may be chosen from several alternatives. For example, in the foraging animal the global aim is to satisfy hunger and a strategy will be planned and carried out until this is achieved. The behaviours in each strategy have their own aims, e.g., the aim of searching for food is to locate it. When this is achieved the next behaviour, getting to the food, can start. Here the aim is to obtain physical contact with the food. Thus, the aim is maintained in each time frame as the planes of this level are run through. The global aim is achieved when the aims in all time frames have been achieved and then a new global aim may arise.

The achievement of each aim must be signalled by internal and external signals resulting from the activity of the animal or from changes in the environment. These signals can be interpreted in terms of achieving the aim only if the plan in each time frame contains internal representations with which these inputs can be compared. In the example of searching for food, the animal must have some internal representation of food in order to recognise it when it finds it; food is located, and the aim achieved, when sensory inputs match the internal representation. These internal representations may be memories set up by past experience or programmes built into the nervous system.

The change from one element in a time frame to the next, e.g., from preparing food to eating it, could be thought of simply as a chain of linked responses, each response being triggered by the sensory input resulting from the previous activity. This, however, does not explain the variability in behaviour seen particularly in field studies (e.g., Rost and Honegger, 1987) where, even in what have been regarded as "fixed" behaviour patterns, one response does not predictably lead to the next. Our description, in which each process derives from a complex interaction in various time frames, predicts such variability.

2.2 A Model for Motor Organization Based on Time Frames

The time frame/content concept can be used to describe the organization of the motor system. The basic time scale in which the nervous systems of both invertebrates and vertebrates operate is on the order of milliseconds (synaptic and action potentials). Movements take hundreds of milliseconds and behaviour operates at the level of seconds or longer (Fig. 1). The motor system must therefore always select its immediate output within a longer-lasting context.

Applying the time frame organization in Fig. 2b to the motor system (Altman & Kien 1991, in prep.), the cone represents the organization of neuronal operations in different temporal levels. The operations in all time frames show, from top to bottom, a decrease in the time scale or prospective duration of the contents which they select and organize. The most important feature of this model is that the different levels all function concurrently and influence each other.

The same two processes operate at every level to organize the large variety of options and motor programmes stored in the memory or to generate new programmes. The CHOOSE process selects the plan for each level. The COMPARE process measures the difference between incoming sensory inputs with the internal representations of the plan of each level. Neither CHOOSE nor COMPARE is an intentional process; these terms are used here to describe the process by which the motor system arrives at one of its many possible output states at any instant.

CHOOSE is the selection and maintenance of both the plan and its internal representations and encompasses the functions of selection, initiation, and, by continual repetition, maintenance of behaviour. These functions are all decision processes. COMPARE tests for consistency and monitors achievement of the aim by measuring the difference between inputs and internal representations. The CHOOSE process continues its operation as long as COMPARE reveals a difference; the aim is achieved when there is no difference and CHOOSE can select the next step in the plan or set up a new plan. COMPARE also tests that inputs remain consistent with experience, i.e., that they do not differ too greatly, both qualitatively and quantitatively from internal representations of the likely consequences of each operation. If an inconsistency does arise, the differences may be so large that CHOOSE cannot maintain the selected plan; this could mean danger, a change in the external world or a mistake, all of which may

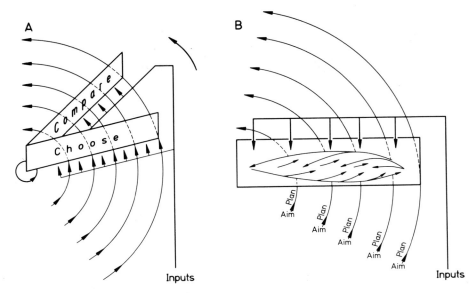

Fig. 3A, B. Operations in the time frame model. The time frames are represented as concentric circles as in Fig. 2a. (**A**) There are two operations in each level, CHOOSE and COMPARE, which are circularly connected. All inputs feed into both CHOOSE and COMPARE. CHOOSE selects and maintains the plan and the internal representations at each level. The difference between the chosen internal representations and the inputs is measured by COMPARE whose outputs are re-entered into CHOOSE. Inputs to the CHOOSE and COMPARE processes are: all sensory information (external and internal world); motivations; motor and experimental memories evoked by sensory inputs and by the activity of CHOOSE and COMPARE; and similarly evoked internal representations including representations of the immediate consequences of actions. (**B**) An expanded schema of one of the operations to emphasize that, due to interactions between levels, each process can be viewed as one operation across all time frames

require large changes at all levels. Note that it is not the comparison which determines whether a difference in inputs is significant or not; it is the effect that the difference has on the CHOOSE process.

The operations in the time frame model, summarized in Fig. 3 are: (i) using all available inputs to CHOOSE an output; (ii) feeding the output of the CHOOSE process at each level into those in all other levels; (iii) comparing the external inputs with the internal representations of what CHOOSE has chosen to be achieved at each level; and (iv) feeding the results of the comparison back to the CHOOSE process of this and other levels. The operations within one level and between levels are essentially circular and the system continually updates itself. The choices in the longer time frames continually influence those in the shorter time frames. Similarly, activity in the lower levels is continually fed back to CHOOSE in upper levels and so will be able to influence the maintenance of the longer lasting choices, perhaps changing them or bringing the whole operation to a stop. That is, the outcome of the CHOOSE process at every level influences the context for the others. There is, thus, one multi-layered CHOOSE process. This also applies to the COMPARE function.

In summary, the time frame model can be described as a multi-layered nested system where inputs – sensory inputs, memory inputs and inputs from the system itself – are simultaneously organized and intergrated in different time frames by a CHOOSE process which derives outputs in each time frame. This time frame model is a descriptive approach to understanding the functional organization of the motor system. The next step is to see how this model can be expressed in neuronal terms.

2.3 The Time Frame Model As a Neuronal Network

A detailed analysis of insect motor systems (Kien, 1983, 1990a, b; Kien and Altman, 1991b; Altman and Kien, 1987a, b, 1989–1991) and comparisons with vertebrate systems (Kien and Altman, 1991a) shows that these systems have 3 major organizational principles:

  (i) large groups or ensembles or neurons work in concert, their activity being best described by their across-fibre-pattern (AFP);
 (ii) the output of each local network within the system is determined by the consensus between the AFP's of the inputs to the local network and the AFP of the neurons in the local network;
(iii) the local networks are all connected to each other either directly or indirectly.

These interconnections ensure that the each network is fed back the results of its effects on the other networks – the property of re-entrance. Thus, functions do not reside in individual networks but rather in the loops formed by their inter-connection. We have suggested (Altman and Kien, 1989; Kien and Altman, 1991b) that these properties can best be modelled by the Hopfield (1982) network design (Fig. 4). The clipped-synapse binary networks we have used range from symmetrical to anti-symmetrical and may receive changing input from sources external to the network (Kien et al., 1991).

Decision-making is also modelled by such networks. In motor systems decision-making is the process of selecting motor outputs appropriate to the input situation and we have shown that this is a distributed process (Altman and Kien, 1987b, 1989; Kien and Altman, 1991b). Hopfield networks model this process in that they tend to relax into a constant or stably oscillating state, an attractor, and this itself is the decision-making process (Kien and Altman, 1991b).

Both the decisions made in motor systems and the process whereby a Hopfield net relaxes into an attractor are the equivalents of the CHOOSE and COMPARE processes of the time frame model. In Fig. 4 all inputs are fed into the elements of the network and, if there is a consensus between inputs and the

Fig. 4. Basic layout of a clipped-synapse binary network based on the Hopfield (1982) network. External inputs feed into "neurons" (I to III) each of which also feeds its output to the other neurons as described by a 3 × 3 connection matrix for the network. Connections between neurons can range from symmetrical through noncorrelated (asymmetrical) to anti-symmetrical, and the strength of connection may lie between −1 and 1. Neurons may be modelled as analogue amplifiers or as binary units with non-linear input functions and a set firing threshold

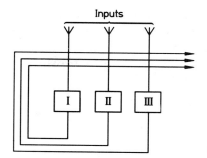

activity of these elements, the network reaches a particular stable state or attractor which is the choice made; this is the CHOOSE process. COMPARE is part of this CHOOSE process. The sensory inputs and the internal representations are all fed to the same network elements where their effects are summed. This is essentially a comparison. The choice is maintained as long as the network remains in this stable state. On the other hand, changes in the inputs can move the network away from its present state; if the move is small the existing choice or plan may be modified, and if the change is large the network will reach a quite new state and a new choice will have been made.

If the CHOOSE process of each time frame and the whole motor system can both be described as Hopfield networks, then the nested interconnections between layers must also form a Hopfield network. One way of nesting the levels in each other is shown in Fig. 5. In this network, the outputs of the subnets represent the aims and plans of the different time frames. New external inputs shift each subnet to a new state which provides new inputs to the other subnets. These change their activity and re-enter the change to the other subnets. No subnet can reach a stable state before it has adjusted to the activity of all the others, which means that the network as a whole must settle to a new state. The CHOOSE process, thus, occurs at all subnets at the same time and is one process. A consequence of this type of operation is that the exact details of movements can never be planned from the start in only the shorter time frames but result from the operation of the whole network.

We envisage that, although all levels work with the same basic time units (milliseconds), the inputs from the upper levels or longer time frames to the lower levels are organized on a longer time scale than the working of the lower levels (e.g., thousands of milliseconds) and that the inputs from lower to upper levels change at a shorter time scale (e.g., milliseconds). That is, inputs from upper levels provide a certain definition of possible output states or framework for the operation of the lower levels, allowing the lower levels to work through their plans without shifting the upper levels out of their long term plan state. On the other hand, as all levels operate on a millisecond basis, large or new inputs from the lower to the upper levels can rapidly shift all levels to a new state.

The functioning of the model can again be illustrated by a hungry animal searching for food. Hunger, signalled by internal receptors, shifts the whole system to a new attractor in which new aims and plans for each frame are generated, e.g., from the food-getting strategy to the motor output to the muscles. The new inputs resulting from the animal's movements keep shifting the system to new states resulting in new movements; there is no plan of the exact flow of movements. These new states lie in the same attractor area as long as the hunger signal maintains the global aim of getting food. When food has been obtained and the aim fulfilled, the new input configuration will shift the whole network into a quite new attractor which gives the aims and plans for new activity.

2.4 Simulation of the Proposed Neuronal Network

The network is Fig. 5 has been simulated (Kien et al., 1991 in prep.) with 4 to 16 neurons with set firing thresholds and spins of $+1,0$. The nesting of the subnets as in Fig. 5 is given by introducing two different delays; each neuron

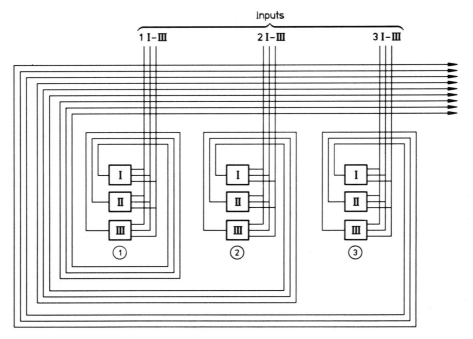

Fig. 5. Network for modelling nested motor organization in 3 time frames. Three basic nets (1 to 3) as in Fig. 4 form a nested arrangement

Fig. 6. One delay matrix used in simulating the network in Fig. 5. Nine neurons arranged in 3 blocks (1–3). Nesting is achieved by structuring the delays between the neurons; here neurons in the same block are connected with a short delay ($d1$), neurons in different blocks have a longer delay ($d2$) in their connections. The delay structure increases the cycle length of network output by several orders of magnitude. However, the particular delay-symmetric arrangement shown here is not necessary; equivalent outputs are obtained where $d1$ and $d2$ are randomly distributed (see Kien et al. 1991)

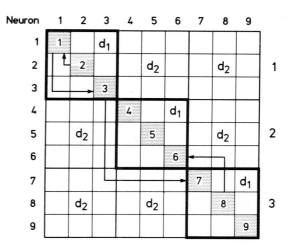

has short delay connections to 1 to 3 neurons, the remaining connections have the longer delay (Fig. 6). This delay structure results in the production of extremely long, stable cycles; delays of less than 5 time units can result in cycles many thousands of units long with complex local patterning within these cycles. These long time cycles are also created when the two delays are randomly distributed and so an ordered distribution of delays as in Fig. 6 is not necessary. The simulations show that a single network with delay structure can produce outputs organized at different time scales, i.e., with temporal characteristics as would be expected from nested time frames. It is, thus, in principle possible for such a network to bridge the gap between milliseconds and seconds.

Application of an oscillating external input to the network can result in complex but stable cycles which need not be related to the input oscillation. Gradual changes in input intensity can result in abrupt transitions to new oscillatory states or they may show no effect. Only at extremely high input amplitudes does the input clearly drive the system in a one-to-one fashion. This property of input-induced fast transitions makes the network ideally suited for modelling motor systems, as this is exactly what we would predict from our descriptive time frame model.

3 Memory and Motor Planning

Remembering and planning involve the same sort of decisions – choosing which memory to activate can be viewed as a similar process to choosing the motor output. Previously, memory has been treated simply as an input to the motor organizing network, but as memory plays a vital role in motor planning, and therefore in the time frame model, it is necessary to examine it in more detail.

Memory and learning of some form are found probably in all animals; this is not surprising because most organisms show some form of long term adaptation to changes in their world. In particular, an animal that is capable of learning from experience and of using its experience to influence new courses of action requires an effective memory. It must be able to retrieve information on past courses of action, the environments in which these actions occurred and the results of the actions as well as value judgements on the efficacy of the actions; these are the data that constitute "experience" (Kien, 1990c).

Experience is required for planning to achieve an aim in the future; memories associated with the aim form a context for planning and setting up the internal representations with which the inputs are compared. This applies down to the level of movements; memory and internal representations are involved in motor programmes like saccadic eye movements (Berthoz and Droulez, 1989) or locust flight (Möhl, 1988) and adapting posture to environmental conditions in insects (Horridge, 1964; Hoyle, 1979). In the time frame model these memories feed into the CHOOSE process at all levels of the network.

3.1 Models of Memory

Many neuronal network models for memory have already been proposed (see Amit, 1989a); which model is most useful here? As it is unlikely that a nervous system uses two quite different algorithms, one for memory processes and another for motor organization, the simplest approach is to try the same algorithm for memory as proposed for the motor system.

Hopfield networks have already been used to model memory storage and retrieval (see Amit, 1989a, b). Their function can be visualized by considering the memory network as a 2-dimensional array of interconnected neurons, and the activation of the neurons as the third dimension. Each possible output state of the network is given by a particular 3-dimensional pattern or activity landscape. Each memory trace is represented by one of the landscapes. It is a property of these networks that an input, which contains part of a particular stored activity landscape or is very similar to it, activates the network to produce the whole landscape. This is content-addressing and associative recall. The network is capable of generalization if inputs which are similar to, but not the same as a memory, also evoke this memory. The number of memories that can be stored rises proportionally with the number of neurons in the network, there being no theoretical limit on the number of elements involved (Gardner, 1988).

The activation or retrieval of memories within these networks is stable; once an input has pushed the network into an attractor and so activated a memory, the network will remain in this attractor until new inputs push it out (Horner, 1988). Continually changing inputs to the network retrieve a flow of memories or associations whose nature depends on the mixture of external and re-entered memory inputs. Both change continually, and to some extent independently, as the world changes and the animal behaves. This means that the sequence of outputs produced by such a network may not necessarily be predictable.

This general network type has been proposed for memory in vertebrates (Amit, 1989b). Similarly, a small Hopfield network has been used as a memory bank in the computer model LIMAX, a detailed simulation of taste memory and categorisation in the mollusc Limax (Gelperin, 1986). In this model, each taste is a pattern of activity in the inputs which activates one stable system state. This becomes the memory that can be retrieved by content-addressing by a similar input. As expected, the network can carry several taste memory patterns simultaneously without mixing them.

The LIMAX type paradigm can be used for memory in the model I am proposing but the network requires considerable enlargement and elaboration as the memory in my model must store the whole experience of the animal. The same form of generalization and associated categorisation can be used but the number of parameters involved in both may increase with the size of the network.

3.2 Memory in the Model for Motor Planning

Memory could be incorporated in the model either by a separate Hopfield network or by simulating memory storage with the configuration coupling strengths in the motor network. If both motor and memory functions were combined in the same network, the network would either become fixed at some stage and no longer learn or remember, or, if the network remained modifiable, the original memories would be lost or modified in the course of the animal's life. This does not appear to happen. Further evidence for the separation of motor and memory systems is given by the Korsakoff syndrome in which patients can no longer recall the past or lay down new memories but are capable of learning new motor skills (Kolb and Whiteshaw, 1985). Thus, the memory and the motor organizing networks must be modelled separately although the coupling strengths in the motor organizing networks may be set by learning processes. Separation of the two nets has the advantage that the memory store stays stable regardless of what the motor organizing system does. Furthermore, loose coupling of Hopfield networks can increase the categorising power of the whole system (Fassnacht, 1990; G. Pöppel, K. Bauer, pers. comm.).

In the time frame motor model the memory receives the same inputs as the motor networks and these activate associated memories which are fed to the various time frames of the motor network. Each time frame requires memories appropriate for setting the context of its plan and for its internal representations. This raises the question of whether it is necessary for the right memory to get to the right place, and if so, how? Is the memory itself also organized in time frames?

Memories are not just of single features but are of sensory and motor experiences embedded in a situational and motivational context that also includes the life history of the animal, e.g., the memory of a new location includes the context of coming to the location as well as local sensory details of the place. This means that what is to be remembered is itself structured in time frames, long lasting contexts and local details. I hypothesize that the same time frames are represented in the operation of the memory network as in the motor organizing network and that the memory network is temporally structured in the same way as the motor network. The time frame nature of the inputs to the memory from behaviour and the world combined with the computational structure of the memory network produce the different time scales in the activity pattern when the memory is recalled.

Such a time frame model of memory can be incorporated in the motor model by linking a second Hopfield network system to the inputs in Fig. 5; the model now consists of two interconnected Hopfield networks which receive the same sensory inputs (Fig. 7). The arrangement on the left is based on a symmetry of design with the motor networks and feeds the output of each nested memory subnet into the equivalent level subnet in the motor system. A second possibility is shown on the right of Fig. 7. Here the output of the memory is distributed over the whole motor network. The two possibilities may be equivalent; our computer simulations have shown that, due to the complete interconnections

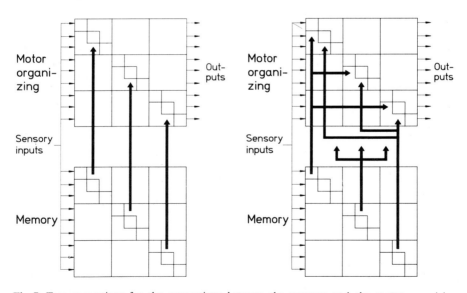

Fig. 7. Two suggestions for the connections between the memory and the motor organizing networks. For simplicity the same network and delay matrix as in Fig. 6 are used to represent each network. The same sensory inputs are fed to both networks. On the left, each block in the memory network, representing each time level, feeds into the corresponding level of the motor organizing network. On the right, the output of each block or level is distributed to all levels of the motor organizing network. It is possible that both arrangements are equivalent. Inputs from the motor organizing networks to the memory are not shown but are assumed to be similar in arrangement

between the elements, the network output does not critically depend on the exact location of an input (Kien et al., 1991a).

The transfer of information from the memory to the motor system raises a further problem. If the memory network is large and each memory trace occupies the whole network, then each memory forms an extremely complex activity landscape all of which must be precisely transferred to the motor networks. Precise patterns of activity across a large group of neurons can be transferred elsewhere only by an equally large read-out mechanism. Therefore some data reduction mechanism must be postulated to process down the memory outputs without losing essential information.

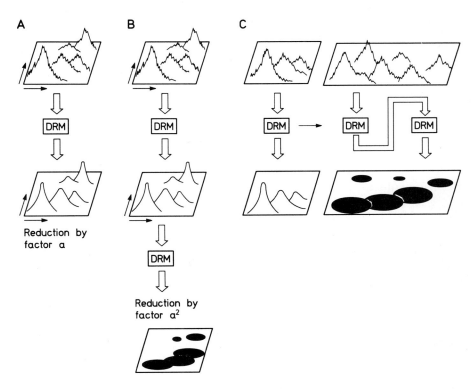

Fig. 8 A,B,C. How re-iterative data reduction can lead to an exponential reduction and to "neural symbols" (**A**) the memory store is represented as a 2-dimensional neuronal array and the activity of the components is graphed in the 3rd dimension. Each memory, when activated, forms a particular activity landscape. This landscape can be passed through a data reduction mechanism (DRM) to smooth it. (**B**) If the smoothed landscape is again passed through a DRM the landscape may be reduced to a 2-dimensional array, for example, delineating the extent of activity over a certain level in the smoothed landscape. This is now a "neural symbol" for the original landscape which it can reactivate in the memory store by content-addressing. (**C**) Increase in brain size leads both to enlargement of the array and to replication of the DRMs. Serially coupling these leads to the production of neural symbols as in B. (After Kien, 1990c)

This need for data reduction must have been exacerbated as the size of the memory networks increased during evolution. The advantage of a large memory store is that the more data that are stored, the more flexible the motor output may be, but to be useful it must be possible to retrieve and transfer these data easily and quickly. The larger the data banks are, the longer these processes may take, resulting in ineffectiveness of the whole system. Therefore as memory stores increase in size there must be a development of data reduction mechanisms to maintain the usefulness of the memory in other parts of the system (Kien, 1990c).

The data reduction mechanisms must not reduce the stored memories. A logical postulate, therefore, is that only the output of the memory is reduced before transfer to other parts of the system; memories would be processed only on retrieval. The simplest possibility is to smooth the memory landscapes down to their essential forms similar to electronic filtering to remove small signals or fluctuations (Fig. 8A). This reduction could be achieved by a higher firing threshold in the output neurons from the memory to the other nets or by a small amount of convergence in the memory output. Both mechanisms can easily be included in the model and computer simulated. The binary elements in the simulated networks already include a firing threshold (Kien et al., 1991).

3.3 Conclusion – the Animal Model for Planning and Remembering

The neuronal network model for remembering and planning, schematized in Fig. 7, is based on the time frame organization of behaviour in which longer term aims and strategies form a context or frame in which the instantaneous motor output is formed. The time frames are formed in the network model by the delay structure within the memory and motor networks which introduces extremely long and complex, stable output cycles. The memory and motor networks interact to create a world of internal representations with which all incoming sensory inputs are compared; matches and/or consistency in the comparison influence the choice of motor outputs. The organization schematized in Fig. 7 predicts that biological motor and memory systems each consist of subfields with multiple interconnections and that there are extensive linkages between both systems. This has been found in vertebrates; many structures are involved in memory and these are multiply connected with each other and with the many structures involved in motor programming (Mishkin and Appenzeller, 1987; Brooks, 1986). Similarly, insects also show a variety of structures involved in memory storage in the brain (Menzel, 1983) and some in the ventral nerve cord which are also involved in motor activity (Hoyle, 1979). The model also predicts that memory is organized in time frames and, indeed, a variety of different memory forms with different time scales is known to exist (Squire, 1987). Thus the model in Fig. 7 is proposed as a general model for both vertebrates and invertebrates.

4 Memory and Motor Planning in Humans

Can the model for remembering and motor planning be extended to humans? Humans differ from all other animals in their extended ability to symbolize (Premack, 1976) and in their ability to organize, store and reproduce long sequences in serial order (Calvin, 1991). These new abilities allow humans to plan their motor activity and organize it syntactically as in manipulating tools (Holloway, 1969). Any model of the human motor system must include these new abilities. However, a model for motor planning in humans must derive from the general animal paradigm if it is to be plausible in evolutionary terms. Indeed, comparison of temporal segmentation of hand movements in chimpanzees and humans (Schöttner, 1990) suggests that there is a quantitative rather than a qualitative difference between human and ape motor systems. Thus, the uniquely human abilities to symbolize extensively and order serially must derive from simple re-organization of those already present. I will show how simple elaborations of the animal paradigm in Sects. 2 and 3, compatible with a possible scenario for human evolution, may lead to these new abilities.

4.1 Re-Organization of Data Reduction May Lead to Evolution of Symbolization

I have argued elsewhere (Kien, 1990c) that the re-organization of the hominid brain became necessary because its increasing size placed increasing pressure on its data management or data reduction systems. A further development in data reduction from that in the model in Sect. 3 (Fig. 8A) to a process unique to hominids could have been achieved by re-iterating or re-entering this smoothing process as shown in Fig. 8B (Kien, 1990c). As the brain enlarged, the neural machinery for the data reduction already present must also have enlarged, most probably by the replication of existing structures (Allman, 1987). The simplest result of replication would be a multiplication of many channels in parallel (Fig. 8C). If the replicated reduction structures are coupled serially instead of in parallel, then the reduced output is reduced a second time resulting in an exponential increase in reduction. Thus, re-iteration achieves a great increase in reduction without the need for developing new processes.

The re-organization to serial coupling could be achieved by a different synaptic coupling within or between cortical columns. This is the only sort of "histological" re-organization that is feasible (Jerison, 1982) and we should expect to find it throughout the neocortex as neurological findings indicate that many cortical areas work together for producing memories and recognition (Damasio, 1989).

The re-iteratively reduced landscape can be visualized as a 2-dimensional plot delineating, for example, the location of peaks in the original memory trace when it is retrieved (Fig. 8B, C). This pattern acts as a "neural symbol" for the original landscape, representing it outside the memory and maintaining the original landscape in the memory through content-addressing when re-entered into the memory. Thus, the symbolization will not result in loss of the original data which remains stored within the memory. This symbolization process is dynamic, the "neural symbol" representing the output of the memory at a given instant. Thus, the neural symbol of a memory will change as the memory trace expands by adding new associations.

4.2 Use of Neural Symbols

Using neural symbols has many advantages. They would be small and convenient and could easily be fed into systems outside the memory that require information about large numbers of experiences or associations. Neural symbols could be held in a small short-term memory or buffer for immediate reference and their small size would permit operations which would otherwise require enormous working capacity, setting up strings, i.e., sequences reproducibly characterised by the serial order of their components and storing relationships.

That symbols can be used to generate strings is possibly their greatest advantage. The ability to order serially is necessary for the generation of both intentional manipulation and language. Movements involved in tool making must be serially organized according to rules dictated by the material and the intended structure. These rules can be considered syntactic and are comparable to the syntax of language (Holloway, 1969).

Symbolization is a prerequisite for serial ordering. If non-symbolized data from the memory were used for serial ordering, the working capacity would have to be very large or only very short strings could be processed. Indeed, only 3 items can be stored in serial order by pigeons whereas monkeys can store at least 5 (d'Amato & Colombo, 1988). The use of symbols, on the other hand, reduces the necessary working capacity and so a working memory buffer could store longer strings, as humans can.

Symbolizing relationships results in an enormous increase not only in the data quantities but also in the type of data that can be stored and worked with. Data may be retrieved not only in the category in which they were perceived (e.g., good tasting apple) but also in a number of other categories expressing their relationship to incoming and also to past data (e.g., as food which relieves low blood sugar). Relationships of relationships can also be derived (e.g., $A = x \cdot B$, $C = y \cdot D$, $x > y$). Humans can do this whereas chimpanzees cannot and this ability is one of the major differences between humans and chimpanzees (Premack, 1976). Relationships can be used to structure the perceptual world in new ways and so the development of symbolization expands the complexity of the perceived world.

4.3 Re-Organization of Human Memory

The memory storing capacity of the hominid cortex must also have increased in the course of hominid evolution. It is likely that memory underwent some form of re-organization, like the data reducing mechanisms, rather than just expansion.

A re-organization could lead to a maximal exploitation of symbolization resulting in increased context-dependence and associative capacities, and new abilities like forming strings, i.e., sequences in fixed serial order. The sequences of outputs produced by memory and motor networks form strings only when their order becomes fixed and can be stored, learned or reproduced in whole or in part. String formation is, thus, a property of the memory rather than of the motor systems.

4.3.1 Elaboration of Context-Dependence in the Memory

In humans small lesions in the left temporal cortex can disrupt the ability to name animate but not inanimate objects (McCarthy and Warrington, 1988). This finding led to a suggestion that, in the first stage of processing as new categories of stimuli are learned, a new neuronal field is opened in each sensory modality in which the stimulus is learned; later memories in this category and modality are also stored there. The overall category specificity is derived from the relative contributions of each of these fields to the total computational process (Warrington and McCarthy, 1987; McCarthy and Warrington, 1988). Damasio (1990) has suggested that these relationships defining the relative contributions are also stored in a further field.

This expansion of human memory and its division into groups of fields may itself give rise to new properties. For example, theoretical considerations show that subdividing a network into loosely coupled fields greatly enhances its categorising power (see Sect. 3.3; Fassnacht, 1990). Modelling human categorical fields by substituting many equivalent and parallel networks for the single memory network in Fig. 7 allows us to test what further properties can be derived from this change. The existence of parallel fields suggests that memories in different categories could be activated and processed simultaneously. Loosely linking the various category groups via weak interconnections may result in the function of each memory field depending on what all the other fields are doing; that is, linking the parallel fields may create context-dependence. This context-dependence would allow an input to evoke different associations depending on the momentary state of the whole memory system. For example, an input may be weakly associated with several categories; if one group of fields becomes active, due to other stimuli, then the re-entrance of the memory from these fields could now evoke it in the other fields. That is, even though memory has replicated into parallel networks the memory process is still distributed over the whole system. Context-dependence means that all categories are searched for associations with a memory input but the momentary context will influence the complex of memories activated. This can be a creative source of variation.

4.3.2 Elaboration of the Memory to Hold Serial Order

Storing strings is a case where reproducibility, rather than context-dependence, is important. How both long-term memory and working buffers have become modified to hold serial order is unknown. Calvin (1991) has suggested that serial ordering evolved from the organization of buffers required in fast and precise motor timing and I have suggested that symbolizing is an important prerequisite simply on the grounds of capacity. However, once long-term and short-term working memory can both store symbols in serial order, then they can be used to produce new strings of associations which can lead to planning without action and setting up and remembering plans for future actions.

This elaboration of using a buffer to store serially ordered symbolized sequences enhances the benefits of a context-dependent recall in that sequences of new associations, set up through context-dependence, can now be retained and ordered. Together context-dependence and serial ordering may lead to the ability to make new associations or openness (Hockett, 1961), and to creativity in human thought.

4.4 Use of Symbols in the Motor System

Does the motor system symbolize its output like the memory networks? Converting the output of the motor networks to neural symbols for transfer to the memory to elicit associated information provides the same advantages as does symbolizing memory outputs. Furthermore, symbolizing the output of the motor system would be necessary for the generation of motor strings, i.e., motor sequences of fixed order. Strings of symbols of the motor output stored in a buffer may allow the setting up of new motor programmes or courses of action with or without acting them out.

4.5 Conclusion – the Model for Human Memory and Motor Organizing Systems

The model for human memory and motor organizing systems is derived from the animal model in Fig. 7 but contains three new features necessary to model new human abilities. The type of changes incorporated in the model are based on biological evidence. Firstly, the data reduction mechanisms which must be present in all large animal systems, are expanded by replication and re-iterative connection. Re-iterative connection enhances reduction exponentially and leads to the dynamic symbolization of the output of a network necessary if the output of a large network is to be transferred to other systems. Symbolization is a prerequisite for deriving relations of relations and for storing long sequences or strings. The symbolization step in the model lies at the outputs of both motor and memory networks.

The second modification is the expansion of the memory by replication of the original network into multiple, parallel and loosely linked networks each of whose outputs is also symbolized. This expansion into multiple, parallel fields greatly enhances categorizing power, whereas the loose linkage leads to context-dependence and creativity. The third change is the modification of both long-term memory and short-term working buffers to store symbols in serial order.

The increase in power due to symbolization and enhancement of the memory in the model are both achieved by replicating given structures and changing their interconnections – a change from parallel to serial connections in data reduction, weak interconnections demarcating parallel memory fields. Replicating existing structures and changing their interconnections may also be a general way of re-organizing neuronal systems to achieve new abilities. It is important to note that no new integrational principles are required in this derivation of the human model from the animal model and that the same principles still apply; functions are carried out by, and distributed through, loops of interconnected neuronal ensembles working in consensus as exemplified by the context-dependence of the human model.

5 Neural Symbolization, Motor Organization and Language

The sequences of memories or motor plans in symbol form produced by the model for human motor and memory systems are nonverbal. It is important now to examine the relationship between neural symbols and words. If the motor organizing network can organize sequences of non-motor symbols, can it organize sequences of words and so subserve language?

5.1 Relation of Words to Neural Symbols

Words are initially experienced as a sensory-motor event within a situational context and so they will be remembered and symbolized like other experiences. A child learns to associate heard sounds with other sensory-motor events, then during its babbling phase lays down motor memories of vocalization and learns to associate these with the previously learned sounds; this leads to a store of memories of words (acoustic/vocal) and the sensory-motor memories derived from the context (situation, object, action) in which the words are used. These are stored like any other memory and symbolized in the same way.

This would mean, in the model in Sect. 4, that as continued use associates words and context the total memory landscape for the context would come to include the word for it. Use of the word now activates the total memory (word and other experiences) by content-addressing. If the memory trace is transferred out of the memory store it will be symbolized and so the neural symbol may contain a mixture of the experience and the word. Thus, neural symbols are not words, they are dynamic and depend on the output of the network at the time.

The binding of a word to experience does not, however, result in language ability; language also requires syntax. For example, a dog, on hearing the word "walk", may pick up its lead and go to the door but it will do this regardless of the situational and syntactic context in which the word was used. Syntax involves the use of the same words in different ways and so we must be able to associate a word with different experiences or different words with the same experiences, an important characteristic of human language.

The model must, therefore, be elaborated further to achieve this loose coupling of words and experiences. Neurological findings suggest that memories of words are laid down in fields other than those storing the experience associated with the words (McCarthy and Warrington, 1988; Damasio, 1990). Thus, the model of memory must now include separate and parallel lexical and experiential fields. These may be derived by further replication similar to that resulting in parallel fields for experiential categories. If lexical and experiential fields are only loosely linked like the linkage of the experiential fields, then one word field may activate different experiential fields depending on the context and vice versa. If each field symbolizes its own output, then some neural symbols would be those of words whereas others would be of the associated experiences.

5.2 Attaching Meaning to Symbols and Words

The memory model suggests that content-addressing enables each input to activate the total collection of related memories which give the total previous experience of the person with these or similar inputs, i.e., the context of the input. In the model, these memories stay activated in the memory store as long as the input remains and during this time a neural symbol may be transferred to other systems. I propose that this also applies to the brain and that, although we are not conscious of this context, it is precisely this context which gives the feeling of "meaning" or "core meaning" (Warrington and McCarthy, 1987) to an input, or which forms the "concept" of what the input is. That is, words derive their meaning from the collection of memories they evoke, their context. A simple illustration is the word "table": it is difficult to define rigorously what a table is, to formulate the "concept" of table, for there are exceptions for every definition. But we still all know what is meant when the word is used. The model suggests that the core meaning of "table" is the total collection of experiences activated when this word inputs to the memory. This collection grows with experience to include not only situations, objects and action but also related words, e.g., chair. This formulation of a biological basis of "meaning" derived from the model is equivalent to that of Warrington and McCarthy (1987; McCarthy and Warrington, 1988) and Damasio (1990) who, based on data from patients with temporal and parietal lesions, suggest that the specified interactions of memories stored in different categorial fields lead to meaning.

A further development of meaning is achieved when serial order is used to derive new meaning, i.e., the development of syntax. Calvin (1991) rates this as the greatest jump in the evolution of language but Boehm (1988) provides evidence for simple syntax in chimpanzee vocalization. Again, the evolution appears to be quantitative rather than qualitiative. However, no mechanism for change in meaning by serial ordering has yet been proposed.

5.3 Use of the Motor Organizing System for Choosing and Planning of Strings of Words

The development of symbolization allows the human motor organizing system to produce sequences of symbols which can be stored in serial order in the memory. A further use of the motor organizing system could be to organize non-motor memories and create new non-motor symbol strings in the same way that it can organize new motor programmes. That is, assuming the motor system underwent the same sort of expansion as the memory, then it could have replicated the motor organizing nets, now using one to organize neural symbols of words just as the other could be used to organize symbols of actions.

The argument for the connection between motor and language organization has been given in detail by MacNeilage (1987). Concentrating on the form or the serial organization of language, he describes two levels of organization in language: the phonological level, in which consonants and vowels are organized into syllables to form meaning units (morphemes); and the meaning, or morpho-syntactic, level in which morphemes are organized into sequences in terms of syntactic rules. MacNeilage argues that both levels have a frame/content organization. At the phonological level the sounds have no meaning on their own. Meaning derives from the way they are combined; consonantal and vocalic content elements are inserted into syllabic frames. Similarly at the morpho-syntactic level, content words are inserted into a grammatical frame.

MacNeilage has compared the frame/content organization of language with that of the hands holding and manipulating food. It is also directly comparable to that of the time frame model for motor organization in Sect. 2. In a language utterance, the frame for what is to be spoken, the global aim, is the nonverbal concept to be expressed. This provides the frame for the strategy level which organizes the general mode of expression. This provides the frame for the choice of phrases – the behaviour level. The chosen phrase provides the frame for the individual words at the routine level which is the frame for the individual syllables – the movement level. These provide the frame for the vocal motor outputs. A CHOOSE process selects the appropriate output at each level. As in motor organization, the inputs to this CHOOSE process include inputs from the memory, here giving lexical information, and information from previously learned programmes, here syntax, as well as inputs from sensory-motor systems informing on the process of speaking. These influence the CHOOSE process in the same way that learned strategies, behaviours or motor routines influence the CHOOSE process in motor organization.

As in behaviour, the word sequence will not be planned from the start. The global aim will push the network into one state which will result in a fast sequence of sound outputs forming a syllable and sequences of syllables making a word.

The rules of syntax, together with the global aim and remembered associations with the word just chosen will push the network into new states resulting in a coherent flow of words.

Thus the networks for speech in the brain and in the model could be organized in the same way as those organizing body movements and behaviour. They can produce sequences of word neural symbols, which stored in the short-term working memory, make it possible to think in words without speaking. Both speech and motor networks can also organize non-word neural symbols producing flows of nonverbal memories, which in the case of the language network would be syntactically organized.

The model presented in Sects. 4 and 5 is not a model for deriving natural language but shows how developments of the motor and memory systems could lead to the development of language. Expansion of the memory system into multiple parallel subfields provides the basis for loosely linked storage of words and experiences. Expansion and new connection of the necessary data reduction steps leads to the development of symbolization which is a basic prerequisite for the formation of meta-relationships and serially ordered sequences. The parallel expansion of the motor organizing networks can lead to the development of one such network for organizing symbols according to syntactic rules while the other network organizes actions according to experiential rules as before. All these innovations are derived from very simple expansions and very slight re-organization of nonhuman systems and they lay the foundation for human language ability.

There are still many major problems – how are internal representations related to memory landscapes, what is the exact nature of the neural symbols, how does the memory store serial order, what sort of rules does syntax have, how is understanding language related to producing it? These remain unsolved in spite of many attempts to answer them (e.g., see Fanselow, 1990 for a discussion of the biological basis of syntax). Perhaps rather than trying to solve these problems by analysing linguistic models or processes, it may be more useful to compare neurophysiological and neuroanatomical data with systems models such as the one I have presented here, and to try to derive the answers from the emergent system properties.

6 Biological Evolution and Information Processing – the Limitations of Models

The time frame description of behaviour and our model for motor organization are a first attempt to show how long term contexts can influence momentary outputs. Each muscle contraction is regarded as being chosen within the frame of the current motor routines, behaviours and long term strategies, and so the model consists of a temporal frame/content organization with the shorter time frames nested in the longer. This nesting of different temporal levels can be achieved in a neuronal network model by incorporating a temporal structure consisting here of delays in the connections between neurons. The model of the memory system is similarly organized. Importantly, activity in all time frames in both motor and memory systems can affect all others. This property leads to a circular non-hierarchical model in which all choice or decision processes are distributed throughout the whole system. This model is useful throughout the animal kingdom because the problem of producing an instantaneous motor output within long term contexts, of remembering and planning, is a general one. It applies just as much to choice of words in human language as to co-ordination of motor outputs in invertebrates.

Does the nested time frame model fit with predictions from the six hypotheses for the evolution of information processing systems (IPS) that form the background for this book? There are some similarities but also basic differences. The elaboration of the time frame model in animals to a model for remembering and planning actions and words in humans fits the predictions of Hypotheses 3 to 6 quite well: I have proposed that the human IPS resulted from a re-organization made necessary by increase in brain size (Hypothesis No. 4); that the reorganization utilises previously existing components (No. 4) and previous methods of information processing (No. 6); that it consists in part of an expansion and differentiation into multiple subsystems with specialized functions (No. 5). Furthermore, this re-organization occurred in a relatively short period of evolutionary history, maximally circa 2.5 million years (No. 3). The model may disagree with Hypothesis 2. A major feature of the time frame model is that all levels communicate with and influence each other continuously. Although this communication is within one IPS as predicted in Hypothesis 2, an expanded analogy of the time frame approach to other systems may suggest that, in contrast to Hypothesis 2, there is more exchange between systems.

The greatest difference, and this is a basic difference, is with Hypothesis 1. The time frame model, like the biological systems it models, does not transmit or receive information either in the semantic or pragmatic sense (Riemann-Kurtz, 1990) and so is semantically closed and self-referential (Roth, 1989). The model and biological systems convert the signals received from the outside world to information internally as the signals acquire significance only within the system. For example, the present is used only to confirm or reject the use of a remembered world for guiding actions, the memories having acquired significance through contextual and emotional associations within the system. This constructionist view of brain function, which also applies to the model, has been convincingly defended by Roth (1985, 1989) and disagrees completely with Hypothesis 1 which predicts that all information systems are semantically open. This basic difference shows that, although living and nonliving systems share certain features, it may not be possible to extrapolate too closely between them.

It is important to examine the validity and political implications of Haefner's (1988) application of a model for the natural evolution of information processing systems to artificial technical systems. Based on very generalized ideas of evolution from cosmology to biological evolution he predicts that information processing technology will inevitably expand to include genetic manipulation of humans, direct biological-electronic linkages between humans and computers, long-term governments to control use of information and, finally, that we will find "Nirvana", "omnipresence", "omniscience" and "liberation from matter" in total networking of information systems. We must remember the limits of scientific models: they are based on both a reduced view of the world and a specific, very narrow set of rules for their argumentation; they are Gedanken experiments to explain our world *post facto* and their very complexity limits their usefulness for predicting the future course of events. Indeed, models specifically designed for complex systems such as chaotic systems show unpredictability as a basic property. A lack of inevitability is also a basic tenet of evolution theory. It is, therefore, irresponsible both to suggest that sociotechnical developments are inevitable and to use a scientific model, here of natural evolution, to substantiate this claim (Haefner, 1988). This common form of legitimization is a strategy to avoid ethical and moral questions. These, however, are questions that must be asked here, particularly in the light of the misuse of the very similar concepts of eugenics to legitimize the Holocaust.

Acknowledgements. Supported by SFB-4 Project H2 of the Deutsche Forschungsgemeinschaft. Many thanks to Dr. Jennifer Altman for the vital role she has played in the development of the models in part I, for all support and for all critical discussions.

References

Allman J (1987) Maps in context: some analogies between visual cortical and genetic maps. In: Vaina LM (ed), Matters of Intelligence, Riedel Dordrecht, pp 369–393

Altman JS, Kien J (1987a) Functional organisation of the suboesophageal ganglion in arthropods. In: Gupta AP (ed), Arthropod brain: its evolution, development, structure and functions, Wiley, New York, pp 265–301

Altman JS, Kien J (1987b) A model for decision making in the insect nervous system. In: Ali MA (ed), Nervous systems in invertebrates, Plenum, New York, pp 621–643

Altman JS, Kien J (1989) New models in motor control, Neural Computation 1, pp 173–183

Altman JS, Kien J (1990) Highlighting Aplysia's networks, Trends in Neurosci 13, pp 81–82

Altman JS, Kien J (1991) A model for the context-dependent selection and organisation of motor outputs, (in prep)

D'Amato MR, Colombo M (1988) Representation of serial order in monkeys (Cebus apella), J Exp Psychol 2, pp 131–139

Amit DJ (1989a) Modelling brain function, Cambridge University Press, New York

Amit DJ (1989b) Attractor neural networks and biological reality: associative memory and learning. In: Intelligent Autonomous Systems, Univ of Amsterdam, Amsterdam

Berthoz A, Droulez J (1989) The concept of dynamic memory in sensorimotor control. In: Humphrey DR, Freund H-J (eds), Motor control: concepts and Issues. Dahlem Konferenzen, Wiley, Chichester

Boehm C (1988) Proceedings of the NATO ASI on the origin of language, Cortina, Italy, in Language Origin: a multidisciplinary approach. Wind J, Chiarell B, Bichakjian B, Nocentini A, Jonker H (eds) Kluwer Academic Publ, Dordrecht (in press)

Brooks VB (1986) How does the limbic system assist motor learning? A limbic comparator hypothesis, Brain Behav Ecol **29**, pp 29–53

Calvin W (1991) A brain for all seasons: climate and intelligence from the ice age to the greenhouse era, Bantam, New York

Damasio A (1989) The brain binds entities and events by multiregional activation from convergence zones, Neural Computation **1**, pp 123–132

Damasio AR (1990) Category-related recognition defects as a clue to the neural substrates of knowledge, Trends in Neurosci **13**, pp 95–98

Fanselow G (1990) Zur biologischen Autonomie der Grammatik. In: Suchland P (ed), Akten des dritten Jena Symposium: Biologische und Soziologische Grundlagen der Sprache, Wiss Beiträge, F Schiller Universität, Jena (in press)

Fassnacht C (1990) Über strukturierte Neuronale Netzwerke mit asymmetrischer Verdünnung. Diplom Thesis, Physics, University of Göttingen

Gardner E (1988) Optimal storage properties in neural network models, J Physics A: Mathematics and General **21**, p 271

Gelperin A (1986) Complex associative learning in small neuronal networks, Trends in Neurosci **9**, pp 323–328

Haefner K (1988) Evolution der Informationsverarbeitung, Report for Fachbereich Mathematik/ Informatik, Universit of Bremen

Hockett CF (1960) The origin of speech, in Human language and animal communication, Scientific American **203**, pp 88–111

Holloway RL (1969) Culture: a human domain. Curr Anthropol, **10**, 395–412

Hopfield J (1982) Neural networks and physical systems with emergent selective computational abilities, Proc Natl Acad Sci USA **79**, p 2554

Horner H (1988) Spingläser und Hirngespinste, Phys **B1 44**, pp 29–33

Horridge GA (1964) The electrophysiological approach to learning in isolatable ganglia, Animal Behav Suppl **1**, pp 163–182

Hoyle G (1979) Mechanisms of simple motor learning, Trends in Neurosci **2**, pp 153–159

Jerison H (1982) Problems with Piaget and pallia, Behav Brain Sci **5**, pp 284–287

Kien J (1975) Neuronal mechanisms subserving directional selectivity in the locust optomotor system, J Comp Physiol **102**, pp 337–355

Kien J (1983) The intitiation and maintenance of walking in the locust: an alternative to the command concept, Proc Roy Soc Lond **219**, pp 137–174

Kien J (1990a) Neuronal activity during spontaneous walking. I Stopping and starting, Comp Biochem Physiol A **95**, pp 607–621

Kien J (1990b) Neuronal activity during spontaneous walking. II Correlations with stepping, Comp Biochem Physiol A **95**, pp 623–638

Kien J (1991) The need for data reduction may have paved the way for the evolution of language ability in humans, J Hum Evol, **20**, 157–165

Kien J, Altman JS (1991a) Neural substrates of locomotion in vertebrates and arthropods, Comp Biochem Physiol A, (in prep)

Kien J, Altman JS (1991b) Deciding what to do next: a model for selection and maintenance of motor programmes. In: Winlow W, Kien J, McCrohan C (eds), Neurobiology of motor programme selection: new approaches to the problem of behavioural choice, Manchester Univ Press, Manchester, pp 147–169

Kien J, Nützel K, Altman JS, Cycle lengths in neural networks with delays and the problem of time hierarchies in biological systems, (in prep.)

Kien J, Winlow W, McCrohan C (eds) (1992) Neurobiology of motor programme selection: new approaches to the problem of behavioural choice, Manchester University Press, Manchester

Kolb B, Whiteshaw IQ (1985) Fundamentals of human neuropsychology, 2nd ed, Freeman, New York

MacNeilage PF (1987) The evolution of hemispheric specialisation for manual function and language. In: Wise S (ed), Recent explorations of the brain's emergent properties, Wiley, New York, pp 285–305

McCarthy RA, Warrington EK (1988) Evidence for modality-specific meaning in the brain, Nature **334**, pp 428–430

Meglitsch P (1972) Invertebrate Zoology, Oxford University Press, New York

Menzel R, Bicker G (1987) Plasticity in neuronal circuits and assemblies of invertebrates. In: Changeux J-P, Konishi M (eds), Neural and molecular bases of learning, Springer, Berlin, 433–471

Mishkin M, Appenzeller T (1987) The anatomy of memory, Sci Am **256**, pp 80–89

Möhl B (1988) Short-term learning during flight control in Locusta migratoria, J Comp Physiol A **163**, pp 93–101

Premack D (1976) Mechanisms of intelligence: preconditions for language, Annals New York Acad Sci **280**, pp 544–561

Riemann-Kurtz U (1990) Aspects of information and information processing for the project Evolution of Information Processing, Bremen (preprint)

Rost R, Honegger HW (1987) The timing of premating and mating behaviour in a field population of the cricket Gryllus campestris L, Behav Ecol Sociobiol **21**, pp 279–289

Roth G (1985) Die Selbstreferentialität des Gehirns und die Prinzipien der Gestaltwahrnehmung, Gestalt Theory **7**, pp 228–244

Roth G (1989) Konstruktivität des Gehirns und Konstruktivität der Wahrnehmung – ein notwendiger Zusammenhang?, paper presented at DFG Symposium, Kognition und Gehirn, Göttingen

Schöttner B (1990) Zeitliche Muster in Bewegungen von Schimpansen (Pan Froglodytes). Diplom Thesis, Faculty of Biology, University of Regensburg

Squire LR (1987) Memory and Brain, Oxford Univ Press, Oxford

Warrington EK, McCarthy RA (1987) Categories of knowledge: further fractionations and an attempted integration, Brain **110**, pp 1273–1296

Nature and Origin of Biological and Social Information

Vilmos Csányi

1 Preliminary Notes on Information 257
2 The Replicative Component-System Model 258
3 Origin of Organizational Information and Origin of Time 261
4 Evolution: The Accumulation of Replicative Information 264
5 Ontogenesis 266
6 Animal Mind as a Component-System 270
7 Linguistic Models of the World: The Rise of Conceptual Thought 271
8 Culture and Ideas 273
9 Evolution of Ideas in Mass Society 275
10 Closing Thoughts 276
References 277

1 Preliminary Notes on Information

The concept of information has not been created by biologists, nevertheless there are important areas in biology where it is widely used without any epistemological solicitude. The briefest of excursions into the subject of molecular biology, which is concerned with probing the secrets of life, reveals that the subject is entirely dominated by the language and techniques of data and data processing. The two sorts of scientist who can really understand each other are molecular biologists and computer programmers. Based upon this success there are many ventures, mainly in the fields of popular sciences, which generalize the findings of molecular biology and try to base all biological sciences on the concept of information. These attempts have provoked the objections of biologists who are interested in development (Oyama 1985) or other traditional fields of biology. As far as I am concerned I am sure that a proper concept of information can be worked out for the biologist and this paper is a preliminary endeavour.

The first problem concerning the use of the information concept arises from its double epistemological nature. Information in the usual sense of "knowledge" or description is perfectly clear and understood; the dilemma arises if it treated as a pure biological phenomenon which *acts* in the biological systems. These two aspects of the information concept were brilliantly separated by Kampis (1987a, b, 1990). Kampis distinguishes *effect* and *knowledge*. Knowledge is a *description* that has meaning only for the observer, while the effect – the final source of knowledge, according to Kampis – is the specific way a given component functions in a given system. Based on effect and knowledge, Kampis defines two kinds of information. The structural arrangement responsible for the effect of a given component and the effect itself are expressed in the concept of *referential information*. The term "referential" points to the fact that for expression of an effect by a given component the whole system is necessary. *Nonreferential information* describes both the effect and the structure responsible for it. If we take for example a simple molecule as CO_2 we might have a *description* of it in a chemical formula, which could represent the arrangement of its constituent atoms, but of course this description can never replace carbon dioxide in its molecular actions. Its very molecular entity is necessary to carry out any chemical or physical action, so its *referential information* is entirely embodied in *itself* and in other compounds with which it reacts, and cannot be retrieved or translated by any informational devices into some other acting form. Our main concern is therefore the appearance and evolution of active referential information.

2 The Replicative Component-System Model

The considerations related to information are treated here in the context of the *replicative component-system* model which has been worked out in detail earlier (Csányi, 1978, 1980–1982, 1985, 1987a, b, 1988b, 1989a; Kampis, 1986, 1987a, b, 1990; Csányi and Kampis, 1987). In the description of the component-system we use the term information essentially as nonreferential information.

The *components* constituting a component-system are built of atoms as building blocks, and the structure of components must accurately be given on the basis of the interactions between the atoms. It is characteristic of the component-system that it occupies physical space; its components are assembled and disassembled continuously by the effect of the energy flux through the system, which thermodynamically is an "open" system. The assembling and disassembling processes of the components create a network of interactions defining the system that we call *organization*, which is an arrangement of interactions in time. Biological systems, like cells, organisms and ecosystems are typical component-systems, which are characterised by their components and by their special organization. Components are molecular structures at the lowest level and sub-component-systems at any higher levels of organization. There are two main features of the biological organization, namely maintenance and reproduction. *Maintenance* or self-production takes place through the continuous production and decay of the components at all levels. A biological entity can be regarded as a functionally closed network of components and component-producing processes, which partially or totally produces the same network again. In the process of *reproduction*, the continuous creation of components of a biological entity in a closed functional network can also be shown. This means, regarding organization, that the two processes, namely self-production and reproduction, belong to the same category. Through analysing the processes of maintenance and reproduction, it can be demonstrated that their common essential process is copying, that is, *replication* (Csányi, 1978, 1985, 1989a; Csányi and Kampis, 1985; Kampis and Csányi, 1987; Kampis, 1990).

Replication is a copying process; a constructor produces a copy (replica) of a component of a given subsystem or system. Two forms of replication are distinguished. *Temporal replication* is defined as the system's continuous renewal in time by the sequential and functional renewal of the system components while the unity and identity of the system are maintained. *Spatial replication* is identical to reproduction. The system produces its own replica, which becomes separated from it in space; From one unit two units are formed. In these processes the component or subsystem to be copied contains in itself the information needed for copying. It is obvious, however, that for the essentials of the process of replication it does not matter at all whether the copy is that of a separated entity or the constructor itself, nor does it matter whether the information necessary for the copying process appears separated in a special memory device (as is mostly the case with replication of proteins and nucleic acids), or is dispersed and found distributed in the system. The fact of replication lies not in the specific implementation but in the functional operation.

Functional operation of biological replication is called *replicative organization*. It is also characteristic of biological systems that functionally more or less closed replication networks include sub-networks; The components include subcomponents like organisms, cells and molecules. *Organizational levels* develop this way (see Csányi, 1978, 1982, 1989a, b).

From an informational point of view the description of the interactions between the components of different organizational levels can be considered as algorithms specific to the participating components. In a general model embracing all levels, the effects calculated from the algorithms of the lower levels seem to be accidental in the event context of the higher levels. On the other hand, the algorithms of the higher level emerge as special *constraints* exerted on the components of the lower level. A cell or an organism of a higher order can be defined but through a simultaneous description of several organizational levels (Csányi, 1985). On the molecular level these systems can be described by their dynamics, expressed in physical and chemical equations; However, these equations provide no information at all on the specific constraints that exist at the higher levels, or on the nature of the specific *architecture* that exists at the higher level, in which the processes of the lower levels will operate. The description of the architecture developed at a higher level – its algorithm – is a feature of the higher level, and in this sense the "laws" of higher levels cannot be reduced into the laws of lower levels. This does not mean, of course, that the phenomena of higher levels cannot be *causally* traced back to the laws of lower levels.

The definition of biological *function* can be derived from the organizational levels. We can define function as specific constraints created by the higher organizational levels. For example, the codons (nucleotide triplets) of DNA have function, namely they assign amino acids to the structures of proteins. The essential part of that interaction is not some kind of chemical affinity, but a role (effect) in the process-network within the cell that is based on the chemical character but remains in fact independent of it. The function can always be fixed as some description in the components and events contexts of the lower level that is specific to the given system.

The function of components of a replicative system can be formulated by means of either a general or a specific interpretation. As for the general interpretation, components of a replicative system are considered to be the entities which, in the course of their interaction, take part in the replication. Thus it is clear that components have a replicative function, for they promote the replication of both the system and themselves. This means their *general function* is to participate, as components, in the replication. On the other hand, if we observe the interactions between the components, we can always reveal in them the *specific mechanisms* of the components affecting the probability of each other's genesis and survival as a substantive manifestation of the replicative function.

The component-systems in which components are assembled and disassembled, but where no organization has yet developed between the components, are called *zero-systems* (Csányi, 1978, 1989a).

The concept of function usually also raises the question of teleology. In the model of the component systems *existence* is the final "goal" and "cause". All functions including the general replicative function can be deduced from simple "being". The operation of the algorithms of functions at the lower levels are both the consequence and evidence of the existence of higher organizational levels and do not need further explanation.

We wish to refer briefly to the relationship between complexity and order emerging in component systems (Kampis and Csányi, 1987c). In the processes that take place in thermodynamically open systems, a growing complexity on the one hand, and a growing order in certain parts of the system on the other, can be simultaneously observed. In a thermodynamically closed system the processes always advance in the direction of growing disorder and complexity, while in an open system the energy flowing through the system is able to create order in certain parts of the system with an increase of entropy in the environment as a natural consequence (Prigogine et al., 1972). It is essential, however, that the complexity created in a chemical system always be *unorganized*, which means that its emergence is sufficiently explained by the thermodynamics, while the understanding of the emergence of an *organized* complexity also needs the description of the *constraints* developed on the higher organizational level.

Constraints developing in the course of organization will also contain some order in most cases, e.g., a large number of certain types of complex molecules, or a certain spatial arrangement of molecules, etc. Thus, the organization of complexity and order are connected in a particular mode and it is just that *mode* that expresses organization.

The replicative systems emerge in a suitable zero-system if an *organizing agent* appears in it. The organizing agent can be a "minimal system" or, as it has been named earlier, an *autogenetic system precursor* (AGSP), (Csányi and Kampis, 1985), which meets the following criteria:

(i) It contains at least one component-producing cycle
(ii) At least one component of that cycle can be excited by means of the energy flux through the system.

It was shown that if an appropriate zero-system develops, AGSP may spontaneously appear in it, to the effect that the *organization of complexity* starts immediately. However simple the AGSP as a replicative system is, it has the capability of picking up few components, and that results in the appearance of new functions in the system, as it is only through some function that a component can join in a replicative system. At the beginning only a few components can enter the cycle of replication. As the system becomes more and more complex however, the possible number of integrated components increases and the complexity of the system can develop at an accelerating pace. This process we have called *autogenesis* (Csányi and Kampis, 1985). Due to the high inaccuracy of the replication, the initial period of replicative information is called the stage of *non-identical replication*; as time passes a certain cooperation and functional differentiation start between the components containing replicative information, and "communities" of components appear, which we call *super-cycles* (Csányi, 1989a). As time passes the replicative coordination of super-cycles begins as well, and the fidelity of replication increases. A separating sub-system develops whose members are separated by their participation in the common replication. This process is equivalent to the functional closure of the network of components having replicative function, and is called *convergence* and *compartmentalization* of the replicative information.

Similar processes also take place at the level of compartments. The different compartments are able to influence the probability of genesis and survival of others, thus their replication is gradually getting concerted; with a continued increase in the accuracy of replication, "compartments of compartments" appear, which means that *a new organizational level* emerges. The compartments become *components* of this new level of organization. The building up of compartments goes on in succession until the whole system begins to behave as one final compartment, that is, one integral replicative unit. At this point, the stage of autogenesis of the given system has come to an end, and its undisturbed existence continues while its environment remains unchanged.

3 Origin of Organizational Information and Origin of Time

Let us consider in more detail the *de novo* genesis of the replicative system (Csányi, 1987, 1989a). In the course of the *origin of life* the zero-system was an open chemical system existing in a continuous energy flux, and in it molecular components were assembling and disassembling *at random*. This system continuously generates molecular complexity, using the energy flowing through the system, but this complexity is unorganized; thus the upper limit of the complexity is determined by the thermodynamic parameters of the system. Description of such a system is possible only by general parameters of composition, temperature, pressure, etc. The nonreferential information which such a description contains, concerns only the components and some statistical features of the components involved. Referential information is embodied in the components, and only in the components. Nonreferential information is a description of the components that is a list of their various properties. This list could never be exhaustive because we define properties from known interactions and these are necessarily finite and limited by our observations. In fact, the list of properties of any kind of a component is infinite; this is inherent in the nature of the interactions of matter, which are infinite and immeasurable (Bunge, 1963). Properties other than those included in our description we call *hidden properties*, and they play a rather important role in the evolution of information.

The state of the zero-system can be changed only by changing the thermodynamic parameters of the system. We might call this kind of information both non referential and referential as *component-information* (I_{comp}); hidden properties appear only in referential information but by definition they are excluded from nonreferential information. An important feature of such a system is that *time* does not appear among its parameters of state. As components assemble and decompose at random, with a system that is big enough, the structure of the system shows only statistical oscillations; component-information is time-*independent*.

If in such a zero-system the simplest possible AGSP appears, we can speak of a *creative act*, because in the cycle of the AGSP using the components already present but unorganized up to that moment, a *new entity* that is separable and characteristic in its organization, has appeared. Conditions for the emergence and existence of completely recycling networks have been analysed in detail (Csányi, 1989a). If a reaction network includes at least two bi-molecular reactions, complete chemical reproduction may spontaneously emerge (King, 1982).

A consequence of the creative act is the emergence of *time* as a parameter. Even the simplest AGSP is a *cyclic process*, and as such it can be characterised by the cyclical period on the basis of our definition. This time signals the emergence of a new kind of being for an observer. For this reason I will call it *time of the first kind*. This notion of time is essentially identical to what we call "time" in physics. In cyclical processes interactions of the components experience time as a *sequence of events*, and therefore a description of them must contain information concerning time of the first kind. Therefore, nonreferential information concerning the AGSP is more than the known features of the components and some statistical parameters: it must contain the *description of the organization* of the AGSP, however simple it might be. Referential information is also increased by the specific relationships of the components of the AGSP to each other and to the AGSP as a unity.

Description of organization involves description of interactions in time sequences; this kind of description provides *organizational-information* (I_{org}).

The kinds of the components and organization constitute a special feature which we call *identity*. A zero-system has no similar feature.

Chemical reproduction is realised through the spontaneous reactions of its molecules, and the existence of the whole cycle depends on *reaction rates*; therefore a chemical cycle in itself is not a component system according to our earlier definition because its constituent compounds have no function and are therefore not real components. There are no controlling structures and processes in a simple chemical cycle that would make it a "system" independent of reaction rates, which could transform reproduction to replication. But chemical cycles gradually undergo strong effects of selection. Compounds that through their interactions lead the network toward stability carry *functions*, as they have been defined in our model, and therefore these compounds become real *components*, and the organization of a chemical reproductive network may convert from reproduction to *replication*.

In the course of its functioning, a chemical network may gain new components, as has been explained earlier. The exit from or entry to the reproductive cycle of any component is also a *creative act*: it changes the structure of the given system; it can influence the function of different components or the behaviour of the system as a whole; and it increases the complexity of the replicative system. The creative acts result in real genesis. Let us consider that from the definition of component systems, all new components bring an essential change to the *identity* of the system. As the component-system is not initially given but is determined by the components, it is a consequence of the organization of the components that the appearance of a new component may lead to unpredictable, *emergent* features. The most important of these is also related to time. The identity-change events that take place as a result of successive creative acts make the history of the given system. The history as such can be characterised by time, namely in what Bergson (1907) called *duration*, rather than as an analogy of cyclical period. This time we will call *time of the second kind*. It is basically different from time of the first kind. Time of the first kind can be infinitely divided, and all of its domains can be construed. Time of the second kind can also be divided, but only according to the occurrence of the *events* constituting the history. Time of the first kind that elapses between two pairs of events – whatever the number of intervening cycles may be – has no importance from the viewpoint of the history and the functioning of the system.

Time of the second kind can also be described as a *sequence of creative acts*, from which it will be immediately clear that this time is system-specific also, and that it does not have much to do with the notion of time used in physics. Time of the second kind is closely connected with the complexity and information of replicative systems.

Both kinds of time are the consequence of the *organization* of the replicative system. An elementary act of organization is recurrence, the closed cycle, and as soon as the latter appears time of the first kind appears as well, as a private time of the recursive cycle, and if recursion makes a sudden leap at an external effect (such as the exit or entry of a new component – a creative act) time of the second kind appears. As a consequence it is not time that creates the replicative systems; on the contrary, replicative systems produce time by force of their existence, and time in both forms is a specific organizational parameter of a given system.

Existence of reproductive chemical system depends on reaction rates. Replicative systems, within limits, are rather independent of the velocity of interactions of their components. For example, when a simple substance, a vitamin or an amino acid, is removed from a bacterial cell's environment, its growth immediately stops. But the cell's reaction network can resist breakdown and the cell can live for a long time because while "fasting" various controlling mechanisms are activated and the whole network of the cell's metabolism is channelled into a special kind of "parking state". So a bacterial cell's system is able to modify and reorganize the behaviour of its own reaction network and defend itself to a limited extent against environmental effects.

The maintenance of a network of reactions with controlling systems is based on coordination of the various reaction rates by a separate class of components which could control the reactions by their presence or absence, and themselves participate in the reactions only temporarily. Catalytic proteins are such compounds and their existence requires a *separate description* of them in the form of DNA, which acts as a final controlling device (Pattee, 1977). Emergence of a description in the active system as referential information can be denoted as *symbolic-information* (I_{symb}).

4 Evolution: The Accumulation of Replicative Information

The various kinds of information have different roles in the origin of living structures. An explanation for I_{comp} is outside biology; possibilities of interactions between atoms and molecules are given by the physical and chemical properties of the elements and do not need biological explanations. More or less the same is true for I_{org}; organizations of chemical networks can be understood on the basis of chemistry, with the important note that features of chemical networks provide explanations for *time*, both physical and historical. Nevertheless, the basic shift from the realm of chemistry to the realm of biology occurred when *control mechanisms of time* emerged in the form of I_{symb}. I_{comp} and I_{org} as referential information are active forces which act instantly; their effect can not be delayed or suspended. Those structures whose existence is based upon an organized time series of chemical reactions are extremely labile and limited in complexity. Life started with time-control, with the accumulation of I_{symb}. Any proper description which, in the form of referential information, could influence the processes of a system is a suspended action. This action starts when the information is processed.

Coded structure of an enzyme in DNA is such an I_{symb}: it is a prescription for future action, an algorithm to be performed. A highly developed replicative system like a bacterial cell is an active chemical network with thousands of little "time-capsules", the *genes*, which are messages for actions in the organized future. This ceaseless struggle with time has raised the complexity of living systems to an extremely high level.

But not only time of the first kind influences replicative systems. All replicative systems strive for an equilibrium in which *time of the second kind* is just static, and only time of the first kind passes in its organized way. Replicative organization itself is a conservative agent in the process of evolution because it is a recursive process. A creative act, which changes the replicative system, occurs when recursion is slightly disturbed by external, accidental forces; a *new identity* emerges and a further elementary unit is added to the time of the second kind. The conservatism of replication provides an *integrative force* for the events of the second time and leads to the accumulation of the various forms of information, and to *biological evolution.*

In this interpretation, evolution is thus closely connected to identity. Every elementary change is a *change in identity*, in the course of which the preceding being is annihilated in a creative act, and a new being appears at the same time. Applying the methods of natural science does not allow the appearance of the creative acts of evolution to be followed, but the same is true for everyday perception. The mechanism of our thinking is of this nature; it tends to endow systems essentially in continuous change with the same identity. The development of human personality is a trivial example, as is any case of biological ontogeny. Of course, in our own way of thinking, this feature of our logical processes does not change the validity of the aforementioned ideas.

Creativity of the evolutionary process has at least three main pinnacles.

The first is the replicative system itself. As soon as the most primitive organizing forces emerge in the form of an AGSP further changes of the evolving system are strongly selected for. Even the simplest organization behaves as a tremendously powerful *selection screen*: it only admits changes which do not alter the replicative nature of the organization of the system, and among them those which are able to join to the system functionally (Riedl, 1978). As the complexity of the system increases, it becomes a more and more specific screen, although in parallel to the increase in complexity the degree of freedom of changes also increases. This organizationally maintained kind of selection is termed *replicative selection* (Csányi, 1989a). The only negative effect of replicative selection is its elimination of changes that risk or inhibit replication – that is, division of cells or reproduction of higher organisms. Compared to this negative effect, its *creativity* has a much greater importance. Replicative selection permits every change that does not inhibit replication and the conservative nature of replication helps to maintain and accumulate these changes in the course of evolution. There is an enormous variety and richness of forms or organisms, and all these are capable of self-replication. The creative nature of replicative selection in evolutionary biology can be shown satisfactorily.

The second source of creativity is the outside *environment* whose impact could alter the identity of the replicative system from one moment to the next, and which also acts as a selective device through Darwinian natural selection.

The third very important source of creativity is based upon the *hidden properties of the components*. In a component-system the referential information content of the individual components (I_{comp}) is infinite, and only a small part of this information is used to create organization and I_{org}; therefore, in cases of introducing new components to the system, some new properties of the components, hidden information, will show up which will influence the organization of the component-system in an absolutely unpredictable way.

The organizational information (I_{org}) is carried by given properties of components, while the new organization may recall other, formerly hidden properties. A complex molecule has an almost infinite capacity to interact with other such molecules, but a given organization needs only a small number of interactions to carry out a particular component's function; all other properties remain hidden and are excluded from I_{comp}, but if further creative acts bring new molecular components into the system, they might unexpectedly fit into a new organization based upon the hidden properties and elevate these into I_{org}. Such creativity is not restricted to the molecular level: frogs, for example, have webbed fingers which help them to swim more efficiently, but some species living on jungle trees use them for gliding. Frogs originating in water evolved webbed fingers to solve problems of survival in water, but certain hidden properties of the webbed fingers allowed them to evolve into a gliding device when appropriate environmental conditions appeared.

The three sources of creativity mutually support and even amplify the effects of each other. As a consequence, there is a very limited likelihood of predicting the events of evolution. Within limits, we can make some predictions concerning the general direction of changes, but there is no way to predict the properties of the new identities created in creative acts and in the future evolutionary process.

5 Ontogenesis

Multicellular organisms and their antogeny appeared after a rather long time in the course of biological evolution. Ontogeny starting from the zygote takes place between the previous event of the second kind of time of the species and the next one; in the first kind of time, in fact, this means a very short pause in the second kind of time. Nevertheless, cyclical events of ontogeny in themselves make a series of events very similar to happenings of the second kind of time.

Considering ontogeny and information, our most significant problem is the question of identity, because terms to I_{org} and I_{symb} were deduced from systemic identity. Is the zygote identical to the multicellular organism developing from it? If it is, on what criteria? The zygote of a giraffe and the adult animal are different not only in their volume, organization, actions and environment, but in almost all characteristics. Where did the additional organizational and symbolic information come from?

One of the reasons we can regard zygote and adult as identical is a historical one. The adult animal developed from the zygote in the course of a process having a more or less determined sequence. The other reason may be that certain components are identical in them.

There is a very close connection between ontogenetic and evolutionary histories. A very simplified relationship is expressed in the recapitulation "law" of Haeckel. It seems as if the organism developing repeated the path the given species had already followed during phylogenesis in the form of mutations. This is only partly true, however, as certain mutations shortened development, created shortcuts and new interactions, etc. Ontogenesis reflects phylogenetic events only very roughly. But it reflects them in a way, because all ontogenetic events have to have an independent phylogenetic significance, as only thus could they get through the screens of the different selective mechanisms.

While the genome of single cell organisms with its one-dimensional sequence reflects the phylogenetic second-kind time, ontogenesis takes place in the first time-dimension in which the series of events starting from the zygote's activity is quasi replaying the phylogenetic history, imitating it by smaller or larger differences, and producing in the end the adult organism which fits into the second time-dimension of phylogenesis.

There is much debate on where the excess information of the developed organism came from because the zygote as well as the somatic cells contain basically the same volume of DNA (Oyama, 1985). On the basis of the component system model it is easy to answer this question.

There is a gradual transformation of I_{comp} of the zygote DNA to $I_{org} + I_{symb}$ during ontogenesis. DNA has an enormous number of hidden properties (at least they are hidden for the zygote) which manifest themselves within the continuously changing component-system of the embryo. The repeated replications of the zygote produce a huge number of redundant genomes which, in their manifestations and interactions with other components of the developing organism and with the environment, significantly differ from each other due to their selective activity. That is, the genomes of the somatic cells bear the marks of events like gene-activations, inhibitions, etc. having taken place in the time-dimension of ontogenesis, the same way as the original genome bears the history of mutations having taken place in the phylogenetic time-dimension. The genome expression changes revealed during ontogenesis are perfect analogies of phylogenetic *mutations*. The series of ontogenic "mutations" also compiles information, the result of which, namely the multicellular organism, represents a *new organizational level*, and the information of its functions has developed and accumulated in the course of the ontogenetic process during the ontogenic history.

The mutations during phylogenesis of multicellular organisms make it possible for the genome, through the effects of internal circumstances created earlier by itself in the course of ontogenesis, to perform certain new creative acts. Thus ontogenesis is a *programmed genesis*. In phylogenesis creative acts are unpredictable and accidental. In ontogenesis creative acts can be predicted with rather high accuracy, because they are based on *hidden properties* of the *already existing* DNA component. This is a decisive difference between the two processes. In both processes the creative act takes place through the effects of the external environment, but the largest part of the forces that move ontogenesis are of phylogenetic origin and have been interiorized in the course of phylogenesis.

The role that creative acts play in ontogenesis is clearly reflected in a few examples related to animal behaviour. Studying birds' song learning is interesting not only for understanding its ethological function, but also because it seems to be a good neurological model of *human language learning*. Marler and Peters (1977) followed a full ontogeny of song learning with a rather large group of swamp sparrows up to adulthood. Living in a laboratory, the birds could hear song sparrows and swamp sparrows singing, and all songs heard or produced by them were recorded. The processing of this huge mass of data offers a very interesting insight into the process of learning to sing. Song learning seems to be some kind of selection process; young birds copy many sorts of sounds and syllables, and they form many kinds of songs which, at the beginning of their development only resemble the song specific to the species, but they also contain many differences too. It has been revealed that birds sing at least five times as many songs, on average, in the course of their development, as those they maintain in the end in their song repertoires, continuously refining and perfectioning them. Only those few that get into the final set of songs fit the marks characteristic to the species in every way. This phenomenon shows a clear case in which organizational information is created from given components by selection: I_{org} originates from I_{comp}.

The demonstration of these *mechanisms of selection* is interesting because we have presumed for a long time that some kind of selection must have been functioning in the collection and storage of memory-traces, as was developed earlier in the model of replicative memory (Csányi, 1978, 1981, 1988a). There was very little direct evidence connected to behaviour. As to the brain development of higher species, the idea has long prevailed that the brain fully develops at the early stage of life, neurons stop dividing, and in adulthood it is only in the *fine-structure* of the animal brain, perhaps only at the organizational level of cell that change takes place. To everyone's surprise it has turned out that in the brain of the male canary the nuclei of neurons of the brain connected with singing are not static in adulthood (Nottebohm, 1988). After the reproductive period the volume of brain nuclei and the number of neurons therein decreases to almost half that of the earlier level. This neural change exactly coincides with a decrease in the disposition to sing. The songs of the canary become uncertain again for several months, even ceasing for a few weeks in the middle of the reproduction period. Despite the instability of the old songs the canary is able to learn new ones, especially when the brain nuclei begin to grow again. By the next reproduction period the nuclei regain their original volume, songs become complete and stable again, the male even sings new ones, while he forgets a few old ones.

It has been successfully proven that the basis of this behaviour is related to the differentiation of brain neurons. Neurons originating from other areas of the brain and not yet fully differentiated *multiply and transmigrate* to the territory of brain nuclei connected to canary song; There they differentiate into functionally active interneurons and fit into the network of existing neurons. At the same time a part of the old neurons die. Biochemical and behavioural data both prove that the canary brain is continuously *"re-building"* itself at the pace of the reproduction cycles, and that re-building is in close relation to song learning and forgetting. It is an extremely important observation that, by means of measurement at higher precision, it could be shown that the re-building activity of the canary brain is not exclusively limited to areas connected to song learning. To a much smaller extent though, similar processes can be detected everywhere in the cortex. Nottebohm (1988) supposes that in the canary brain only those neurons which are not connected with learning are *permanent*; that is, those that control *inherited behaviour patterns* or the innumerable neural controlling processes of animal organisms. These neurons comprise the major part of the neural network. The special interneurons, on the other hand, which are connected with learning – and not only song learning – are continuously dying and coming into being. After their emergence they undergo a process of differentiation in the course of which memory traces are somehow stored in them. When they die the animal forgets a part of its learned behaviour.

Our own experiments with fish demonstrate that, in the course of the fine adjustment of brain structure with the environment, a functional model emerges which helps the animal predict the behaviour of other beings in its environment, and it uses these predictions in the interest of its own survival (Csányi, 1986). The paradise fish is instinctively attracted to objects which have two eye-spots and are moving. These key stimuli provoke an active exploration in paradise fish and start a process of learning (Csányi, 1985a, b). If no unpleasant event affects the animal during the exploration, it will ignore, in encounters that follow, the object or living creature seen earlier. However, if that creature had chased it or caused trouble in any other way, then in another encounter the paradise fish shows an active avoiding reaction. The learned reactions can still be triggered with success after several months (Csányi et al., 1989). Amidst natural conditions many memories related to predators and peaceful creatures are stored in the brain of the paradise fish, which also activate the appropriate sequences of behaviour by activating the memory traces. The neural system evolved in this way works functionally as a model of environmental events (Csányi, 1986, 1988a, 1989a).

These examples show that the history of creative acts during the ontogenesis of behaviour transmit the smallest changes in the environment to the organism and continue in the adult animal. Ontogenesis, lasting until death, remains sensitive to external conditions during the whole life of the animal. The process of ontogenesis and that of evolution do not differ essentially (Csányi, 1989a). The brain structure creating the song repertoire of the male canary is a particular system that is different in each reproduction period; structural changes can be interpreted as a historical series of external events, thus ontogenic history is essentially an evolutionary process. The final structure of the paradise fish brain is determined by creative acts of environmental effects, just like the genome is, in another historical time-scale. The time changes that take place in the second-kind time of the species make the first-kind time manifested in the ontogenesis a relative second-kind time again, for one time-cycle at least, for the life-period of one single individual. In doing so, phylogenesis creates a model, a special and ephemeral form of itself represented in ontogenesis.

6 Animal Mind As a Component-System

In earlier studies (Csányi, 1980, 1982, 1987b, 1988a, 1989a) ontogenic development of the animal mind was considered as an evolutionary process and was termed *neural evolution*, which also can be treated in the framework of the replicative model, but here I will only give a short summary.

The essence of neural evolution is the assumption that the basis of animal memory is the interaction of replicative components formed by neural interconnections. From this assumption follows the development of a new type of evolutionary system.

The most important biological function of the animal brain and memory is the construction of a *dynamic model of the environment*, the continuous maintenance and operation of the model, and the use of the data obtained by this operation for predictions in the interest of the survival and reproduction of the animal. This model-constructing ability of the animal brain is regarded as *consciousness* (MacKay, 1951–52).

The modelling activity of the brain manifests itself in the development of higher organization above the level of neurons. Emergence of structures consisting of intercellular connections which would correspond to the stimulus, the drive, the response or even the expected results of possible responses is assumed. The basic units of these associative structures are the **concepts**. The confluence of concepts form higher active neural networks that determine the behaviour of the animal in a given situation. Concepts formed by interactions of neurons are considered to be components of the neural system which are continuously built up and dismantled. The replication of concepts takes place under the continuous controlling effect of various motivational mechanisms, and of the environment. Concepts having adaptive value for an animal get into the multiplication cycle more often, and are therefore produced in greater numbers than irrelevant ones or those that are harmful. The essential process of learning is the selection of concepts generated by the brain.

Three-part functional structures can be regarded as units of the concepts. Percepts originating from external stimuli in the environment, through animal perception, play the part of a "*key*" which is the basic connection between the inner and the outer worlds. Sooner or later the stimuli are always followed by some action of the animal. Neural structures which organize animal *actions* form the third part of elementary units of the neuronal models. Nevertheless, there is no direct connection between percepts and actions. The same stimuli could activate different actions, depending on previous experience. But actions also depend on the current inner state of the animal, whether it is hungry, or thirsty, for example. The functional elements that link percepts and actions are called *"reference structures"*. The triadic "key-reference structure-action" units can be combined, and the reference structures and their complicated compositions are the very models of the environment (Csányi, 1988a).

The three-part units of these models are called *concepts*. A concept can be considered as a neural controller unit of a behavioural act or thought in the animal mind. The general capacity of the higher structures, which can be built from the concepts, is rather limited in the animal brain because each individual can form its models only from its own limited experiences. Each and every model made by animal brains is highly particularised in this respect.

Various data suggest (Csányi, 1982) that, in the course of ontogenesis, the animal brain goes through phases of neural autogenesis, non-identical replication and identical replication, and the phenomena of each phase (functional differentiation, rise of supercycles, compartmentalization, convergence) can also be observed. The human brain, which develops in the course of embryogenesis, represents a zero-system capable of neural evolution. This zero-system can be considered as a *cognitive space*, in which the physical components corresponding to concepts are continuously synthesized and decomposed. In this process, a decisive and guiding role is played by various motivational mechanisms and external stimuli.

Our earlier terms of information can also be applied here. Building blocks of the concepts can be regarded as components of a component-system and their I_{comp} is the basic information content of the brain's models. When organized structures of the concepts emerge both I_{org} and I_{symb} appear, and later become more and more important during animal evolution. The creativity of the animal brain manifests itself as an ability to create such organization in concepts which are able to represent the events, entities and interactions of the outer environment. This creativity is based upon the hidden properties of the neuronal building blocks of the concepts and also on the selection process which acts during learning. In less-developed brains the possible organizations are rather limited, but in higher animals like dogs, apes, dolphins and man the concept structures become highly complex and their information contents are finely tuned to the environmental structures.

7 Linguistic Models of the World: The Rise of Conceptual Thought

In humans the replication of concepts is connected with the use of *language*. With the aid of language, very accurate replication of complex concept structures and their storage for as long as a lifetime becomes possible.

The evolution of the linguistic competence of man resulted in a fundamentally new model-making mechanism. By *naming* something, a key which has only a very loose connection with percepts and actions arises. The "word-referential structure-action" segmented units could be combined not only through experiences, but also through grammatical rules and this results in the very complicated super-structures of *conceptual thought*. A linguistic concept can be regarded as an utterance or human thought. The linguistic concept could also activate actions, but primary experience is no more a prerequisite for these actions than it was in the case of the animals. Mental super-structures built up from linguistic concepts also reflect experiences, of course, and therefore they can be regarded as models of the outer world-much more so, as it is clear that animal-type concepts and linguistic concepts combine together to create a world model in man. In the evolution of man the symbolic information content of the brain's models plays the most important role. Animals are capable of thinking in their own ways but according to our knowledge only man uses *descriptions* in his thinking. Descriptions have double functions. First, as descriptions they are nonreferential information about something observed, reflections and representations of the outer reality; on the other hand, they are active entities in the human brain, and thus they have referential information connected to the human world. Their "meaning" is connected to this second function. *Meaning* is an active property of a description bound to the whole system by which the description was made. The concept of meaning is not of much use in discussing lower biological systems because in these symbolic information content is not especially high. One notable exception is the genetic code for protein structure. Meaning of the genetic code lies in the system's individuality. Specific DNA codes can act only in given systems, as their actions are bound to the individual organization of the system. Of course, meaning and identity are also closely connected in human thought. Conceptual thoughts have meaning only in such human systems where they can influence actions and these systems are always individualistic in a systemic sense.

A further important consequence of linguistic competence is the opening up of the closed inner world of the individual. By linguistic communication individuals are able to exchange parts or the whole of their models in the form of linguistic concepts. Through this process the survival of the models becomes independent of the individuals carrying of them. An independent evolution of the models can commence in this way. This is called cultural evolution (Csányi, 1978, 1982, 1989a).

The most important consequence of language competence is the ability of an individual *to transfer the concept-components of his own memory (by copying them with varying fidelity) into the memory space of another individual through language*. This copying process is replication, that is, the propagation of the concept-components in a physical sense. As a further consequence, the evolution of concepts is not restricted any longer by the longevity of the individual. Individuals living in a linguistic community join their memory spaces and create a common replicative space many orders of magnitude larger than the individual ones.

With this new mode of concept-replication a new evolutionary process, *culture*, commences.

8 Culture and Ideas

It is worth comparing animal and linguistic concepts in an example. It is a well-known observation that in the cooperative groups of certain higher mammals, like the wolf, individuals use hunting tactics in which they are very attentive to each other's actions. Each pack member positions itself in such a way that the appearance of a fellow member at the right moment and in the right place is clearly supposed. This significant form of cooperation can exist because of the high similarity of the models each wolf's brain. The individuals have the capacity to identify themselves with fellow members, and they can figure out the next actions of the latter. That is, they have appropriate models of the behaviour of others. This kind of cooperation occurs *without the exchange* of plans, intentions or thoughts. Wolves can communicate only certain parameters of their inner state, like hunting or aggressive moods.

Members of a human hunting group with linguistic communication are not only able to predict the behavioural actions of their fellow members, but can also divide the common tasks among themselves, can make plans and then assign individuals a particular action (Eibl-Eibesfeldt, 1982). In this way, concepts existing in the individual brains become parts of a higher collective structure which then determines the aim and the exact route to other achievements. We call this higher organization of individual concepts an *idea*.

Concepts comprising an idea are not selected at random, instead they form a functionally organized set which makes the performances purposeful and possible. It is not even important for every member of the hunting group to know everything about the task or roles of the others. It is enough for the leader to know the main programme, but even for him it is unnecessary to learn the finer details. Ideas can be organized hierarchically and the whole is available only in the whole group whose activity is regulated by the idea.

Individual concepts existing in the brains of the group members can be functionally combined only by a specific self-organization of the ideas. Only ideas that contain those and only those concepts suitable for achieving the given goal can act and accomplish something.

Besides linguistic competence, it is the rule-following ability of our species that made ideas possible. Most of the concepts of an idea are simple behavioural rules, which are the elements of the collective action. Language itself can be described by a series of elementary rule-following behavioural actions.

Many different goals arise in the life of a group. Therefore, very strong selection is exerted on the formation of the ideas connected to these goals, and those concepts which are unsuitable for a given goal are selected out. Suitable concepts are memorized by the members of the group and these are connected by strong functional bonds. Survival values of the ideas are much stronger than those of the concepts because of the higher organization of the ideas. Brain models of the environment made from animal concepts also have a survival value if the individual has the appropriate experiences during its life and can make the right model. However, ideas built up from the linguistic concepts above the individual level are useful controlling agents which carry both the experiences of the acting and contributing individuals, and the experiences of their ancestors as well. Over the generations, concepts which are unsuitable for realising the given goal for some reason or other are selected out from the idea pool. Therefore, ideas are organized by the collective experience of the group and they tremendously enhance the effectiveness of the group.

A classification of the ideas has not yet been worked out, but if we define an idea as an organized set of concepts necessary for achieving a given goal, it is obvious that even a primitive group society had many different ideas simultaneously. For hunting, fishing, for the defence of the group from predators or from other groups, they needed different ideas. It is also clear that the different ideas could not be entirely independent of each other because there are many common behaviour elements of these actions, and it is also necessary to coordinate the maintenance and expression of the various ideas. The seed of this coordination is the group as an entity itself. The ideas organize the maintenance and survival of the group which in turn must harmonize the ideas. Various organizing ideas, such as creation myths, legends, and religions emerged. Values and norms appeared.

An important organizational principle of the ideas has to be shown. Both human groups and ideas are *population* entities, that is, they consist of individual components which are bound into some higher functional organization. Components of the group are fellow members, people who are born in or immigrate into and are accepted by the group. Therefore, the group can be defined only stochastically. At a given moment all members of a group can be counted only artificially and without certainty. For example, we can consider the newborns, who have not yet acquired the ideas of the group and whose life might be short as members, or we can logically exclude them by restricting memberships to adults. It is also important that components of the group are continuously changing by death, birth, immigration and emigration. The group consists of people as components, but the group is an entity formed above the people, and the definition of its identity is theoretically a difficult problem which I wish only to mention here.

If we examine ideas closely, their populational nature also shows up. Components of ideas are concepts, both linguistic and nonlinguistic, which exist physically in people's brains. Ideas are formed from concepts by a functional organization at the group level. Because concepts are entities of the nervous system, and they correspond to rather elementary units of behaviour or thought, it is impossible to determine the exact concept constituents of an idea.

If, for example, we study an ancient craft, like arrow-making, it is clear that this craft follows certain rules concerning the selection and preparation of raw materials, and in the sequence of the technological procedures. It cannot be said that only one sequence or set of concepts results in appropriate products; small deviations are allowed during the whole technological process, and therefore the idea of arrow-making is also only a stochastically characterised entity. It is a stochastic set of concepts, members of which can fluctuate very much without perceptably changing the effectiveness of the idea.

It is extremely important for our further discussions to emphasize that the stochastic entity of an idea is bound to the physical structure of another stochastically determined entity, the group. Therefore, the action and the effect of an idea is fundamentally influenced by this double stochastic characteristic.

Human creativity is based upon the informational nature of ideas. Their components, brain concepts, are in themselves enormously complex entities which have a great many hidden properties and are highly variable. Idea organizations which connect the concepts into a higher acting structure contain both organizational and symbolic information which has its origin in the hidden properties of the concept components, themselves originating from the hidden properties of neurons. Thus, organizational and symbolic information transcend the inherent nature of matter through to symbolic human culture. Its unlimited creativity comes from the deepest organizational layers of the physical world.

9 Evolution of Ideas in Mass Society

Ideas organized around group identity, and people carrying these ideas, formed a complete unity. A given group was a set of particular people and at the same time it was also a set of particular ideas, these two components forming the culture of the group. Mass societies have appeared during the last couple of thousand years of cultural evolution, and they have fundamentally changed the relationships among groups and ideas. Ethological traits are unsuitable as bonding forces for groups of hundreds of thousands of people, and later for the many millions of people of modern societies. This is so in spite of the continuous, uninterrupted actions of these factors, because it is well known that a human being can develop strong personal bonds only in groups not exceeding the limit of 60–80 people. Bonding forces for larger groups like armies, states, parties or major religions are provided completely by ideas (Csányi, 1990).

Emergence of mass societies also promoted the *competition* and selection of ideas that has been at a very low level in group societies. Idea competition led to new forms of self-organization. Ideas in group societies served practical purposes (technique and technologies), or they supported group identity by means of myths, legends and primitive religions. Exchanges of ideas between societies were only occasional and ineffective, mostly because group societies are rather closed systems. In modern mass societies, the number of potential idea carriers is enormous and the acquisition of ideas can occur by new means other than the traditional group framework. Mass communication makes it possible for ideas to encompass people who do not compose a real group in the ethological sense.

Survival of the ideas of group societies was ensured by early learning and socialization, and by extremely strong tradition. Ideas were acquired in childhood. Ideas of mass society compete for the adult members of the society. This change has important evolutionary consequences in the structure and organization of present-day ideas. New kinds of components appeared in the set of linguistic concepts which form the various ideas. These new concepts promote the spreading of the given idea and help it compete. The missionary role of some modern religions is a good example; such a role was impossible in the culture of a group society. The creation myth of a group society need not be logical or convincing at all, because every member of the group believes it, it is taught by the elders, and there is no reason to question any part of the myth. Ideas in mass society are continuously challenged. The creation myth must be supported by science and practice. Logical overall explanations are needed; moreover, we want the possibility of choice among alternatives.

Ideas in mass society can organize groups of immense size, but only if there are such concepts among its constituents which are suitable for maintaining large groups. This leads to the appearance of propaganda and ideology as the main tools of the organizers of the ideas.

As we have said, the major advantage of idea-organized groups is the emergence of idea competition, which is the basis of the development of modern science and also of the modern welfare state. A disadvantage comes from another direction. Group society has ideas in the form of values and norms which guarantee the undisturbed cooperation or at least tolerance of the various ideas. In mass societies this harmonizing effect of values and norms ceases to exist. Ideas appear as a selectable entity. The individual must choose ideas from an enormous pool to construct his or her personality, but there is no comprehensive idea framework accepted by everybody, which could serve his or her unambiguous selection.

Alienation of modern man, his insecurity and his loss of values can be explained by idea competition.

Perhaps this is just a temporary stage, because today new ideas are being formed which have a global nature. The unity of humanity, global peace, global protection of nature and other ideas emerge, which most probably will form a network of organizing ideas and create a harmonious environment for others. They will also result in a closure of the idea world again. As the new global-system identity emerges, new meanings of the ideas will be set for: *survival of humanity*.

10 Closing Thoughts

Information of biological systems originates from the hidden properties of the physical components, and eventually from the atoms of the elements. Chemical interactions of the atoms and molecules could create a chemical zero-system and initiate autogenetic processes, which inevitably lead to the creation of organizational and symbolic information. Creativity of the autogenetic process lies in the hidden properties of the atoms, which manifest themselves in creative acts and form new interactions. Hidden properties of the components at any organizational level are responsible for the emergent organizations.

Hidden properties of nerve cells, of concepts and ideas *create* the variability and complexity of information in the human world.

Creative information is responsible for all *actions* of the biological and social realms.

The final meaning of biosocial information is *being*.

References

Bergson H (1907) L'Évolution Créative, Paris

Bunge M (1963) The Myth of Simplicity, Prentice-Hall, Englewood, Cliffs, NJ

Cavalli-Sforza LL, Feldman MW (1981) Cultural Transmission and Evolution Princeton Univ Press, New Jersey

Csányi V (1978) Az evolució általános elmélete, Fizikai Szemle **28**, pp 401–443

Csányi V (1980) The General Theory of Evolution, Acta Biol Hung Acad Sci **31**, pp 409–434

Csányi V (1981) General Theory of Evolution, Soc Gen Syst Res **6**, pp 73–95

Csányi V (1982) General Theory of Evolution, Publ House Hung Acad Sci, Budapest

Csányi V (1985a) Autogenesis: Evolution of Selforganizing Systems. In: Aubin J-P, Saari D, Sigmund K (eds) Dynamics of Macrosystems, Conf Proc, Laxenburg, Austria 1984; Lecture Notes in Economics and Mathematical Systems No. 257, Springer, Berlin, pp 253–267

Csányi V (1985b) Ethological Analysis of predator Avoidance by the Paradise Fish (Macropodus opercularis) I Recognition and Learning of predators, Behaviour **92**, pp 227–240

Csányi V (1985c) Ethological Analysis of predator Avoidance by the Paradise Fish (Macropodus opercularis) II. Key Stimuli in Avoidance Learning, Anim Learn Behav **14**, pp 101–109

Csányi V (1986) How is the Brain Modelling the Environment? A Case Study by the Paradise Fish. In: Montalenti G, Tecce G (eds) Variability and Behavioral Evolution, Proc Accademia Nazionale dei Lincei, Rome 1983, Quaderno No. 259, pp 142–157

Csányi V (1987a) The Replicative Model of Evolution: A General Theory, World Futures **23**, pp 31–65

Csányi V (1987b) The Replicative Evolutionary Model of Animal and Human Minds, World Futures **24**(3), pp 174–214

Csányi V (1988a) Contribution of the Genetic and Neural Memory to Animal Intelligence. In: Jerison H, Jerison (eds), Intelligence and Evolutionary Biology, Springer, Berlin, pp 299–318

Csányi V (1988b) Il modello replicativo dele'evoluzione biologica e culturale. In: Ceruti M, Laszlo E (eds), Physics: abitare la terra, Feltrinelli, Milan, pp 249–260

Csányi V (1989a) Evolutionary Systems and Society a General Theory, Duke Univ Press, Durham, p 304

Csányi V (1989b) Origin of complexity and organizational levels during evolution. In: Wake DB, Roth G (eds), Complex Organismal Functions: Integration and Evolution in Vertebrates, Wiley & Sons, pp 349–360

Csányi V (1990) Ethology, Power, Possessions: A System Theoretical Study of the Hungarian Transition, World Futures **29**, pp 107–122

Csányi V, Kampis Gy (1985) Autogenesis: Evolution of Replicative Systems, J Theor Biol **114**, pp 303–321

Csányi V, Kampis Gy (1987) Modelling Society; Dynamical Replicative Systems, Cybernetics and Systems **18**, pp 233–249

Csányi V, Csizmadia G, Miklósi Å (1989) Long-term Memory and Recognition of Another Species in the Paradise Fish, Anim Behav **37**, pp 908–911

Eibl-Eibesfeldt I (1979) Human ethology: Concepts and Implications for Sciences of Man, Behav Brain Sci **2**, pp 1–57

Eibl-Eibesfeldt I (1982) Warfare, Man's Indoctrinability and Group Selection, Z Tierpsychol **60**, pp 177–198

Ellul J (1965) The Technological Society, Jonathan Cape, London

Kampis Gy (1986) Biological Information as a System Description. In: Cybernetics and Systems 1986 (ed Trappl R) Reidel D, Dordrecht, pp 36–42

Kampis Gy (1987a) Some Problems of System Descriptions, I Function, Int J Gen Syst **13**, pp 143–156

Kampis Gy (1987b) Some Problems of System Descriptions, II. Information, Int J Gen Syst **13**, pp 157–171

Kampis Gy (1990) Self-Modifying Systems in Biology and Cognitive Sciences: A New Framework for Dynamics, Information, and Complexity, Pergamon Press, Oxford

Kampis Gy, Csányi V (1987a) Replication in abstract and natural systems, BioSystems **20**, pp 143–152

Kampis Gy, Csányi V (1987b) A Computer Model of Autogenesis, Kybernetes **16**, pp 169–181

Kampis Gy, Csányi V (1987c) Notes on Order and Complexity, J Theor Biol **124**, pp 111–121

King GA (1982) Recycling, Reproduction and Life Origins, Biosystems **15**, pp 87–89

Lee R (1969) I'Kung Bushmen Subsistence: an Input–output Analysis. In: Vayda P (ed) Environment and Cultural Behaviour, Natural History Press, Garden City, pp 47–49

Mackay DM (1951–52) Mindlike Behaviour of Artifacts, Brit J Phil Sci **2**, pp 105–121

Mumford L (1976) The Myth of the Machine, Harcourt, Brace, Jovanovich, New York

Marler P, Peters S (1977) Selective vocal learning in a sparrow, Science **198**, pp 519–521

Nottebohm F (1988) Hormonal regulation of synapses and cell number in the adult canary brain and its relevance to the theories of long term memory storage. In: Lakoski JM, Perez-polo JR Rassin DK (eds) Neural control of Reproductive Function, Liss

Oyama S (1985) The Ontogeny of Information Cambridge Univ Press, Cambridge, England

Passingham R (1982) The Human Primate, Freeman, Oxford

Pattee HH (1977) Dyambic and Linguistic Modes of Complex Systems, Int Gen Syst **3**, pp 259–266

Prigogine I, Nicolis G, Babloyantz A (1972) Thermodynamics of Evolution, Physics Today **25**, pp 23–28, pp 38–44

Riedl R (1978) Order in Living Organisms, Wiley, New York

V The Evolution of Information Processing Systems at the Social Level

Unitary Trends in Sociocultural Evolution

Ervin Laszlo

Chaos and Bifurcation 282

Stability and Instability 282

Convergence to Higher Levels 284

Interface with Nature 285

Extrapolation from the Trend 285

References 286

The new sciences of systems and complexity – cybernetics, information and communications theory, chaos theory, dynamical systems theory, nonequilibrium thermodynamics, and general systems and general evolution theory – have made great progress in the detailed understanding of the laws and dynamics that govern the evolution of complex systems, regardless of whether they are physical, biological, ecological or human. The new insights contradict the facile views that we have reached the end of history, and that the future is made by mere "bricolage." The course of history has a logic of its own; a logic that is not rigidly predetermined and yet not the plaything of chance. This is a logic that governs not only the evolution of human societies in history but also the evolution of life on earth, and even the evolution of the cosmos in the observable universe.

The manifestations of the evolutionary logic discovered in the new sciences are more than "mega", from the Greek "megas"; they are "giga", meaning "gigas", that is, giant. The gigatrends of evolution hold good in all major spheres of investigation. Scientists are discovering them because they pay increasing attention to evolutionary processes. The cosmologists find laws of cosmic evolution: patterns and regularities that govern the creation, configuration and ultimate annihilation of baryonic matter in space and time, from the super-burst of the Big Band to the final evaporation of black holes. The life scientists discover evolution in the more than four-billion-year time span during which man emerged from the protocell, and the self-regulating system of the biosphere evolved from the interaction of organisms and environments. The social scientists find that the social, economic and ecological systems formed by human beings evolve in the course of recorded history, following processes that are just as fundamental and irreversible as evolution in other fields of experience. Even the previously encapsulated human scientists have awakened to the fact that mankind is embedded in the biosphere, and may be evolving toward a consciousness-impregnated noosphere.

Chaos and Bifurcation

The gigatrend underlying evolution in the various realms is irreversible, chaotic and nonlinear. Evolution takes place through periodic outbreaks of chaos, in a process known as "bifurcation". Bifurcations occur when systems are destabilized in their milieu: they then shift from one set of "attractors" to others. Depending on whether the shift is smooth and continuous, entirely abrupt or just sudden, chaos scientists speak of "subtle", "explosive" or "catastrophic" bifurcations. All three are the subject of extensive investigation.

The new discipline of nonequilibrium thermodynamics applies the mathematical simulations of chaos theories to real-world systems. The relevant kind of system is that which exists, like a vortex in a stream of water, in a state of dynamic nonequilibrium within an enduring energy flow. The condition of nonequilibrium signifies the presence of energy concentrations and chemical gradients, hence the condition under which dynamical systems can perform work. When changes that occur within the systems or in their milieu upset the dynamical balance between system and environment, a chaotic state comes about. In terms of dynamical systems models, point and periodic attractors give way to chaotic ones. If and when the system regains its dynamical balance with the environment, its unstable chaotic attractors yield to a new set of more stable point or periodic attractors.

The description of the evolutionary trajectory of dynamical systems as irreversible, chaotic and nonlinear fits the observed course of development in history as well as in nature. This is because human societies, like biological species and entire ecologies, are dynamical nonequilibrium systems immersed in the enduring flow of energy that streams from the sun to the surface of our planet.

Stability and Instability

The evolutionary gigatrend that underlies the long-term development of human communities from the stone-age tribes of the Paleolithic era to the techno-industrial societies of modern times manifests a constant dynamical interplay between stability and instability. Traditional societies are relatively stable: they are built into their milieu by powerful myths of reverence for nature and natural cycles. But as innovations change traditional relations between man and nature, the myths are revised and societies expand in population and resource use. In time, they reach the limits of the resources available in their environment. This leads to structural instabilities. The instabilities are overcome either by adopting practices and technologies that make do with, or else stretch, the existing resource base, or by acquiring, or emigrating to, outside territories.

Major instabilities bring in transitional crises. The reins of power change hands, systems of law and order are overthrown, and new movements and ideas surface and gain momentum. When the causes of the instability have been removed or at least mitigated, order is reestablished and the crisis of transition gives way to a new era of stability. Because some variety of improved technique or technology is usually involved in removing the causes of instability, the new order tends to be more complex, and its procedures more structured and sophisticated, than those of the foregoing one.

Technological innovation plays a specific role in this process. In the long run it creates access to a wider range of energies and brings about a more intense utilization of the energies that are actually accessed. In the earliest societies of prehistory, the main source of energy was the caloric content of the food ingested by our distant ancestors. Some one-and-a-half or two million years ago the energy of fire was added, and about eight thousand years ago the energy of draft animals. Further sources of energy were tapped through technological innovations throughout recorded history – for example, the power of moving streams by water mills, and of moving air by windmills. The exploitation of the existing energy sources was steadily improved by an array of inventions using levers, pumps, clockworks and similar mechanical devices. The discovery of the steam engine created access to the energy of superheated water and brought in the industrial era. When boilers came to be fuelled with coal rather than with wood, further increments were achieved in the gross use of energy. In the last century the steam engine began to be replaced by the internal combustion engine, and coal, oil and water were the main fuels used to generate electrical power. Further increments in gross energy use were achieved in recent decades with the installation of nuclear power reactors as the mainstays of extensive power grids in many industrialized regions.

In the transition from the first to the second industrial revolution, innovation concerned not the gross quantity of the energy that was used, but the density of the effectively exploited energy flux. The new breakthroughs enhanced energy efficiency by creating technologies that constantly did more with less. These technologies extended not only the power of human muscle and human sense organs, but also the power of the human brain. With its capacities to process specifically patterned energy – that is, information – the second industrial revolution created another quantum leap in the effectiveness and efficiency of energy use.

Thanks to the growing input of technological innovation, contemporary societies set forth the historical progression toward the exploitation of ever more energies with ever more efficiency. High-energy societies perform much work and support a more complex social structure. They are further from thermodynamic equilibrium than any human society in recorded history. As a consequence the gigatrend in history, as in biological evolution, takes evolving systems from low-energy to high-energy (high negative entropy) states, and from simpler to more complex structures. The trend unfolds through the nonlinear alternation of stability and instability, with chaos interspersing the more stable periods.

Convergence to Higher Levels

Societies, as they move further from equilibrium and become more dynamic, develop correspondingly complex forms of organization. The progressive complexification of the structures of society illustrates another aspect of the evolutionary gigatrend: that of inter-system convergence toward successively higher organizational levels.

In politics, convergence means a growing similarity and ultimately the full uniformity of the converging parts. In the context of evolution, however, convergence has a different meaning: it stands for the progressive coordination of the diverse functions of at first relatively independent, and then increasingly interlocking, dynamical systems. In this sense convergence is a universal feature of evolution. In nature, chemically reactive atoms interlock in multiatomic molecules, and simpler chemical molecules build more complex polymers and in some cases organic macromolecules. The simpler forms of life, based directly on systems of organic molecules, interlock in more elaborate cellular lifeforms, and single cells can evolve coordinate functions in loosely integrated colonial and later in more fully integrated multicellular organisms. Organisms interlock in intra-species social structures as well as in inter-species ecologies. All living systems converge ultimately in the self-regulating system of the global biosphere.

In the context of history, convergence brings together tribes, clans, villages and provinces in more extensive, complex and diversified social, economic and political systems. The archaic empires of China and India incorporated and coordinated villages and regional communities in subcontinental administrative structures; the classical Roman Empire was built of numerous city-states, regions and provinces under the common aegis of Pax Romana; the colonial empires of Europe consisted not only of villages, towns and provinces in the mother countries but also of strings of overseas colonies. Even if there are no longer empires and colonial powers, contemporary societies still display the consequences of the process of historical convergence. Every modern state consists of metropolitan centres and rural areas with villages and towns, and federated states consist of politically and socioeconomically integrated states, republics, provinces or cantons, administered by regional governments overseen and coordinated by the central powers.

In the middle of the 20th century the gigatrend of inter-system convergence was temporarily reversed. This was a consequence of the dissolution of the colonial world of European nation-states without a new set of relationships to take its place. The result was a proliferation of societies organized on the principle of nation-states. Since the majority of the newly created nation-states could not achieve a real measure of independence, pressures have been building for the integration of these structures in broader international structures. Thus today we are in an epoch of rectification. The process of covergence has been relaunched with the creation of subregional and regional communities and associations of interdependent national states. The creation of the new Europe is a clear expression of inter-nation convergence in the post-modern epoch.

Interface with Nature

Growing complexity in the structure of increasingly dynamic societies creates a closer and more critical coupling between societies and their physical environment. The coupling involves all raw materials and natural resources essential to fuelling the new technologies and sustaining the large populations served by them, and is the most crucial in regard to the flow of energy through the system. No matter how efficient a form of technology is, it creates some form of thermal pollution. If further large quantities of energy were produced, further increments of heat would be discharged into the biosphere and a critical threshold would be reached. The thermal balance of the biosphere is highly sensitive to human impact, and its dynamics are nonlinear. Just as a small additional discharge of chemical waste into the sea can bring about a sudden eruption of algae, in the same way a little more output of waste heat could rapidly shift atmospheric heat balances into new and humanly unfavourable equilibria.

The unfolding of the gigatrend has accelerated throughout history, and in the span of the last century it has done so at an exponential rate. Societal size has been growing everywhere as well as structural complexity; technologies have become increasingly broad-based and, through their newly-won information processing capacities, extremely energy efficient; national societies have been brought together within a variety of economic, social and military alliances; and man and nature have become infinitely more closely and vulnerably coupled.

Extrapolation from the Trend

The evolutionary gigatrend permits an extrapolation to be made into the future. The matter is not as simple as it may seem, however. At first sight it may appear that human societies will be larger in size, greater in complexity and more intensely integrated both with each other and with their physical milieu. A linear extrapolation suggests that states will be multinational, and their network globally extended. The new social, economic and political units will emerge from the ongoing convergence of today's national states and national or regional economies. On their various levels and forms of organization, future societies will incorporate in coordinated economic, social, and political structures villages, farming communities, urban neighborhoods, townships, districts, provinces, national and federated states, and continental or subcontinental associations and federations. Future societies will be more stable than societies today. Evolution, as we have seen, is not necessarily the infinite acceleration of bifurcations: periods of instability alternate with epochs of stability, and a conscious mastery of the process could effectively extend the latter and reduce the former.

A closer scrutiny of the trend casts doubts on the validity of such linear extrapolations. The gigatrend is not only nonlinear, it is also nondeterministic.

This makes extrapolations difficult: we are dealing with probabilities and not with certainties. The future could bring genuine novelty, including the novelty of human extinction. The gigatrend is only the envelope within which there can be numerous deviations. In today's highly stressed world, a negative deviation could have serious consequences. If it is in the form of a world war, or of a further important degradation of the environment, the planet would become uninhabitable for aeons to come.

An ultimate catastrophe of this kind would be something new on the world scene: previously, evolution has always led to the disappearance of some of the systems, but never all. Almost 99 per cent of the biological species that have emerged in the course of the last three billion years have become extinct, and a large proportion of the culturally specific human groups and societies that arose in human history have likewise vanished. But some species and societies have always survived, and then evolved. Today, however, biological extinction outdistances evolution by a factor of a million to one – this is the proportion of the number of species that are currently becoming extinct, compared with the number of new species that are still emerging. If Homo were to continue his short-sighted interventions in the ecosystems of the biosphere, he would flatten diversity to the point of catastrophic vulnerability. And if he were to commit the ultimate folly of a major war, he would not only become extinct himself, but would take with him all species more evolved than insects and grass.

The fact that a trend or pattern has held good in the past is not a guarantee that it will hold good in the future. We can be assured in this regard only if we can derive the trend from processes that are themselves enduring. This appears to be the case in regard to the evolutionary gigatrend. Its foundations in contemporary scientific theories justify the expectation that from this trend extrapolations into the future are justified. Although bifurcations can always produce unforeseen outcomes, including those that are catastrophic for the bifurcating systems, any dynamic process which permits evolving systems to persist in, or find, dynamic steady-state solutions to critical instabilities and chaotic states is likely to exhibit characteristics conformant to the evolutionary gigatrend. The alternative to evolution continues to be extinction, but the alternatives that permit evolution tend to be situated within the parameters described by the fundamental laws of the evolutionary process.

References

Laszlo E (1969) System, Structure, and Experience. New York, London: Gordon and Breach
Laszlo E (1972, 1991) The Systems View of the World. New York: George Braziller; revised edition in preparation. In German: Ganzheitlich Denken. Rosenheim: Horizonte
Laszlo E (1972, 1973, 1984) Introduction to Systems Philosophy. New York, London: Gordon and Breach; revised edition: New York: Harper Torchbooks; reprinted: Gordon and Breach
Laszlo E (1974) A Strategy for the Future. New York: George Braziller

Laszlo E (1983) Systems Science and World Order. Oxford: Pergamon Press
Laszlo E (1987) Evolution: The Grand Synthesis. Boston, London: New Science Library, Shambhala Publications. In German: Evolution: Die Neue Synthese. Wien: Europa
Laszlo E (1989, 1991) The Age of Bifurcation. New York, London: Gordon and Breach. In German: Global Denken. Rosenheim: Horizonte; München: Goldmann Verlag
Pattee H (ed) (1973) Hierarchy Theory: The Challenge of Complex Systems. New York: George Braziller. (Laszlo E (ed) The International Library of Systems Theory and Philosophy)
Weiss P (1971) Hierarchically Organized Systems in Theory and Practice. New York: Hafner
Whyte LL, Wilson AG, Wilson D (eds) (1969) Hierarchical Structures. New York: Elsevier

The Replicative Model of the Evolution of Business Organization

Mika Pantzar and Vilmos Csányi

1   Introduction   288

2   The Replicative Model of Evolutionary Processes   290
2.1 Dynamics of Replicative Systems and Replicative Information   292
2.2 The Replicative Model of Cultural Evolution   294
2.3 Evolution of Sociocultural Systems   295

3   The Replicative Model of the Evolution of Business Organization   296
3.1 The Replicative Nature of Business Organization   296
3.2 Business Organizations as Replicative Networks:
     Convergence and Compartmentalization   298
3.3 Origin, Evolution and Maturation of Business Organizations   299

References   302

1 Introduction

A remarkable period of transition in business organization, which we call the formative period of American business enterprise manifested itself in the United States of America in the latter part of the nineteenth century. The most obvious indication of change was the appearance of large integrated enterprises; administrative coordination began to assume the role of impersonal markets, and management was detached from ownership.

Alfred Chandler describes the background of the transition as follows:

> "Modern business enterprise, as defined throughout this study, was the organizational response to fundamental changes in process of production and distribution made possible by the availability of new sources of energy and by the increasing application of scientific knowledge to industrial technology. The coming of the railroad and telegraph and the perfection of new high-volume processes in the production of food, oil, rubber, glass, chemicals, machinery, and metals made possible a historically unprecedented volume of production. The rapidly expanding population resulting from a high birth rate, a falling death rate, and massive immigration and a high and rising per capita income helped to assure continuing and expanding markets for such production. Changes in transportation, communication, and demand brought a revolution in the processes of distribution. And where the new mass marketers had difficulty in handling the output of the new processes of production, the manufacturers integrated mass production with mass

distribution. The result was the giant industrial enterprise that remains today the most powerful privately owned and managed institutions in modern market economies" (Chandler, 1977: 376).

Let us summarize some aspects of the maturation of big business:

 (i) In the beginning of a business organization, exogenous factors – chance, inventions – dominate. This might potentiate pioneer advantage and growth. In the final stage, endogenous inertial path-dependency matters most (c.f. Arthur, 1988; David, 1988). The managerial problems of maturing business change from problems of creation (invention) to problems of maintenance and sustainability. The history of the American railroad exemplifies the shift of focus from issues of exogenous "mutations" (inventions) to problems of internal coordination and, finally, to questions of external coordination and politics (Beniger, 1986; Chandler, 1977; Pantzar, 1990).

 (ii) In the beginning of the maturation process, changes in cultural determinants and behaviour direct structure-formation. In the final stage it is structure that both restricts and enables routine acts through increasingly fixed cognitive maps. Scientific management and the dominance of the T-Ford exemplify both changes in conditioning factors and the blind alleys of over-specialization (c.f. Koestler, 1967: 219).

(iii) Different phases of maturation in different organizational and developmental levels feed each other positively. For instance, a perfected railroad system was needed before firms with massive integrated flows of production and distribution could emerge.

(iv) In time business organizations that survive reproduce themselves better and better; they become more predictable and dependable. What was in the beginning an inner directed attempt to find workable combinations of inputs will be transformed in time to coherent and predictable behaviour dictated by the external environment (emergent combinations).

Here we concentrate on two questions:

1. How do economic systems of interdependence (organizations of organizations of organizations) evolve?
2. To what extent are economic, historically evolved hierarchic configurations within firms (organizational structure), between firms, and between industries analogous to hierarchies of nature?

We presume that the tendency of ever-increasing complexification of structures in modern business organizations and economies and the shift toward hierarchic organization structures illuminated by Alfred Chandler are compatible with what we call "the general evolution viewpoint". It will be claimed that the replicative theory of evolution (Csányi, 1980, 1985, 1987, 1989, 1990; Pantzar, 1991) provides a useful theoretical basis for understanding the origin, formation and perfection of business organization.

Evolutionary (or ecological) theories of economics have only partially adopted concepts of modern evolution theory. Evolutionary views, such as Alchian and

Demsetz's (1972), Nelson and Winter's (1982), or the perspective of organizational ecology (cf. Aldrich et al., 1986; Carroll, 1988; Hannan, Freeman, 1989) concentrate largely on selective forces and homeostatic feedback mechanisms. According to these authors, market competition is analogous to biological competition and firms must pass a survival test imposed by the market. The approach adopted in this paper is a further step toward modern evolution theories.

We see evolution to imply more than only the existence of selective forces. The replicative model of evolution gives a rich description of the unfolding processes[1] that generate history, change and permanence. We are interested in processes in which the consequences of economic choices continuously transform or maintain the mental and institutional structures and goals that determine choices (cf. Pantzar, 1988).

What makes the general evolution research context relevant for the explanation of the maturation of business organization? The deep-seated tendencies of routinization (of action), rationalization (of organizations and technology) and disciplination (in the cultural sphere) of ageing organizations do (Pantzar, Csanyi, 1990). These tendencies are analogous to tendencies of natural evolution. In evolutionary terms, an organization's replicative quality seems to increase when ageing. In evolutionary terms, a business organization's increasing capability to reproduce itself and survive is our concern here. In addition, the evolutionary point of view is more complete. It can provide organization theory and economics with explanations that emphasize the radical, divergent processes of self-organization. The current theory of economic organizations still seems to follow the classics in its emphasis on convergent processes at the expense of divergent processes (Fombrun, 1986). We need a theory that explains the interplay between divergent and convergent processes and the replicative model of evolution is one candidate.

2 The Replicative Model of Evolutionary Processes

The laws of a general evolutionary process inherent in nature were first formulated by Spencer (1862). A replicative model of evolutionary processes encompassing both natural and social evolution has been based on the recognition of

---

[1] Actually, evolution is the history of unfolding of complexity, or differentiated order. However, evolution is not simply "building up". Jantch (1980: 75) put it as follows: "Unfolding is not the same as building up. The latter emphasizes structures and describes the emergence of hierarchical levels by the joining of systems 'from the bottom up'. Unfolding, in contrast, implies the interweaving of processes which lead simultaneously to phenomena of structuration at different hierarchical levels. Evolution acts in the sense of simultaneous and interdependent structuration of the macro- and the micro-world. Complexity emerges from the interpenetration of processes of differentiation and integration, processes running 'from the top down' and 'from the bottom up' at the same time and which shape the hierarchical levels from both sides. Microevolution (such as the emergent forms of biological life) itself generates the macroscopic conditions for its continuity and macroevolution itself generates the microscopic autocatalytic elements which keep its processes running."

the various biological and social entities as *systems*, and the recognition of the common role of replication in both the maintenance and the origin of such systems. Studies have been published in several papers (Csányi, 1980–1982, 1985, 1987–1990; Csányi, Kampis, 1985, 1987; Kampis, 1986, 1987a, 1987b, 1990; Kampis, Csányi, 1987a–1987c; Pantzar, Csányi, 1990).

If we look at any of the natural entities on Earth that are organized in some way, we find that these entities embody various components. A single-celled organism is built up from various molecules as components. A higher organism, such as a plant or an animal, consists of various cells. Ecosystems are composed of organisms of various species. Human societies also contain components such as humans, artefacts, and other living beings. We may call these entities *component systems*. These systems can be characterised by their *components*, their *processes* and their *organizations*.

Examining the various components of entities like cells, organisms, societies, one finds that these entities exist by self-maintenance or *self-production*, that is by the continuous renewal of their components. The components are assembled and decomposed in various *processes* at the expense of a continuous energy flow going through the system. The cell for example, can be considered a functionally closed network of molecules and molecule-producing processes that continually produces the same network of components and processes. The same is true of organisms or ecosystems. Production of the same components and subcomponents by a functionally closed network of processes is also the basic mechanism of *reproduction*, which is the other characteristic trait of these entities. It can be found in cells, organisms, ecosystems and societies.

As far as *organization* goes, all biological and cultural entities are characterised by the mutual interactions of their components. Various molecules, components of a cell, are interacting chemically; the cells, components of an organism, are influencing each other's life processes. People, artefacts, and ideas, the components of a society, are also mutually interacting.

In general, the components of a natural entity participate in the maintenance of the whole system through their interactions. The particular mechanisms of the components' interactions are varied, but there is a common aspect; through their interactions, the components always influence *the probability of each other's genesis*. Each of these interactions can be rendered to particular components and named *functions*. A digestive enzyme in a cell, for example, splits the structure of other proteins (components), i.e. it decreases the probability of their existence. A predator taking its prey also decreases the probability of existence and genesis (of progeny) of the prey animals. The existence of cars increases the probability of genesis of gas stations and repair shops in society.

This common aspect of the various functions can be expressed as a general replicative function by which they promote both the system's and their own maintenance. By analysing self-production and reproduction one can show that the underlying principle of organization in both cases is the formation of a network of the various components with special functions resulting in copying or *replication* of the very components of these networks and the replication of the whole system.

In the functional organization of the cells, molecular components are copied repeatedly. In an organism the cell components are originated by cell divisions, that is by copying the cell components. Therefore this kind of organization is called *replicative organization* and the general function by which each component joins the replicative organization is called *replicative function*.

One can work out a common general model of natural systems based on these considerations. The basic general process in this model system is replication. Replication is generally considered a synonym for copying, where a constructor produces a copy of a component or a given system. To do so, the constructor needs a description, the information necessary for this copying process. The essence of replication is the function of copying regardless of the particular mechanisms of storage and retrieval of the information for copying. It does not matter whether this information is stored separately (as is true of DNA for example) or distributed in the whole system (as in society). Neither does it matter whether a separate object, a component, or the constructor itself is copied. The nature of copying lies in the functional organization, not in the particular mechanisms. As a result of our argument about self-production and reproduction, the concept of replication will be used in two senses.

We define *temporal replication* as the continuous renewal of the system in time. This is the uninterrupted existence of the system that is manifested via the sequential and functional renewal of the components of the system. The components are assembled and decomposed, but the unity and identity of the system and its organization are always maintained.

The definition of *spatial replication* is identical with that of reproduction. The system produces its own copy, which becomes separated from it in space. From one system or unit, two units are formed. Spatial replication also proceeds by copying components of the system, as in the case of replication in time, but in doubling numbers, while the original organization of the system remains unchanged.

2.1 Dynamics of Replicative Systems and Replicative Information

A system of components that has not developed functions yet is considered a *zero-system*. A zero-system lacks organization. A precondition for starting a self-organizing process in a zero-system is the presence of a least set of components that can replicate and that fulfills the following criteria:

 (i) It contains at least one cycle of component-producing processes.
(ii) At least one of the components participating in this cycle can be excited by the energy flux flowing through the system.

Such minimal sets of components will be called autogenetic system-precursors (AGSP).

It has been suggested for conceptual and empirical reasons that the functional informational content of a zero-system containing an appropriate AGSP will increase in time. This process is called *autogenesis*. During autogenesis, as time advances, an increasing proportion of functional information becomes replicative information. This can appear only as an extension of AGSP, that is, additional replicative cycles appear that are interconnected with the AGSP. These cycles are called *supercycles*. As time passes, replicative coordination of various super-cycles develops, and the fidelity of replication increases. We can speak of some kind of functional differentiation and cooperation that results in the formation of communities of simultaneously replicating components, that is subsystems, called *compartments*. Their components are separated from others by their participation in co-replication. The emergence of compartments is equivalent to the organizational, i.e. functional, closure of the network of component-producing processes and components having a replicative function. This succession of events is called the *compartmentalization and convergence of replicative information*.

Components and compartments embodying them may co-exist with different levels of replicative fidelity. For a while components may replicate with relatively high accuracy, but the compartments formed by them do so with a low level of fidelity. As time passes, the fidelity of replication of both may increase and perhaps a next level of organization, a "compartment of compartments" is formed, which also replicates with an increasing fidelity.

A system may contain several different kinds of compartments, which are all replicative units with diverse fidelity. Among these, interrelationships develop and, as a result, their replication becomes coordinated. Gradually, the *whole system* will start replicating as a final replicative unity. In the autogenetic process, the organization of the system and of its parts changes due to the functions of the emerging new components. Thus autogenesis is possible only while the state of identical replication has not yet been achieved. In that state, the system has become functionally closed and its replication in time continues as long as the environment does not change. There are no further organizational changes initiated by organizational causes because new functions cannot originate. In the state of identical replication, the system is an autonomous, self-maintaining unity, a network of components and component-producing processes that, through the functional interaction of the components, produces exactly the same network that produced it. Its organization is almost closed and cyclic. Its input and output are subordinated to its replication, but through them, its existence depends on the invariance of the environment.

A replicative system is thus characterised by a functionally closed replicative network that embodies various replicative sub-networks which form *organizational levels*. At the organismic level the *organisms*, for example, are autonomous networks of components and component-producing processes which are replicating in time, i.e. the organisms are component systems. Organisms are also components of higher organizational levels like *ecosystems* or the whole *biosphere*. The components of the organism are cells, which are, in turn, autonomous replicative networks of molecules, components of the lowest level of organization. Atoms are the elementary building units of the biological system.

2.2 The Replicative Model of Cultural Evolution

The replicative model enables the whole human society to be studied as a single system. It provides an explanation for the origin, the organizational levels and future changes of culture. Human society has all the necessary characteristics of a component-system. Its components are complex structures, as there are several organizational levels below culture. Its basic components belong to three classes: living *organisms*, *artefacts* and *ideas*.

  (i) The category of living organisms includes all biological organisms that contribute to social interactions in a human society.
 (ii) All formed, processed objects that have been changed by human intervention are regarded as artefacts.
(iii) Ideas are mental representations of the smallest, intelligently definable actions or thoughts determined by physical factors of the nervous system that can be communicated, copied or formed as artefacts or performed as social acts. They belong to three categories: social, material and mental ideas.

Ideas, artefacts and people are the building blocks and components of higher structures, which have been called second-order organisms by Cavalli-Sforza and Feldman (1981), technical complexes by Ellul (1965) and *cultural machines* by Mumford (1967). The components of human societies are continuously created and destroyed because of the energy flowing through the cultural systems. The creation of components involves the action and contribution of other components. We use the concept of *creative space* to indicate the complexity of the creative process.

The creative space is an abstract space in which a representation of each component involved in the production of other components is given. So we speak of technical space, i.e. the creative space of artefacts, of cultural space connected with the creation of the various political and social ideas, and of biological space that creates living beings.

It can be shown that *creative spaces have a replicative organization*, because the information represented in the components created feeds back to the process of its creation. The information is of replicative type. Cars, for example, are created in factories, where the information for production is in blueprints. Information transfer from blueprint to the artefact is a seemingly one-directional process. But if we examine the entire creative space of artefacts, it appears that the manufactured cars which are used by the society influence the car-producing process and the design of cars in many ways. The most common process of design is copying. The designer copies, and sometimes recombines information of the reliable parts of the previous production cycles. Even if the designer invents something, it may have originated from the process of creating a different artefact.

A further important feature of the creative space is that various kinds of components are acting together, so that each component of the cultural system can *influence the probability of genesis of some others*, that is, the components have replicative functions. Thus the creative spaces·form a functionally and organizationally united replicative system. In a more detailed analysis it has been found that the basic processes of autogenesis such as compartmentalization and convergence of replicative information are also characteristics of sociocultural systems.

2.3 The Evolution of Sociocultural Systems

It has been shown above that one of the preconditions of autogenesis is the emergence of an autogenetic system precursor, which organizes the construction of special constraints on the dynamics of the organizational level(s) already in existence.

The emergence of cultural man is usually attributed to a single characteristic, for example, language, or common hunting to mention just two. Recent comparative evolutionary studies have shown that these simplifying explanations are inadequate, because these traits depend on one another. Human *proto-cultural groups* were formed on the biological bases of sociability and certain properties of the brain of the proto-cultural man (Eibl-Eibesfeldt, 1979, 1982). The *biological proto-cultural organization was the system-precursor of cultural evolution* and the organizer of the autogenesis of human culture.

It was the *autogenesis of ideas* in these proto-cultural groups that initiated cultural evolution, promoted co-evolutionary biological changes and resulted in *group-societies*. Group-societies are replicative units not only in their biological aspects, but also in their cultural aspects. Members of group-societies required cultural complexes, including social and mental ideas, and objects to satisfy their basic needs. Social ideas shaped the organization of the group, mental ideas included useful knowledge and techniques, and artefacts were made for ritual or practical purposes. The ideas, artefacts and social connections of the group replicated from generation to generation within the limits of fidelity provided by linguistic communication.

The group-society is an organization controlled by cultural constraints and built upon the dynamics of the biological level of human existence. The replicative competitive selection of components have played an important role in the evolution of the group-society.

The structure of group-society satisfies the conditions for being a *replicative machine*. Its components make up a functional network which reproduces an almost identical network during its operation. The group-society, as a replicative machine, lacks inputs and outputs. It only creates the conditions for its own existence but it cannot produce a surplus to be traded, exchanged or unevenly distributed. It is organizationally closed, stable and in equilibrium with its environment. The emergence of this organization took several million years of human evolution.

These most developed compartments of cultural autogenesis could have provided building blocks for an even higher level organization. Humans began to learn to organize special sub-groups permanently cooperating for a given aim and accepting discipline, i.e. *creative cultural machines* emerged. This organization did not simply mean that more people were available for a certain activity, but also that participants were willing to submit to the necessary training and discipline, and to accept leaders who could organize the whole action. People within the replicative machines could cooperate and produce surplus to be used for purposes beyond mere existence. Therefore creative cultural machines are *functionally open* entities which are very active in forming higher level organizations. The formation of creative cultural machines led to the appearance of a new social organization, which may be called the *village-society*. The formation of a village-society requires permanent dwellings, division of work, and cooperation of large communities.

Inputs and outputs of cultural machines can be joined to a super-organization, and therefore creative cultural machines are also *system-precursors* to *states*, which are compartments of the highest social-organizational level. During the development of states, the creative cultural machines underwent further perfection. New technologies, services and mechanical machines were invented which also embodied complex ideas. Later in the industrial revolution, common machines and people gave birth to cultural machines of a new kind; *modern industry and technology emerged.*

3 The Replicative Model of the Evolution of Business Organization

Early social theorists such as Emile Durkheim, Auguste Comte, Herbert Spencer and Thorstein Veblen used biological metaphors which meant that "if such and such a law is true for animals, then it must be true for societies." We use biological metaphors in a heuristic sense instead. A metaphor is useful only if it suggests new and testable ways of thinking about social phenomena. The general evolution view on which our theoretical view rests does not reduce cultural phenomena directly to the biological realm. However, it reminds us about the historical, and thus biological (evolutionary), roots of human culture. We aim at a consistent picture of the mechanisms of the evolution of business organizations. As modern philosophers of science like Thomas Kuhn, Imre Lakatos, Paul Feyerabend and Richard Rorty have stated, confirmation, verification and falsification of theories corresponding to "reality" is doubtful. Several valid theories explaining the same phenomena might exist. Our model, or interpretative scheme, is one of them. We pay attention to three main features of the large business organization, namely its replicative nature, its origin and maturation.

3.1 The Replicative Nature of Business Organization

The mere existence of a biological system is a manifestation of a replication process; Temporal replication is the continuous renewal of the system over time.

Biological systems are component systems and they are continuously renewed by the replacement of their components. In an ecosystem each generation of animals and plants and other organisms supplies a new, renewed set of similar components. Organismic reproduction is spatial replication, a copying process in space. We are interested in finding mechanisms of temporal and spatial replication in business organizations.

What could the term "replication" mean in the business enterprise context? It refers to the tendency of everyday business practices and related structures to maintain (replicate) themselves, we argue. Participating people, cultural determinants, networks of agents, and the spatial and temporal structural conditions of business are continuously renewed. Not only are people and their relationships in the business life renewed but also the whole system of artefacts they use, process, create and transfer during their business. If we regard a firm as an organizational unity with a given input and output, then the most characteristic feature of this unity is its perseverance in producing its output; The firm is a replicative process. Old employees are dismissed but new ones hired and taught to perform the same functions. In a systemic sense it is a replicative process where one (human) component has been replaced by a similar one that can perform the same function. At the same time, all relationships of the dismissed employee are renewed so that their substitute will fit into the organizational network of the firm and perform the same function and maintain similar relationships as the previous one.

Old furniture or machinery is discarded but new sets are bought to replace and functionally substitute the old ones. That is also a replicative process, which must be understood in a functional sense. A piece of equipment, say a paperweight in an office, is necessary for the activities of the office. When it wears out, it is replaced by a *similar one*. Similarity is determined by the component's function, kind and form, which make the person responsible for replacing it select a similar substitute fulfilling all the functions of the discarded one. This particular paperweight may have been produced elsewhere, but its production is controlled by the demand created by the replicative (substitution) process.

Let us take up socioeconomic interpretations. The possibility of functional differentiation, i.e. division of labour, is dependent on some certainty that all subcomponents replicate and coordinate properly. Obviously the emergence of cooperation demands at least some predictability (Heiner, 1984; Axelrod, 1983) concerning the "rules of the game". Such constancy is based on replication.

Organizational ecology (cf. Carroll, 1988; Hannan, Freeman, 1989) has shown that social selection processes favour organizations and forms that have high levels of reliability of performance and accountability, and that these attributes depend on a capacity to replicate structure with high fidelity. Fidelity of replication is assumed to increase with age.

In evolutionary terms a business enterprise is a replicative network which is closed enough to produce its own components and subcomponents and to embody various replicative sub-networks. Temporal replication of behavioural routines, organizations or innovations means the continuous renewal of these practices and structures. Spatial replication is a similar copying process in space. The adaptation of one company's method of accounting in another business firm exemplifies spatial replication, which may take place by intentional adaptation (imitation) or by selection. In the replicative network of a business firm this process is manifested by ever-increasing standardization of lower-level sub-component networks. This was also what Chandler (1977) maintained: due to more complex input and output flows, firms were forced to develop more structured administrative machineries. Better administrative skills propelled firms to growth cycles. Ultimately, cumulated competitive advantage resulted in the rise of integrated industrial giants with administrative hierarchies of middle and top management.

Finally, it is important to emphasize that replicative fidelity has its limits. In practice, replication seldom occurs with perfect fidelity. Substitute employees or artefacts fulfil their assigned functions only approximately. They have new features that will open up new possibilities for change and evolution. *Incomplete replication* is the creative force in every evolutionary change both in biology and in business. In further work this particular phenomenon deserves special emphasis.

3.2 Business Organizations as Replicative Networks: Convergence and Compartmentalization

Every replicative system can be characterised by an increasing tendency to compartmentalize. Certain sets of components are separated from others by their co-replication. At the same time there is a tendency for all components to participate in the overall replication, in renewal of the whole system, which is termed convergence. Both basic processes can be observed in business.

In the replicative model of business organization the degree of closure of a firm refers to the real (or descriptively useful) borders of a firm and its environment. A firm is always a component in a higher-level network that imposes constraints on lower-level entities, like production inputs. Cycles of cycles, ultra-cycles, are manifested in super-systems (development blocks, national economies, etc.), where sequences of complementarities exist. The hypothesised tendency of higher and higher-level integration and thus coordination is manifested, for instance, by increasingly hierarchic societal structures and division of labour (cf, Dodgshon, 1987).

The emergence of cooperation is one explanatory idea to be taken seriously. A system, here a business organization, can contain many different kinds of compartments which are all replicative units. Interrelationships develop between them until replication becomes coordinated. Different replicating components connect more tightly, and finally the cooperating components form a **new** replicative unit, which in time becomes a part of a larger totality.

In organizational terms a firm represents one level of integration. A firm comprises functional units like production and marketing, and participates in higher level organizations, like industry or a block of firms, which represent higher compartments. The evolution view focuses on such issues as the number of intra-organizational and inter-organizational links and an organization's tendency toward hierarchic structures. It makes conventional limits of organizations problematic by emphasizing the functionally open nature of organizations. Inter- and intra-firm relations between organizations no longer necessarily differ decisively, since they are all replicative component systems. The concepts of resource interdependence (Williamson, 1981, 1985) and firms as coalitions (Alchian, Woodward, 1988) are compatible with this viewpoint.

The evolutionary view studies how firms are related to each other and to macrosystems, and how networks of business relations generate macrostructures and macrostructures select those networks. Firms are each other's environment; one-way relationships between the environment and the system do not exist. In co-evolutionary terms, evolution is characterised by continuous reciprocal responses of two closely interacting species. The organization as an active agent constructs its own objective envionment.

There is no neat separation between cause and effect any more. There is a continuous process in which an organization (organizational level) evolves to solve an instantaneous problem that was set by the organization itself. When organizations construct (intentionally or non-intentionally) an environment they generate the conditions of their survival and replication. In other words it is both the object of selection and the creator of the conditions of that selection.

3.3 Origin, Evolution and Maturation of Business Organizations

In the course of history, kinship-society configurations defined the social existence of humans for the longest time. They also influenced human biological evolution, and we can say that species-specific characteristics of humans are in harmony mostly with any *kinship-society* as a superstructure. The texture of the kinship-society is based on family relations, and individual development is in harmony with the slow alteration of the social structure. If the socialization of the individual born into the kinship-society is perfect, he completely accepts the conditions that exist in the society, and when he becomes an adult he is given the opportunity to climb the social ladder. Values and norms change very slowly, generation by generation, if at all. The harmony of biological factors and a given social structure explain the fact that certain signs of kinship-societies prevailed even in *mass-societies*. In the semi-mass-societies predating the industrial revolution, biological family relationships were an essential part of the social structure. The relationships that are new compared to those of the kinship-societies appeared in fact because family relations could no longer organize the society as a whole. Religion, ownership and the state gave rise to new forms of dependency, which can be interpreted as social extensions of family relationships. In a society of relatives, an individual's existence depends on a network of given family relations. Feudal society is a typical example. The liege lord is at the top of a social group which includes many people other than his relatives, but all relationships are of a family character (for example vassals get certain protection in exchange for their services) with one important exception: the shift of generation is not followed by a change in the seigniory, since vassals do not become liege lords after a time. Feudal relations are quasi-infantilizing for a major part of society, defining for ever an individual's place in the social order of ranks. Like mass society, it is simply too large to utilize childhood socialization to the full in order to conserve a given social structure, and there will be individuals who reject the superstructure of the society.

We think our remarks are valid for the early social organizations of artisans and traders. They also apply for individual entrepreneurs of the early capitalist societies, who acted in a pseudo-kinship network. Although kinship relationships gave way to the social relationships of the production process, trade and business, entrepreneurs' activities were dependent on a network of personal relationships.

It is only after industrial development takes off that the idea of social autonomy emerges. Autonomy results in the appearance of individuals and groups that reject the given family and dependence conditions of society, and organize around an independent objective or idea. An autonomous individual can choose from different connections, ideas and social options. He or she is free from kinship-type dependencies. The autonomous group unifies individuals pursuing similar ideas and aims. The acceptance of autonomous ideas implies a refusal of monolithic ideologies that organize society.

Social autonomy has accelerated *modern industrial development*. The society based on family relationships can only change very slowly, and the appearance of autonomous groups dramatically accelerated the change in social structure, first in the field of production, as autonomous production units, enterprises, developed. There are several factors involved: the autonomous entrepreneurs' groups recruit their members with no regard for social constraints; they recruit those who, due to their personal knowledge and skill, are able to fulfil a given task, to follow a given target, or to implement a given idea. If the society shows tolerance towards the autonomous groups, competition between these groups can begin, and the ones unable to achieve their goal will disperse and promptly yield to other groups of a new composition. Since the lifetime of autonomous groups is usually much shorter than a generation, and selection also takes place, the groups' development will also accelerate. The organizers and executives of a given enterprise create a well-defined organization, which faces the given social conditions, and if it is successful, it will survive and replicate prosperously. If it fails, it will disintegrate without causing too much harm to its members, who will take part in new autonomous groups and set up new enterprises. A recombination and selection of organizing ideas is taking place this way. This is the very process of organizational evolution.

Of course, not only enterprises, but also other social organizations, such as editorial boards, associations and parties can function autonomously. It is worth noting that the autonomous individual, who can create autonomous groups, is the result of socialization. Modern industrial societies invest significant energy in the development of individual autonomy and the freedom to create or dissolve autonomous groups in adult life. Jurisdiction and the political system as a whole should be fit for putting up with autonomy, but this is only one precondition. The other is the social production of individuals educated to autonomy.

A social system of autonomous individuals and autonomous small enterprises is the *zero-system* for giant business. A small enterprise is an autogenetic system-precursor (AGSP), which can already replicate on a small scale. A small enterprise satisfies all characteristics of the compartmentalization and convergence processes of the replicative model: it can use both human and material resources for new inventions, and during its incomplete replication it can develop entirely new functions and join to new functional networks.

Clearly the timing of the emergence of big business depends on the size of the social system. In a somewhat homogeneous system in which components and AGSPs can freely join and separate, where language and social order are in favour of social autonomy, large business can and will evolve as fast as the autonomous subcomponents emerge. The industrial environment of the early United States of America fulfilled such conditions. It had not been a kinship-society, it promoted social autonomy and its huge size permitted rapid expansion. Schumpeter's faith in the transience of firms (profitable innovations), social classes and economic systems (Schumpeter, 1911/1949: cf. 156), has been proven mistaken. The current evidence concerning modern business adverts to the possibility that Schumpeter's early "theory of transience" should give way to a "theory of permanence" (of giant business). We should focus more properly on those mechanisms that generate, protect or potentiate continuity in the realm of business organization.

In the social science context, the convergence of the replicative model expresses the general trend of societal evolution toward a greater scale of integration, and concurrently greater differentiation, functional specialization and hierarchization. In human society this evolution could be visualized with schemes that present societal evolution in stages of bands, egalitarian tribes, chiefdoms and state-systems. Successive stages show a qualitative jump in societal complexity over the previous stage (Csányi, 1987: 237–240; Dodgshon, 1987: 19). Therefore the maturation of large business is a logical consequence of replicative evolution.

We hope that the descriptions and arguments presented are plausible enough to support our conclusion that the replicative evolutionary model is suitable for further study of the maintenance and future development of large business organizations. Compartmentalization explains the rise of organizational units and their hierarchical levels as well as the vigorous growth of the early stages and the conservative nature of the mature, more complex organizations. A finer analysis certainly would show more exact and detailed descriptions.

References

Alchian AA, Demsetz H (1972) Production, Information Costs, and Economic Organization. American Economic Review, 82, 777–795.

Alchian AA, Woodward S (1988) The Firm is Dead; Long Live the Firm. A Review of Oliver E. Williamson's The Economic Institutions of Capitalsm. Journal of Economic Literature Vol XXVI, March 1988, 65–79

Aldrich HE, Auster E, Staber U, Zimmer Z (1986) Population Perspectives on Organizations. Acta Universitas Upsaliensis, Studia Oeconomicae Negotiorum 25

Arthur B (1988) Self-Reinforcing Mechanisms in Economics. In: Uppsala Anderson P, Arrow K, Pines D (eds) (1988) The Economy as an Evolving Complex System. Addison-Wesley, Reading, Mass

Axelrod R (1984) The Evolution of Cooperation. Basic Books, New York

Beniger J (1986) The Control Revolution, Technological and Economic Origins of the Information Society. Harvard University Press, Cambridge, Mass

Corroll GR (ed.) (1988) Ecological Models of Organizations. Ballinger Publishing Company, Cambridge, Mass

Cavalli-Sforza LL, Feldman MW (1981) Cultural Transmission and Evolution. Princeton Univ Press, Princeton

Chandler AD Jr (1977) The Visible Hand: The Managerial Revolution in American Business. Belknap Press of Harvard University Press, Cambridge, Mass

Csányi V (1980) The General Theory of Evolution. Acta Biol Hung Acad Sci **31**, 409–434

Csányi V (1981) General Theory of Evolution. Soc Gen Syst Res **6**, 73–95

Csányi V (1982) General Theory of Evolution. Publ House Hung Acad Sci, Budapest, 121

Csányi V (1985) Autogenesis: Evolution of Selforganizing Systems. In: Aubin J-P, Saari D, Sigmund K (eds): Dynamics of Macrosystems, Proceedings, Laxenburg, Austria 1984, Lecture Notes in Economics and Mathematical Systems **257**, Springer-Verlag, Berlin, 253–267

Csányi V (1987) The Replicative Model of Evolution: A General Theory. World Future **23**, 31–65

Csányi V (1988) Il modello replicativo dele'evoluzione biologica e culturale. In: Ceruti M, Laszlo E: Physis: Ab itare la terra. Feltrinelli, Milano, 249–260

Csányi V (1989) Evolutionary Systems and Society. A General Theory. Duke Univ Press, Durham, 304

Csányi V (1990) Ethology, Power, Possessions: A System Theoretical Study of the Hungarian Transition. World Futures **29**, 107–122

Csányi V, Kampis G (1985) Autogenesis: Evolution of Replicative Systems. J Theor Biol **114**, 303–321

Csányi V, Kampis G (1987) Modelling Society: Dynamical Replicative Systems. Cybernetics and Systems **18**, 233–249

David PA (1988) Path-dependence: Putting the Past into the Future Economics. Institute for Mathematical Studies in the Social Sciences, Stanford University, Technical Report No 33, August 1988

Dodgshon R (1987) The European Past – Social Evolution and Spatial Order. Critical Human Geography, Macmillan Education, London

Eibl-Eibesfeldt I (1979) Humanethology: Concepts and Implications for Sciences of Man. Behav Brain Sci **2**, 1–57

Eibl-Eibesfeldt I (1982) Warfare, Man's Indoctrinability and Group Selection. Z Tierpsychol **60**, 177–198

Ellul J (1965) The Technological Society. Jonathan Cape, London

Fombrun C (1986) Structural Dynamics Within and Between Organizations. Administrative Science Quarterly **31**, 403–421

Galbraith JK (1967) The New Industrial State. Hamish Hamilton, London

Gigch Van (ed) (1987) Decision Making About Decision Making, Metamodels and Metasystems. Abacus Press, Cambridge, Mass

Hannan, MT, Freeman J (1989) Organizational Ecology. Harvard University Press, Cambridge, Mass

Heiner R (1983) The Origin of Predictable Behavior. American Economic Review, Sept 1983, 560–595

Jantsch E (1980) The Self-Organizing Universe, Scientific and Human Implications of the Emerging Paradigm of Evolution. Pergamon Press, Oxford

Kampis G (1986) Biological Information as a System Description. In: Cybernetics and Systems '86. Riedel D, Dordrecht, 36–42

Kampis G (1987a) Some Problems of System Descriptions I: Function Int J Gen Syst **13**, 143–156

Kampis G (1987b) Some Problems of System Descriptions II: Information Int J Gen Syst **13**, 157–171

Kampis G (1990) Self-Modifying Systems in Biology and Cognitive Sciences: A New Framework for Dynamics, Information, and Complexity. Pergamon Press, Oxford

Kampis G, Csányi V (1987a) Replication in Abstract and Natural Systems. BioSystems **20**, 143–152

Kampis G, Csányi V (1987b) A Computer Model of Autogenesis. Kybernetes **16**, 169–181

Kampis G, Csányi V (1987c) Notes on Order and Complexity. J Theor Biol **124**, 111–121

Koestler A (1967) The Ghost in the Machine. Hutchinson and Company, London

Langlois R (ed.) (1986) Economics as a Process, Essays in the New Institutional Economics. Cambridge University Press, Cambridge

Mumford L (1967) The Myth of the Machine. Harcourt, Brace, Jovanovich, New York

Nelson R, Winter S (1982) An Evolutionary Theory of Economic Change. The Belknap Press of Harvard University Press, Cambridge

Pantzar M (1986) Economic Agent as Changing the Structures and Adapting to the Structures – an Evolutionary Vies on Consumer Choice. Helsinki School of Economics, Studies B-86, Helsinki

Pantzar M (1988) Some Thought Experiments Concerning the Evolution of Complex Societal Structures. Labour Institute for Economic Research, Discussion Paper 70, Helsinki 1988

Pantzar M (1990): Toward an Evolutionary Theory of Hierarchic Organization – The Rise of Big Business as an Exemplary Case. Draft August 1990, forthcoming in Labour Institute for Economic Research, Research Report

Pantzar M (1991) A Replicative Perspective on Evolutionary Dynamics, The Organizing Process of the US Economy Elaborated Through Biological Metaphor. Labour Institute for Economic Research, Research Report 37, Helsinki

Pantzar M, Csányi B (1990) The Replicative Model of the Evolution of the Business Organization. A Case Study: the Rise of Giant Business. Labour Institute for Economic Research, Discussion Papers 98

Pfeffer J, Salancik G (1978) The External Control of Organizations, A Resource Dependence Perspective. Harper and Row, New York

Schumpeter J (1911/1949) The Theory of Economic Development. An Inquiry into Profits, Capital, Credit, Interest, and the Business Cycle. Harvard University Press 1949, third printing, first published in German 1911 (Theorie der wirtschaftlichen Entwicklung)

Schumpeter J (1947) The Creative Response in Economic History. The Journal of Economic History, Vol VII, Nov 1947, 149–159

Spencer H (1862) First Principle. Williams and Norgate

Veblen T (1904) The Theory of Business Enterprise. Social Science Classics Series, Transaction Books, New Brunswick, New Jersey, 1978

Weber M (1978) Economy and Society, An Outline of Interpretative Sociology. In: Roth G, Wittich C (eds) University of California Press, Berkeley

Williamson O (1981) The Modern Corporation: Origins, Evolution, Attributes. Journal of Economic Literature, Vol XIX, Dec 1981, 1537–1568

Williamson O (1985) The Economic Institutions of Capitalism, Firms, Markets, Relational Contracting. The Free Press, New York

VI The Evolution of Information Processing Systems at the Sociotechnical Level

Information Processing at the Sociotechnical Level

Klaus Haefner

Hans Primas:

New ideas are viable if they
lead to new viewpoints which
can be used to define new
contents that allow interesting
divisions of the holistic world.

1 From Social to Sociotechnical Structures 307

2 Technical Information Processing 310

3 Validity of the Hypotheses from the Basic Concept 312

4 Trends for the Future of Sociotechnical Systems 314

5 Consequences 317

1 From Social to Sociotechnical Structures

All developments in societies have been determined heavily by information processing in human brains. This is particularly true for human organization: raising children, living in a family, organizing business, setting up local and regional administrations, building national governments, developing scientific and technical insights, organizing the military, establishing religions, developing culture, etc. For all those activities the human brain has to cope with information.

In the course of our cultural development we have developed and structured an "informational environment", which is the result of modelling our understanding of the relevant material and biological environment. In this informational environment we live as conscious human beings by using it and steadily developing it further.

In this world, education (in its broadest sense) is the main way of allowing human beings to understand the informational environment available and to perceive it for solving individual and societal problems. Humans do not inherit basic knowledge for their complex behaviour, they must learn. Thus, internal and external organization of information is a crucial and essential component of human life. There is no way whatsoever to bypass learning, it must be done if a proper "survival program" is to be acquired. Language helps in the organizing and communicating of information. Until the invention of primitive paintings, writing, and finally scripts, human brains were the sole means of storing representations of the informational environment.

Thus, widespread use of writing and scripts was tremendous progress, a big leap forward. First, it was a method for communicating independently of personal contacts. Second, it allowed the proposing and enforcing of general rules and principles in a society (see for example the ten commandments of Moses). With written information it was possible to establish and store a detailed body of knowledge about our material, biological and human environment without using the brain as the sole storage device. This was the very beginning of information technology. Thus libraries turned out to be an extremely important ingredient in all higher, organized human societies. Third, the drawing of figures and writing of text were important for enlarging the functional capacity of our short-term memory in most areas of cognitive work. Thus, even quite complex procedures could be handled easily (e.g. making calculations for charging taxes or for astronomical observations). "Extended memory" in ancient cultures was as essential as in modern societies.

However, up to the 1950s, complex *processing* of information could only be handled by human brains. The most exciting and informative book was completely useless as long as it was not read and interpreted by a properly trained person. Therefore, before the 1950s, societies and all their activities were determined *in every detail* solely by the action of human information processing. Societies were strictly *social systems* in the sense that their behaviour was in every detail determined by the collective action of human brains, although scripts and pictures had been around as "external memory" to be used by some members of society. In this sense, all human groupings were *social* structures.

This setting has changed dramatically with the invention and the increasing use of *technical information processing*. This allows for an increasing transfer of cognitive processes from the brain to the computer. Automatic calculating was the first process to be technically established because arithmetic operations was the first process to the technically established because arithmetic operations can easily be described and implemented technically. In recent years, however, tens of thousands of quite complex cognitive processes have been organized technically in computers as algorithms or even in the form of heuristics (see the approaches of "Artificial Intelligence").

The combination of technical storage and processing of information with recent developments in technical telecommunications allows for the establishment of sophisticated computerized networks which gather, process, store and transfer information without using human brains at the operational level at all. (Such systems must be designed and implemented by human beings; however, after successful testing they can "work" almost completely on their own.)

Technical information processing is spreading intensively. The information technology industry today turns over some $500 billion a year worldwide with an increase of some 10% per annum. Almost no area of our society has remained untouched by the "computer revolution".

The coexistence of human and technical storing, processing and communicating of information gives rise to completely new structures which can be called *sociotechnical systems*. They are characterised by a strong influence of technical information processing systems, which are deeply embedded into social structures, this integration significantly altering these structures.

In sociotechnical structures, control is in many instances being transferred almost completely from the social to the technical domain. This results in an increasing autonomy for the technical structures which determine social life. Let us reflect on this changing situation in three examples.

(1) *Electricity.* Human life in the highly industrialized countries is strongly determined by the use of electricity. Social life is completely dependent on electricity. However, production and distribution of electricity today is only possible via intensive use of technical information processing for control of demand, distribution and production. In nuclear power plants particularly control computers are the central source of information processing; no group of engineers working solely with their brains could by any means provide the necessary security asked for by society.

Here, a feedback loop is quite clear; we have set up a technical infrastructure which needs electricity and this needs technical information processing. Therefore, society is heavily dependent on information technology in the electricity system. In all highly industrialized countries a long-term shut-down of electricity would result in complete disaster. This makes the system a sociotechnical one. We can no longer separate human and technical information processing, social and technical structures. The "mixture" of both forms a crucial and complex base of our existence.

(2) *Finance.* Computers and telecommunications are used to a great extent in all financial activities of the highly industrialized societies. (In Germany, e.g., it would take some seven million persons to do the daily computational workload of the banking system if it had to be done solely by human brains.) Banking and its integration into financial activities in commerce is forming a complex mix of technical transfer, storage and processing of information. Larger companies and particularly the multinational corporations have almost completely delegated their financial *routines* to computer programs combined with world-wide telecommunications networks.

Thus, most financial systems are national or even world-wide sociotechnical structures. They cannot be handled without information technology. Their failure would result in a complete breakdown of broad commercial activities including the markets for food and basic goods. Today, we rely on this computerized financial system as an essential component of a sociotechnical structure. We can control details, but we cannot make decisions about whether to have or not to have such structures.

(3) *Early Warning Systems (EWS).* NATO and the Warsaw Pact countries have established a highly complex technical information processing system allowing the observation of all flying technical objects, hoping to detect in time every object which might attack their own territory. Some thirty computer centres, many systems operating in aeroplanes and satellites as well as a sophisticated telecommunication structure are integrated into each

system. However, top politicians still have to make decisions about final actions.

Using this approach East and West have been successful in keeping peace by working with the "Launch On Warning" strategy. SDI is now elaborating this concept even further by trying to destroy an attacking object automatically before it reaches its target; it is heavily dependent on technical information processing.

Without EWSs, peace would be much more vulnerable, although the reliability of the systems themselves can be questioned. (Even in a future of little tension between East and West such systems will be essential against military blackmailing of "Third World" countries.) EWSs cannot be replaced by human information processing. And EWSs are sociotechnical systems which cannot be "switched off" without dramatic consequences.

Such a list of examples of sophisticated, critical, sociotechnical systems can be made for many areas, e.g. production (CIM – Computer-integrated manufacture), trading (Just-In-Time), travel services (reservation systems). In all cases, we have established complex mixtures of human and technical information processing. In all cases, world-wide integration is under way.

In addition, we are going to establish an increasing number of interconnections between various sociotechnical systems, forming "Super Systems": electricity, for example, is needed for the use of computers; these are essential for the supply of electricity in a peaceful world protected via Early Warning Systems as well as for stabilizing financial structures; they are needed for supplying goods to markets using Just-In-Time structures; and so on.

In order to understand the future of sociotechnical structures it is necessary to anticipate in some detail the potentials of information technology and its future developments (see Sect. 2). Combining this knowledge with our present understanding of social systems and our (growing) understanding of the evolution of information processing systems (see Sect. 3) makes it possible to anticipate some essential trends of sociotechnical systems for the near future (see Sect. 4). In Sect. 5 we are then going to draw some conclusions for our future situation.

2 Technical Information Processing

Technical information processing allows (by physical means) messages to be processed. Input to **Tech**nical **I**nformation **P**rocessing **S**ystems (techIPS) in most instances comes from human beings or (in an increasing number of cases) from other techIPSs. The output of the techIPS usually goes to a human user, who interprets the messages (e.g. at a terminal) in his or her brain. However, in an increasing number of applications, output from one techIPS goes into another techIPS directly (e.g. process control).

Internally, techIPSs are sophisticated electronic, optical and mechanical structures which utilize a discrete set of binary representations. Looking "from outside", computers combine hardware and software which perform according to procedures pre-defined by human beings. However, they allow some adaptation ("machine learning") to occur and – at least in *some* instances – perform *much* better than biological neural networks. Internal information today is often distributed in complex technical structures which can encompass several techIPSs in conjunction with many human brains.

TechIPSs can be considered an extension of human information processing, functioning, however, with *non*-biological methods. Computers are somehow an "externalization" of the human brain, they are an "addition" to neural processing of external information. In many cases in *small* cognitive domains, these "external brains" in the form of hardware and software are much more powerful than human beings. But computers are orders of magnitude away from handling simultaneously messages from many different cognitive domains. This, however, is done all the time by the brain.

Five Basic Principles are of central importance to an understanding of the "level-evolution" of techIPSs and their future within sociotechnical structures.

First principle: Every information processing activity which can be performed in practice by the human brain and which can be specified in detail ("step-by-step") can be handled safely by a techIPS.

This means, that whenever the human brain has reflected and understood a given process of manipulating information ("algorithm"), this process can be performed safely with information technology giving well-defined results (see e.g. calculators, digital telephone switchboards, or completely automated production machines).

Second principle: Information processing activities which can be performed in practice by the brain, but which *cannot* be analysed or understood in *full* detail, can also be implemented with techIPS. However, in these cases we cannot expect a "safe result" instead we get only an "acceptable result".

This is a statement which has to do with the way human information processing works. In many cases we cannot ensure that our human problem-solving approach will be 100% correct as in e.g. chess playing, choosing personnel, making financial decisions or driving a car. We always make (at least some) mistakes. Therefore, it was meaningful to implement "somewhat better" technical problem solvers ("heuristics"). Technical information processing can indeed be helpful in producing results from vague procedures, which are somewhat faster, more reliable, available all day long, etc. (as in, e.g. chess-playing computers, expert systems in maintenance, computer graphics in arts).

Third principle: From the broad variety of available options for technical information processing allowed for by the first and second principles, only a (small) fraction is really put into practice, namely those approaches which are meaningful either economically or in the military.

This means we do *not* implement all types of technical information processing structures in society. Instead, we select very carefully only a fraction which fit into our economically determined world. Information technology evolves under economic restrictions. It has to "survive" in competition with other structures. Economic selection practically eliminates at an early stage "uneconomic techIPSs" ("survival of the fittest"?). However, they may show up later in the course of evolution when they have become economic (see fifth principle).

Fourth principle: New concepts and innovations for techIPSs (using physical principles and in the future even macromolecular structures determined genetically) allow the volume per unit of performance to be decreased by some 10% a year.

This means that information technology is shrinking steadily in size. In particular, it is becoming more and more possible to construct "intelligent" products, which bring quite sophisticated information processing procedures to the end-user (e.g. camcorders, CAD systems, mobile robots, laptop computers).

Fifth principle: Costs per unit of performance of techIPSs are decreasing by some 10% per year for most components.

Thus, although the third principle stays valid, more and more applications of techIPSs will come into practical existence purely because of the disadvantage of costs for human labour (human information processing) compared with costs for technical information processing. From this principle it is very clear that information technology will spread continuously all over the world into all areas of social life – as long as we stay with economic strategies.

These five principles give an understanding of the rules controlling the level-evolution of technical information processing systems. They have been correct for some fifty years, and it is most likely that they will stay valid in the foreseeable future.

3 Validity of the Hypotheses from the Basic Concept

In the Basic Concept (Chap. I) eight hypotheses were put forward to explain the broad variety of information-processing phenomena in nature, their level-evolution and their mega-evolution. In this section the question is asked whether these hypotheses are valid for techIPSs (as a substructure of sociotechnical systems) as well.

If this could be proved to be correct, then it might be possible to extrapolate the validity of our basic hypotheses *beyond* the year 1991. This might allow more insights into the future of sociotechnical structures.

First Hypothesis: All natural structures continually receive, transmit and process information.

This statement is true for all techIPSs (although they have been formed by human beings). In their basic structure techIPSs are physical information processing systems and thus they always perform physical information processing, e.g.

electromagnetic interactions. At the *operational level* analogue or digital external information is put into information technology either by human beings or via networks from other computers. TechIPSs are programmed to produce continually appropriate output (pragmatic information) which either goes to "users" or into other techIPSs.

Second Hypothesis: IPSs are open, self-organizing systems.

Self-organization takes place in techIPSs at two levels:

(1) the self-organization of a physical structure (hardware) which is, however, trivial vis-à-vis our general understanding of physical systems, and
(2) software in computers allows "self-organization" of hardware and software functions to take place for some task fulfilled within the sociotechnical system. TechIPSs are open systems within social systems. Their information input and output comes from and goes into their "environment".

Third Hypothesis: Information is a system variable which always exists in relation to a particular IPS only.

This is true particularly for techIPSs that work with (digital) impulses as the sole basis for internal information processing. The *functional* information processing system, however, is the human-computer system. Information input into computers comes (directly or indirectly) from human beings and output is used directly or indirectly by people only. Information is indeed only a potential for techIPSs.

Thus, it is difficult to separate techIPSs from social systems. They function only in being embedded into social structures.

Fourth Hypothesis: This is not relevant here.

Fifth Hypothesis: Level-evolution results in further differentiation of the subsystems of a distinct mega-level of an IPS.

Specialization and differentiation in information technology is very obvious. Using the general scheme of the binary von Neumann machine, humans have developed thousands of specialized structures in hardware. In particular, the software approach allowed and allows a high level of diversity of structures as well as further differentiation.

Sixth Hypothesis: This is not relevant here.

Seventh Hypothesis: Basic structures of a new level of IPS come into existence in leaps within a relatively short period.

This hypothesis is true for techIPSs. Although the very early roots of technical information processing can be traced back some 2000 years in human history, actual mega-evolution took place over only some 50 years, starting in the 1930s.

Eighth Hypothesis: New IPS-levels come into existence if earlier structures are "exhausted".

In social systems techIPSs form a new layer of information processing. They allow extrasomatic storage, transfer, and processing of information to occur. However, input–output relations are always dependent on the societal part of the total system. An "independent" techIPS does not exist; they work, and can only work, in reliance on the physical, human and social level of information processing.

Considering these facts one may add to the eighth hypothesis a statement like: "IPS of a higher level need lower-level IPS as essential components." (This is also true, e.g. for biological IPSs; they necessarily need physical IPSs.)

However, the need for a lower-level IPS does not determine the "shape" of the higher-level IPS. A new information processing system *emerges* from an extremely broad variety of possibilities allowed for by the lower level, by "choosing" a particular set of initial conditions from "all possibilities". For techIPSs this is very straightforward; the binary computer is just an *economic* choice of organization of techIPSs. However, it is *not* the only one; it is, e.g. possible to build an "uneconomic" computer which uses a triplet code.

Summing up our arguments about the validity of the relevant basic hypotheses we can clearly state that they are in general true for techIPSs embedded into social systems, although some modifications might be agreed on.

4 Trends for the Future of Sociotechnical Systems

Arguments and considerations put forward in the project "Evolution of Information Processing Systems" show clearly that evolution of information processing is a basic strategy in nature. There have been phases of rapid restructuring ("mega-evolution") and phases of slow unfolding of information processing systems at a given level ("level-evolution"). But there has never been a "total standstill" in the evolution of IPSs.

Therefore, it is *extremely unlikely* that the evolution of information processing systems will be at its end in the year 1991 or in 2000. Instead, new structures within the level-evolution of techIPSs will show up in the near future and entirely new levels of information processing systems will evolve in the distant future!

In the next decades people as individuals and societies as a whole have to work hard on the problem of *organizing sociotechnical systems*. These structures will grow all over the world by the integration of technical information processing into already existing mixtures of humans and machines ("industralized society"). This process will be awfully difficult if we try to master it under the boundary conditions of supporting proper living conditions for more than *10 billion people* in the 21st century.

Today we know more about the world and the universe than ever before in history. We have more information processing capacity on hand (in brains and computers) *than any other generation*. We have some insight into the general framework of the mega-evolution and level-evolution in which our actions for the future should be embedded. Thus, in principle, it should be possible to solve the long list of problems we have on earth: *poverty*, particularly in the "developing" countries; *environmental pollution* all over the earth; *misallocation of national incomes* (see military budgets); *ethical problems in medicine* serving people who are becoming older and older, sicker and sicker; *establishing peace world wide*; etc.

What can be learned from our insights into the evolution of information processing systems in finding solutions to the above problems? *Five trends* have been derived and are put forward in the following. They *do not* solve the great problems but may be considered as an "environment" in which problem-solving is somewhat easier than without this framework. Being "in line" with the trends of evolution might make it less difficult to accomplish our tasks, *whereas actions taken in opposition to the evolutionary trends might be quite difficult or even impossible.*

Trend No. 1: Technical information processing will spread rapidly to all areas of societal life and will take over an increasing proportion of all information processing (level-evolution of the sociotechnical systems).

In sociotechnical systems feedback between technical innovations and their "usefulness" is direct and very fast. In contrast to biological systems it *does not* take many generations to try out the "success" of a "variation". Instead we can and do immediately feed back our experience within sociotechnical systems to reconstruct old or establish new structures.

The steady cost-performance degradation of information technology will be an important factor in the development of sociotechnical systems. It is clear that under conditions of economic competition technical information processing will spread widely. This trend is particularly supported by a world-wide and immediate availability of information processing capacity *without* human learning at the "local level".

However, not only the quantity of technical information processing will increase, but quality as well will improve in the next decades. This means that an increasing proportion of present-day human information processing can and will be transferred to technical systems. The level-evolution of sotecIPSs will be very rapid. (This is in accordance with early phases of other levels of IPSs.)

Trend No. 2: Sociotechnical systems will allow new types of "sociotechnical learning" to take place. Technical storage of data and procedures allow very rapid feedback between experience and the "technical genomes" of sociotechnical systems to occur.

In the evolution of the genIPS it took nature a long time to find out about the "success" or "failure" of new internal information (e.g. mutations and recombinations of the DNA). However, success of data in techIPSs can be checked within very short periods. Thus, new feedback loops for learning on the societal level are set up; they work very efficiently.

In our technical data banks and information systems we are setting up "technical genomes" for societies. At present they are evolving under direct control of human beings ("programming" and "data collection"). However, in the future machine learning will be a new approach for automatically improving "sociotechnical genomes". (Although libraries have already been a type of "social genome", it is clear that with computerized systems we have reached a new level. The storage and direct reusability of *procedures* is a completly new approach.)

Under these conditions "valuable internal information" can be selected very rapidly. The fixing of meaningful knowledge happens within periods of a few years compared to some 100,000 or millions of years in the level-evolution of the genIPSs and neurIPSs.

Trend No. 3: Complexity of sociotechnical systems will increase drastically. They will form complex "organisms".

Telecommunications networks available today (the analogue telephone) and advanced electronic networking in the future (e.g. ISDN, Integrated Services Digital Network) are integrating hundreds of millions of people and millions of computers. We are going to set up a "nervous system" for societies. This will integrate technical and human functions very tightly. With an increasing use of information technology we will get an intensive accumulation of procedures which are stored and used outside the human brain. Thus, we are organizing a gigantic information processing structure which is *beyond* social organization. It is a new complex machinery with its own memory and internal control.

At present there seems to be *little chance* for alternatives to this scenario because of the fifth hypothesis on the evolution of information processing; we are in the phase of intensive level-evolution of the techIPS. This necessarily results in new structures and organizations.

Although today the main workload of information processing in society is still borne by brains, it is quite obvious that by using an increasing amount of technical information processing "old" structures are integrated more and more into the sociotechnical structure in the form of hardware and software. Thus, sociotechnical information processing by itself is a new level of information processing; it is the approach of integrating all structures already established. In particular, social processes, which are a sophisticated combination of various human information processing structures, are "frozen" in the form of technical systems which organize and control societies. Recently developed expert systems are just the very beginning of a new type of transfer of expertise as "social competence" to technical systems.

Up to now sotecIPSs have been composed of relatively loose linkages of human brains and techIPSs. They will, however, increase in their density and complexity. There are various approaches already in sight which will support the integration: genetic manipulation of human beings (particularly cloning of "the brightest"), combining biological with technical neural networking, and "artificial life" which will allow new types of neurIPSs to exist.

Furthermore, the steady "use" of tecIPSs will have an influence on the way of human thinking, thus "bringing closer together" human and technical information processing.

Trend No. 4: The third principle of information technology is conquering the world and its sociotechnical level-evolution more and more. This might be a problem!

For the level-evolution of sotecIPS we have had – up to now – no well-formulated understanding of the driving forces. This is similar to the lack of knowledge at other distinct levels of mega-evolution of IPSs; genIPSs, e.g., "did not know" why and to which goals they evolved (see, however, the notion of the "selfish gene").

At present it looks as if we have only one mechanism for optimizing sociotechnical systems, namely, "doing it economically". We see this in all areas of today's life; whatever is done tries to be in line with economic considerations. However, this is a "restricted view"; what is asked for is a better understanding of the basic forces controlling the level-evolution of sotecIPS.

Trend No. 5: Sociotechnical systems are moving from local and regional IPSs towards global information processing structures. This makes it necessary to control the sociotechnical "organisms" at a completely new level.

In evolution "new procedures" have usually been local for long time before becoming regional or even global. In sotecIPSs we have already reached the second phase; in information technology we use the same approaches and machines world-wide within months or years. Thus, "cultural specialization" becomes a relict of history. The setting up of a world-wide "technical genome" is controlling the level-evolution of sotecIPSs, implemented as a global process. Today and even more in the future the same ideas are, and will be, used with the same programs and the same hardware from South Africa to Japan to Chile.

A special problem is the *lack of competition* for new sociotechnical structures which are growing world-wide. Whereas in evolution structures have always been in competition with other local structures for a long time, we now enter a period in which a *"monon"* slowly occurs, a unique sociotechnical system, "alone" on earth. Local variations within such systems are possible. However, they are relatively restricted compared to the "basic principles" of a unique sotecIPS.

Finally, we have to consider seriously the ethics of a sociotechnical structure. This is a task for politicians, the churches and for sciences. We must find a new approach of "responsibility" for the *entire* sociotechnical system. This is much more than just optimizing its components. Thus, thinking in networks and large contexts seems to be extremely important. But who is going to do this? Humans or sotecIPSs?

5 Consequences

It is highly speculative to draw conclusions for the future of human and social information processing as "lower-level" IPSs embedded in the evolution of the "higher-level" sotecIPS. However, to do this is of high importance for individual human being as well as for societies. Thus, we have to do it as well as we can.

As a guiding principle we have to use Hypothesis 8 of the basic concept on evolution of information processing in this book (see Sect. 1.2 in Chap. I). It states that "higher-level" IPSs are always integrating "lower-level" IPSs. Thus, it is very likely that neurIPSs (including human information processing) and sotecIPSs will serve their task *within* future sotecIPSs. The complete autonomy of local or regional socIPSs is at its end. In the future sotecIPSs will develop according to the principles of level evolution. This always – as has been shown in the basic concept – goes hand in hand with further differentiation and integration.

The level-evolution of technical information processing forces us to think world wide and simultaneously to act locally at a proper level. This is an extremely sophisticated task. In particular, it is necessary to organize all our technical networking in a way that serves the human being. However, the main problem to solve first is an understanding of the principle of sociotechnical structures. Their complexity is *way beyond* any other systems we have on earth. Technical information processing allows simpler structures to exist on the one hand. But it is also a trigger for sociotechnical aggregations of increasing complexity and thus for the need of world-wide thinking.

Studying the evolution of information processing and particularly the new levels of social and technical information processing allows some new insights into our future to be gained. However, dealing only with technology itself will not solve our problems. Instead we have to get an integrated understanding of the various levels of information processing and we have to properly support the development of a new "world-wide organism" which might internally balance the various interests in sociotechnical structures. (This is true today for the industrialized countries; it will become a problem for the developing countries whenever they reach a level where they are extensive users of information technology.)

We must integrate information technology into social structures as a sensible and powerful tool. However, we have to do this by simultaneously establishing proper control over the emerging sociotechnical systems. We must avoid the misuse of technology. This begs for technology assessment at all levels of the implementation.

In education, for example, we have to consider very seriously and support the "typical" capabilities of human information processing which are "beyond" the performance of techIPSs. We have to support creativity, solidarity and communication between human beings as a crucial component of human information processing within sotecIPSs. We have to do everything to get proper complementation of human and technical information processing ("cognitive mobility with information technology").

Setting up a proper organized extrasomatic "technical genome" is a sophisticated task. Therefore, we have to work hard on the organization of the technical representation of our informational environment. It is a task which has to be accomplished by the cooperation of all levels of social organizations. Informational environments must be transparent and usable by every individual. This asks for proper organization of information, proper access techniques and basic training for everyone, making possible access to the "technical genome".

We have to try to anticipate the next steps within level-evolution of sotecIPSs using proper methods. We must develop scenarios of sociotechnical evolution to find a proper way into this development. This means particularly supporting research for the understanding of sociotechnical systems as very complex structures.

It is necessary to establish societal and political organizations which try to establish a proper balance between human and technical information processing in such a way that human life is still meaningful in terms of our traditional goals.

With sotecIPSs there is a chance to set up a completely new "mega-organism" which might guarantee peace on earth because its components are no longer likely to fight one another (as is the case in a stable biological organism). However, this might only be possible if we stop over-population, which seems to be an "inherited cancer" of the new sociotechnical organism.

From Neural Information Processing to Knowledge Technology

Erhard Oeser

1   An Integration Model of Cognitive Science   320

2   Evolution, Function and Performance of the Human Information Processing System   .   322
2.1 Evolution of the humIPS from the Viewpoint of Evolutionary Epistemology   323
2.2 Function and Performance of the humIPS from the Viewpoint of Neuroepistemology .   327

3   The Evolution of Technical Information Processing Systems   331

4   Knowledge Technology: The Future Integration and Co-Evolution of Human
    and Technical Information Processing Systems   336

References   338

1 An Integration Model of Cognitive Science

Since the emancipation of cognitive science as an autonomous discipline an integration model allowing for the representation of the contributions of the individual subject fields has become necessary [1]. But as yet such a stable model does not exist. Perhaps one will never be established, since the perspectives and points of view for such an integration are so different.

In this sense the following proposal from the philosophico-epistemological perspective claims to integrate intentions and experiences from other disciplines as depicted in the figure.

The starting point is the investigation into the human ability to acquire new knowledge, traditionally described in philosophical epistemology from Aristotle to Kant. While Kant was convinced that an empirical experimental psychology was impossible, an empirico-experimental cognitive psychology has developed within the paradigm of information processing, parallel to philosophical epistemology, which is in fact the logic of knowledge. This new discipline provoked the interaction of psychology and computer science and led to *Artificial Intelligence Research* using psychological theories of natural human intelligence for the construction of intelligent computer systems. The development of the computer into a universal symbol-processing system was a precondition for this new ability. At the level of software engineering this change was provoked by the construction of higher-level programming languages.

The development of *computational linguistics* established the connection to linguistics which in turn has been linked to a general theory of signs, that is *semiotics*, with its three dimensions syntactics, semantics and pragmatics.

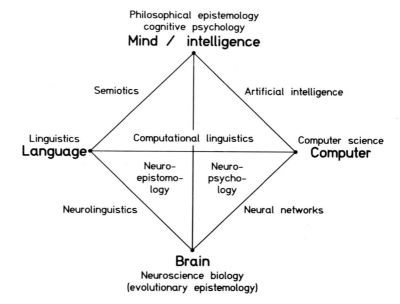

Fig. 1.

At first, the study of brain and thus the neurosciences was excluded from this interdisciplinary triangle of philosophico-psychological theory of cognition, linguistics and computer science, dealing with the human mind, language and the computer as objects of investigation. Due to the language-analytical philosophy of mind the paradigm of serial symbol processing has even been considered as a model of the human mind. It served to become independent of the highly complex structures of the human brain. But this restriction has now been removed in every respect; *neurolinguistics* is now investigating neural structures and mechanisms for explaining organically-based language disorders and thus contributing to a theory of human language acquisition.

In a similar way *neuropsychology* is searching for neural correlates of cognitive human performance, i.e. of perception and thinking. In analogy to these already established disciplines we can speak of neuroepistemology [2] or neurophilosophy [3, 4] searching for correlates of the principles and truth conditions of human knowledge in certain brain functions. This idea had already been conceived in philosophy by Locke with his physiology of human understanding and by Kant with his philosophy of the brain [5]. Today this idea of neurophilosophy is linked to a new approach in computer science, the so-called neural networks. Neuroepistemology is thus the philosophical basis of connectionism.

Symbol processing and connectionism are therefore seen as alternative models of the human mind. But in an integrated cognitive science these distinctions do not constitute an exclusive disjunction. The human mind and its material-organic apparatus are obviously both *a parallel distributed information processing system and a serial symbol processing system.*

Thus it is necessary to search for the relationship between the subsymbolic and the symbolic levels of information processing.

This relationship is constituted by the fact that distributed representation and parallel processing are the prerequisite and precondition of serial symbol processing [6–9]. In this way perception and conceptual thinking or propositional logical reasoning can both be distinguished and linked to each other.

After these general considerations about an integration model of cognitive science we are going to compare in the next two sections the evolution of the human information processing system (humIPS) and that of the technical information processing system (techIPS), and treat sociotechnical integration under the aspect of the co-evolution of the two systems. Concerning the relation between these three domains we shall make the following assumption, which will later be justified and supported by empirical proof: the humIPS is the last step or layer in the evolution of neural information processing systems (neurIPSs). TechIPSs, which originated at a later stage, are not a layer of IPS on their own, but just a side-product of the humIPS which in the beginning was irrelevant to evolution, i.e., it did not have any special selective advantage. But once originated and integrated into human society, this epiphenomenal product has turned into a basic phenomenon [10], since in the sociocultural evolution of man it is a selective advantage to be informed better, faster and more precisely. Even the old evolutionary biologists knew this; de Candolle [11] reinterpreted Darwin's "struggle for existence" as follows: "The struggle for existence is now won by the one who knows, whereas in the barbarian age the most cunning and most brutal have won." Due to the fact that "more knowledge" is a selective advantage, techIPSs are of "Artificial origin" and also have their "Artificial" evolution made by man.

If the humIPS is the "highest" or "last" step of the evolution of information processing systems, this also means that a humIPS is characterised by using other IPSs without being used itself. This also holds for sociotechnical IPSs which do not possess an independent existence superordinated to human individuals but are just the result of coordinated individual free actions of individuals.

This is at the same time a special characteristic of the biological species *homo sapiens* as opposed to all animal societies, the cooperation of which is genetically conditioned, i.e. pre-programmed through the genetical information processing system (genIPS), whereas human cooperation is in fact based on free actions of individuals.

2 Evolution, Function and Performance of the Human Information Processing System

The assumption that the basic functional mechanisms and performances of the humIPS are the result of the evolution of the neurIPS is already supported by an interdisciplinary or transdisciplinary theory: *Evolutionary Epistemology* (EE).

It is a logical consequence of biological evolutionary theory and had been proposed in its basic structure by Darwin. Although there were many fore-runners in the 19th century – some of them biologists such as Häkel, but also physicists such as Mach and Boltzmann – EE was systematically and empirically founded only in the twentieth century by Konrad Lorenz. The empirical foundation is given by comparative ethology, the theoretical basis by the theory of evolution. Due to the latter it is assumed that not only somatic properties such as bones, teeth, hooves, etc., but also fundamental modes of behaviour are genetically conditioned characteristics which are specific for every species, and which have developed in the course of evolution.

This not only holds true for lower animals, but also for man and even his higher knowledge performances. Present-day human knowledge performance and learning ability are based exactly on these standard mechanisms developed in hominid evolution and genetically determined. Without these standard mechanisms the biological species *homo sapiens* would not have survived. Even the point of time when the determination of the genetic determinants of behaviour was finished can be indicated fairly exactly. For the apparatus controlling our behaviour, the central nervous system or the brain of the *homo sapiens*, has not changed since the Neolithic period, as skull casts have shown at least superficially.

This historical fact, i.e. that the evolution of hominids was almost exclusively an evolution of the brain, turns Evolutionary Epistemology, primarily dealing with a comparison of the cognizing apparatus, into *Neuroepistemology* (NE) concerning the humIPS, i.e. into a theory reducing human knowledge per-formance to genetically conditioned structures and functions of the human central nervous system.

2.1 Evolution of the humIPS from the Viewpoint of Evolutionary Epistemology

The peculiarity of the humIPS distinguishing it from all the other new IPS lies in the progressive encephalization of perception. This means that in the evolution of hominids not the structure and resolving power of our sensory organs, i.e. the periphery, are increased, but rather the internal processing, the central nervous system. Thus the somatic evolution of man was primarily an evolution of the central processing system, which rapidly increased, not only in its extension and weight, but especially in the complexity of its macroanatomical and micro-anatomical structures. Whereas the evolution from the first begin-nings of a centralized nervous system to a fully developed brain of the younger mammals or primates has been a process over several hundreds of millions of years, the evolution of the hominid brain cannot be compared in its lapse of time. Within several hundred thousand years, but less than two million years, the volume of the hominid brain has almost tripled. This increase in volume or weight of the brain has primarily happened through the rapid enlargement of the neocortex which dominates the whole brain. The cerebral cortex contains about half of all the neuron cells. This explosive growth of the cerebral cortex turns the brain into a superfluous organ. Its specifically cognitive capabilities by far transcend the original needs of its carrier [12].

As with all organic evolutionary processes, the evolution of the brain in hominids is based on genetic mutations. It is assumed that this was caused by a relatively small change in the genes, which determined the length of development of the brain in such a way that it grows longer and faster than other organs. The phenomenon of neoteny is cited as empirical support for this hypothesis, i.e. the juvenile characteristics of the predecessors of man are preserved in the adult *homo sapiens*. In this sense an adult man is more similar in his facial characteristics (high forehead) to a juvenile ape than to an adult one.

If the structure of the brain contains the condition for its performance [13], this also means that a longer learning phase in man, in comparison to all other organisms, corresponds to a prolonged maturation process of the human brain, which in the postnatal phase mainly consists of neuropile maturation and connectivity. Thus ontogeny or individual development is far more important in man than in animals. But this does not mean that man or humIPSs are emptied of phylogenetically determined, genetically "preprogrammed", behaviour. In addition the genetic determinants of human behaviour are not active from birth onwards, but rather according to a time plan, itself genetically determined, which must not be missed. Thus deprivation of the natural environment in "Kaspar Hauser" situations and the resulting mental retardation do not support behaviourism or environmentalism, but rather indicate the complex relation between "innate" and "acquired" properties. Thus the deprivation of the humIPS of innate standard procedures is not the presupposition of learning. From the viewpoint of ethology and thus from Evolutionary Epistemology learning is rather the capability for completing or changing the innate behavioural program. This can clearly be seen from the example of birds learning to fly. Flying is a highly complex motor interaction of hundreds of muscles, bones, joints, plumage, etc. In extreme cases, as in the case of the swift, the young bird has less than one second in which to learn to fly when leaving the nest. Sverre Sjölander [14] computed how much time would be required in order to carry out all possible combinations of muscle movements, positions and movements of joints until by chance the only right one, fit for flying, is found. The result was that it would have taken the bird about 200 years – eight hours per day – without holidays to find this combination. Thus the bird inevitably must have an innate basic programme enabling it to carry out **a priori**, before any experience, the right, highly complex motor sequence of movements. In addition the bird must have an innate programme for the coordination of the motor sequence of movements by sensory mechanisms and the programmed capability to correctly modify other programmes at once, if this becomes necessary because of individual and new variations.

If we understand evolution on the one hand as a process of transformation (vertical component) and on the other hand as diversification (horizontal component), the human IPS is vertically the most complex of all new IPSs in the transformation process and is distinguished horizontally from all other new IPSs by the fact that the internal processing mechanism is the most advanced, whereas with the periphery, the capacity of the sensory organs is geared towards middle dimensions. Almost all higher animal species outdo man as far as the capacity of individual sensory organs is concerned.

The human capacity for knowledge and learning is not based, technically speaking, on the causal capacity of the sensory organs, which have only a small "spectrum", but rather on the "inference machine". The receptors are, so to speak, only small slots which can do nothing but provide information on certain energetic states of the environment. Everything else is based on the logical processing of elementary information by the internal mechanism of the brain functions.

Based on and in accordance with results of comparative ethology, evolutionary epistemology asserts that this human internal processing mechanism is also a result of evolution in its logical structures.

Similarly to the bird which is genetically programmed to learn to fly in less than a second, man, too, who has to learn to think, has "innate masters" in the form of genetically determined basic programmes, modification and expansion programmes. This is the only way for the human individual who has to start again from scratch to be able to learn the knowledge of his social environment in order to be able to communicate in this society and to perform something. Learning, e.g. of a language, takes place comparatively quickly so that an extremely complex genetically pre-programmed structure has to be assumed. The architecture of this genetic program system is in fact given in the architecture of the central nervous system as investigated by functional neuroanatomy. Thus the above-mentioned assumption, that the living (= functioning) brain is a theory of its environment including its own body or carrier organism, is proved.

The evolution of information processing can thus be seen as an increase in performance in discovering or computing the world, which becomes possible through those programmes and programme architectures which have developed in the evolution of nervous systems. Thus it is a general phenomenon that an increase in structural complexity always results in an increase in system performance. The enormous learning capability of the humIPS is thus not a result of depriving it of all programs, but is due to the fact that a complex architecture of empty programmes ranging from the simplest body reflexes and reactions to the highest mental activities has emerged. In this respect the fundamental conceptual distinction of structural and fluctuating information can be connected with empirical investigations of comparative ethology.

Structural Information is the sum total of all basic programmes, modifying and extension programmes implemented in the structure of the central nervous system. For the individual it exists *a priori*, preceding all experience and is a guarantee for its survival. This structural information, however, emerged *a posteriori* by mutation and selection in phylogeny.

This structural information is, so to speak, organically embodied knowledge of the organism about its environment. But this knowledge cannot inform on the current state and situation of the environment. What is necessary is the *a posteriori* fluctuating information (known in ethology as "instant information") which does not change *a priori* structural information, which is genetically consolidated. Structural information, embodied in the material micro- and macro-anatomical structures of the brain, but not identical to them, is to be seen as templates or schemata serving to categorise current events, i.e. fluctuating information. In this sense this structural *a priori* information has already been anticipated in classical philosophical epistemology from Aristoteles to Kant. This connection to Kant's terminology has been explicitly pointed out in information theory [15] and evolutionary biology [16].

Fluctuating Information is current information which must be processed under the conditions of structural information. It originates not only outside the organism as external signals, but also inside the organism, since for the brain the body is also part of the environment. In addition this fluctuating information also originates within the central nervous system itself, as nonsense stimuli, but in the first place as results from the processing phase, which return as input into the processing phase, as though they were external signals.

This distinction between structural and fluctuating information, which is of fundamental importance to all IPSs, is also the basis for describing the peculiarity of organic information processing. In contrast to classical, symbolic and technical information transmission and processing systems with their separation of the three dimensions of the concept "information", the syntactic, semantic and the pragmatic dimensions, these are always *a priori* linked with each other in organic information processing systems. In the latter case there are no meaningless signals. Biological information always has a "value". This is also true even for the genIPS and all the more for the nervIPS; information is meaningless without evaluation. This means that the mere energetic process is still without information for the organism. Only the energetic process which affects the behaviour of the organism carries information. But this implies that the semantic dimension, i.e. the "assignment of meaning" of an energetic signal is determined by pragmatics. In order to be able to answer the question of how this assignment of meaning happens, the pragmatic aspect must be divided into two components: "first occurrence" and "approval" [17]. Concerning the distinction of structural and fluctuating information this means that an information process can only take place if an event approves the basic structural information as fluctuating information and also adds something new which can then be linked (=approval) because of the already structurally embodied information. Information is what produces information through processing the "first occurrence" or "newness". 100% newness cannot be understood. 100% approval does not provide new information. In general, in all kinds of biological information systems, the optimum of information production must therefore be where there is as much first occurrence as possible and as much approval as necessary [18].

The evolution of the humIPS was evidently such an optimization process, where the approved structural information content of the system quickly increased, and through which the complexity of functions also increased. The principle of the increase in functional complexity can be explained on the basis of comparative neuroethological studies in the following way: a new, complex function often, if not always, emerges through the integration of several, already existing functions which have been efficient as individuals independently of their subsequent integration, and which still function as indispensable parts of a new entity. In some cases the old functions may disappear completely later on or lose their importance [19].

Braitenberg [20] has shown that this principle also works technically by constructing "cybernetic vehicles" for this increase in functional complexity with the purpose of illustrating the evolution of new functions and the increasingly complex crossing of fibre pathways in vertebrate brains. We can assume that this principle was also effective in hominid evolution. The evolution of sensorimotor functions in humIPSs is, for example, a consequence of the integration of organic functions controlled by the brain: the function of the hand and the function of the eye. In contrast, haptic and optic functions work separately and sequentially in other organisms (e.g. snakes). The evolution of the symbol function in humIPSs is due to the integration of the two components communication and representation to form a complex synthetic function, which is not the case in insects, for example, where we can observe communication without any representation.

The fully developed brain of *homo sapiens*, which has not changed since the Neolithic age, thus contains the whole history of the evolution of neurIPSs in its organization, structure and function.

2.2 Function and Performance of the HumIPS
from the Viewpoint of Neuroepistemology

Whereas EE deals with the evolution of the human cognitive apparatus, NE focusses on the function and performance of this apparatus [21]. The point of departure for this consideration is the assumption, especially well supported by palaeoanthropology, that the anatomical organization of the brain with all its structures and functional relations is the result of an evolutionary process lasting one million years. The human brain as a top product of the evolution of neural information processing systems is at the same time the most complex system, concerning both structure and organization as well as function and performance. Opinions on the way this system functions vary accordingly. Depending on which of the various performance measures of humIPSs is taken as a point of reference, basically three points of view can be distinguished:

– The cognitivistic model of symbol processing is based on the analogy between the brain and the computer, with the brain corresponding to the hardware of a digital symbol processing computer and consciousness corresponding to the software. Every kind of information processing in the brain should thus be seen as an operation with symbols, with this internal symbol system itself processing a propositional logical structure. Even simple sensorimotor actions are thus supposed to be based on such a serial symbol processing mechanism.

- The direct realism of ecological perceptual psychology [22] is in contrast to this position. This approach originates from the special field of investigation of movements controlled by the sense of vision (landing operations of pilots). Obviously information processing is based on elementary and more complex invariants generated by sensorimotor feedback processes without having to think of a complicated symbol processing operation. According to this opinion the movements of the organism are driven by the synchronized information flow of elementary and higher invariants.

- Connectionism differs from these two approaches in its direct orientation towards the internal neural processing system of the human brain. In contrast to the symbol processing model, connectionism explicitly starts at the sub-symbolic level. Thus the concept "representation" also changes. Whereas in the symbol processing model the representation by certain signs and sign linkages is similar to a linguistic representation, the process of representation is distributed according to this paradigm over a network of basically "meaningless" non-representational neuron-like processing units. Thus these "neuronal" networks are located at a lower level of information processing than the semantic networks of the symbol processing model. On the other hand they can also explain – in contrast to ecological perceptional psychology – the emergence of higher invariants as a property of neuronal structures based on external constraints of the environment.

Starting from this neuroepistemological approach with the complex performance of the humIPS being based on a gradual evolution of the cognitive apparatus, these apparent alternatives can be seen as emergent functional hierarchies.

These functional hierarchies are due to the neuroanatomical structure of the human brain and they differ in the modality of information processing. The neuroanatomical structure of the human brain has, in spite of its similarity to neurons in its elementary design, several subsystems and levels of integration in its gradual structure. Here an important peculiarity of organismic IPSs becomes obvious, which Leibniz [23] has already pointed out when saying, "The machines of nature, i.e. living bodies are still machines in their smallest parts until infinity." In today's terminology of the theory of neurons this reads, "Every single nerve cell is both a computing unit and a communication apparatus of its own." The brain is not a computer, but rather it consists of billions of computers, each having an internal, highly complex structure and chemoelectrical function, altogether forming a network of interactions, with its overall structure always unknown to us. This modular way of construction also permits, in addition to sequential information processing, massive parallel processing of fluctuating information distributed over many subsystems to be carried out. Furthermore, it is possible to process one and the same piece of information emerging in this system as neuronally distributed representation in several "copies", so to speak, and to unite partial results of this parallel distribution on a higher level of integration. The basis for this highly efficient capability of the human IPS which cannot be reached by connectionist techIP systems either, is the enormous amount of structural information in the form of biological genetic information in organismic matter.

Due to the high structural information content of the humIPS a high variety of programmes is possible at a functional level. The form of these programmes which are typical of the living human brain is different in the various regions of the brain; in those parts of the brain with structural excitation constraints the programmes are rather uniform and rigid. But in other regions of the brain they are rather variable and conditionable, so that we can speak of a "freedom of programmes " (Seitelberger), given under certain structural conditions, as a functional possibility of the humIPS. This freedom of programmes is a peculiarity of the human brain. Basically all programmes are independent of the apparatus used concerning form and content, because they are determined only through the syntactic rules of the signals, their position and their sequence. The more linking possibilities the carrying processing apparatus permits, the more difficult the programmes it can execute. As far as the brain is concerned this means that, given its complexity, we have to assume an incomprehensible variety of possible programmes and thus an enormous performance capacity.

At a quantitative level this can already be seen from the high interconnectivity of neurons, consisting of about 50,000 billion contacts (synapses), even according to conservative estimates.

Thus an exact description of the connection structure and a complete representation of the excitation processes is impossible. What is possible, though, is the global representation of the individual functional hierarchies based on the macro-anatomical structure of the cerebral cortex.

The well-known fact that certain functions can be "located" in the cerebral cortex is not, as had erroneously been assumed in former times, due to a functional specificity of the cerebral parts in question. For in principle all parts of the brain function in the same way. The true cause is that in certain cortical regions information of a certain origin is processed. Those cortical regions, for instance, where certain parts of the information flow from the periphery of the nervous system go in and out are called modal cortical regions, e.g. the optical region, the auditive region or the motor region. By contrast, those cortical regions where information originating from the modal areas is compared and processed in the sense of multimodal computation are called *intermodal* cortical regions.

But multimodal processing of the sensory information is not the last processing step. These results of intermodal information processing are also subject to a further differentiation and construction in order to gain a full comprehension of reality from the totality of information in the sense of cognition and its linguistic representation. This higher level of information processing can be called, according to Seitelberger, "*supramodal function*" and certain cortical areas can be identified as their "workshops". At this supramodal level a functional distribution between the two hemispheres can be observed. This *lateralization* or *specialization of hemispheres*, proved by the famous "split-brain" experiments, does not, however, mean a strict distribution of labour in such a way that for example language as logico-analytical thinking is located or centralized in the left hemisphere, and spatial visualization or intuitive thinking in the right hemisphere. These and the other higher cerebral capabilities are based rather on the cooperation of both hemispheres and need these for their normal functioning. Lateralization is thus an increasing, qualitative extension of the range of action of the cerebral cortex, which apparently is so important from the viewpoint of evolutionary progress that the necessary loss in operational security could be accepted.

Simplifications of split-brain research, such as the "speechlessness" of the right hemisphere or the "blindness" of the left hemisphere, must be avoided under all circumstances, and also the idea that one of the two hemispheres could be separated from the other and formed or designed by the other. Experiments or neuropathological evidence in brain lesions forbid even more (as Eccles did) that one talk of the localization of the self-conscious mind in the language-dominated hemisphere. For severe cases of total removal of the left hemisphere clearly show that these patients still have self-consciousness and that sometimes an even better linguistic capability remains than in the split-brain patients of Sperry with the right hemisphere still being "overshadowed" by the left hemisphere.

This means that there is neither a neuroanatomical spot nor a neurophysiological substrate of consciousness. It is, rather, functional reality linked to the degree of complexity of information processing. The seemingly unambiguous connection, seen by Eccles [24] as being between the "self-conscious mind" and the language-dominated hemisphere, is based on the *modus operandi* necessary for linguistic performance which cannot take place without consciousness. The self-reference necessary for that is another modality of information processing, which is an integrative renewal treatment of all results of information processing of the whole cerebral cortex, including the intermodal and supramodal products of processing. This *metamodal* level of information processing can be assigned to certain cortical areas of the frontal lobe and the temporal lobe. The loss of these cortical areas through lesion or illness results accordingly in the loss of self-reference of experience. But this does not mean that in these metamodal cortical areas self-consciousness is located as "immaterial mind". Self-consciousness is rather a global, unifying function, i.e. a function of second order. In contrast to all other higher functions of the brain this function has a meta-organic character. It is therefore separated from the other levels of neural information processing. This means that the function of consciousness is not a useless knowledge about its own neural processing mechanisms, but in respect to the experience of the environment, it is a considerable extension of the range of action of the body in its environment. This happens through active, purposeful and well-planned observation and screening of its own environment in order to change these mechanisms according to internal plans and ideas.

At this point the sociocultural evolution of man starts as a bifurcation of organico-biotic evolution, characterised by a higher speed of all processes. The artificial evolution of techIPSs operated by man is embedded in this meta-organic sociocultural evolution.

3 The Evolution of Technical Information Processing Systems

"Artificial" evolution of techIPSs has to be distinguished from "natural" evolution of organismic IPSs. In the evolution of both genIPSs and neurIPSs, the two factors mutation and selection are separated. But in artificial evolution, mutation and selection are linked. The classical form of animal husbandry and plant breeding, still without genetic manipulation, has an intermediary position between the two extremes of natural and artificial evolution. Here we can already observe artificial selective breeding instead of natural selection, oriented towards the goals and needs of the breeder. In addition in the development of technical apparatus, natural (accidental) mutation is replaced by planned purposeful invention or change by the producer [25].

The planned and purposeful invention of techIPSs is based on a theory about the processes to be effected by the apparatus to be constructed. This theory must exist, even if it is very rudimentary or faulty. A technical invention never occurs by chance without any theoretical knowledge. But on the other hand there is no complete theory anticipating all processes which can be carried out by apparatus on a theoretical level and indicating how to construct this apparatus, or else there would be no evolution of technology. A supermind à la Laplace knowing all the laws of the real world would be capable of constructing a perfect machine. The evolution of technical tools, apparatus and machines reflects not only the imperfection of human cognitive capabilities, but also their constant capacity for improvement. This particularly also holds true for the evolution of techIPSs, where there is also the process of iteration of structural and fluctuating information, typical of every evolutionary process. In this case structural information is not the result of a natural self-organizing process. Although this information, as an *a posteriori* result in phylogeny through mutation and selection, is a genetic *a priori* for the recent individual, it is rather provided in artificial evolution through the plan of the constructor or system designer of the techIPS. This materially implemented construction plan is the basis of the framework for the real capacity of a technical apparatus. In optical apparatus this structural information has the form of light intensity, focal length, etc. [26]. In transmission systems it has the form of channel capacity and the code, in IP systems it has the form of general programmes (programming languages) and special programmes forming the structural information content as a logical architecture. On the other hand, fluctuating information is what can be received, transmitted and/or processed according to the structural conditions of the apparatus. Thus structural and fluctuating information behave like light intensity and light waves in optical devices, like codes, channel capacities and signals in information transmission systems, like programmes and data in information processing devices.

On the one hand, evolution happens as iteration of structural and fluctuating information in biological IPSs by the change of structural genetic information during the long succession of generations. This change results in the emergence of other kinds of systems, forced by selective pressure to autonomous evolutionary progress. On the other hand the structural information of technical apparatus cannot change on its own. It remains rigid and unchanged without the intervention of the constructor. In technical fields, self-construction or self-reconstruction, as in the case of natural evolution of organisms as phylogeny, is not yet possible. As John von Neumann [27] already suspected, the goal of self-correcting, self-producing and self-developing automata would totally change classical computer architecture. The algorithmic machine would have to change in a pseudo-morphosis towards neurology, which would only be possible by way of thermodynamics. But such early remarks (1956) clearly show that current tendencies in the development towards subsymbolic connectionist neurocomputers, Boltzmann machines, etc. had been planned from the beginning of automation theory onwards. Only such systems, developing themselves by learning, would lead to a really new form of evolution of techIPSs. But if we want to talk about "evolution" not only metaphorically, in the previous phases of development of techIPSs, it has been from the beginning onwards until today only an artificial evolution effected from outside by the human constructor, with the aim of increasing human performance. This means that it is the natural humIPS which produces "prostheses" supporting it for the extension and improvement of its own structurally and materially restricted performance capacity. Just as optical apparatuses are prostheses of the sense of vision, techIPSs are prostheses of the natural humIPS. They substitute, amplify and extend as "brain prostheses" natural human information processing, as it occurs in the central nervous system. This prosthesis-like character of the techIPS has already been pointed out by Leibniz, the first to construct a four-species calculating machine, by saying that because of this development mankind will possess something like a new organ increasing the capacity of mind far more than optical instruments amplify visual acuity and outperforming microscopes and telescopes, just as reason is superior to the sense of vision [28].

But just like all prostheses, techIPSs are also restricted to certain capabilities, i.e. those capabilities which are to a large extent theoretically understood by man himself and thus controlled in their regularities. Therefore, the evolution of techIPSs starts with the construction of calculating machines, since arithmetic operations are those operations which can best be mechanized. Among these arithmetic operations addition was the easiest to implement directly with cranks, cylinders and gears. But the important theoreticians of computer science clearly knew from the beginning that the modern digital computer is a universal information processing system and that it is not restricted to numerical data. Whenever information can be represented in letters and numbers, i.e. "alphanumerically", such universal symbol processing machine systems can be used. With the limitation of the symbol alphabet in the binary system of Leibniz and Boole to only two symbols, not only do we have an easy technical implementation, e.g. a toggle switch which is unambiguously switched on or off, but also the abstract possibility of emulating every other machine with discontinuous states with such a digital computer is provided, though there is the disadvantage that every operation of the imitated machine needs several operations in the imitiating universal machine [29].

Leibniz already knew that, at the level of Boolean algebra, calculation with numbers corresponds to logical reasoning. The difference between formal languages and their embodiment by machines is lost as well. Exactly because of the fact that all other machines can be imitated by the digital universal machine, it is not necessary to build a new machine for each type of information processing operation, but all these operations can be carried out with a digital computer which can be adequately programmed for every application (Turing). In this first phase of artificial evolution of techIPSs, a fundamental difference from natural evolution of the humIPS reveals itself; this artificial evolution starts at the highest supramodal performance level of the humIPS, at the level of regulated and consciously controlled symbol processing. This level of symbol processing is transferred by the natural human neural IPS to a constructed symbol processing machine. What is lost in this process is semantics. The classic computer is a purely syntactic apparatus in its logical construction. All attempts to add semantics to this system are only possible by means of a syntactic model. Computer semantics is thus restricted in this form to the execution of machine commands. There is no direct access to the "meaning" of symbols as knowledge about reality in the sense of the original semantico-concept "information". Whenever a relationship to reality has to be set up, a syntactic model is necessary, with mere sign manipulation sufficing for the computer [30].

All ambitious research programmes dealing with semantic information processing leading to artificial intelligence research have failed due to this limitation, at least as far as their claims are concerned. For the goal was a "General Problem Solver" (GPS) based on the following idea: if we have the elements of a theory of heuristic problem solving as a successful process in most cases but never reliably predictable in man, then it must be possible to simulate these heuristic processes in a digital computer. This idea led to the opinion that intuition, knowledge by cognition and learning will no longer be properties exclusively of the humIPS. Herbert Simon [31] even said that every big high-speed computer can be programmed in such a way that it also has these capabilities.

In this manner the whole of the research for the creation of artificial intelligence has shifted from algorithms to heuristic operations. But the problem is how to find such generally usable heuristics. There are two different answers and consequently two different directions in the evolution of techIPS:

- One direction is the attempt to construct functioning models of human behaviour with the claim that the behaviour of the machine corresponds to human performance. Representatives of this direction are Newell, Shaw, Simon, Feigenbaum, Gelernter, Samuel, etc. [32].
- The other direction is the attempt to build intelligent machines without a preliminary decision as to whether they are similar to man or not. Minsky and MacCarthy are the most important representatives of this direction.

The difference between these two directions can be systematically explained by the concepts "simulation" and "performance", which in turn can be explicated by a classic example from technology. Just like the pioneers of aircraft technology, e.g., the legendary Icarus of classical mythology or Leonardo da Vinci who, by emulating the flight of birds, wanted to fulfil the old dream of mankind being able to fly, the pioneers of AI research wanted to simulate intelligent human behaviour with the computer by observing how people actually solve problems. Herbert Simon and Allen Newell did that by observing what students did in the solution of logical problems. The method they used was to record the oral reports given by test subjects during their solving operations. A program was written following the rules of thumb not always but often successfully used. In this way the program "Logic Theorist" and later the famous program "General Problem Solver" (GPS) were produced.

In contrast to this mode of simulation in which one tries to emulate in a program "how people do it", the mode of performance is characterised by the performance of a program. But this performance can be based on construction principles which are completely different from those which the living organism is following. In aircraft technology fixed wings and propellers were used to achieve the performance of the movable flapping wings of birds without simulating them; in the development of AI the mode of performance has been represented by programs which could "understand" natural language input or solve geometrical analogy problems in a drastically restricted field, e.g. the field of algebraic equations [33]. Although the idea of simulating human thinking in its real process was not important, the line separating the modes of performance and of simulation is – for practical reasons – difficult to draw and to maintain. For, on the one hand, the only way of thinking about how the computer could solve a certain problem is to ask how a human being would do it, on the other hand, a simulation model has performance, i.e. a specific capacity, if it is successful and thus achieving a certain performance.

Last, but not least, both the mode of simulation and the mode of performance can be reduced to a uniform theory, in the same way that the theory of aerodynamics covers the flight of birds and aircraft at the same time. The same can be said about AI research. In this case it is the theory of information processing underlying both modes of production of artificial intelligence. This theory holds that man and computer are only two kinds (species) of one and the same "abstract genus" of information processing system.

This assumption can be interpreted in two ways. In the sense of a "weak functionalism" it just says that certain cognitive functions of man can be formalized with explicit rules, which can also be used for programming a computer in order to achieve adequate intelligent performances. But for the strong form of functionalism, also called "cognitivism", the situation is different. It says that these rules of information processing used for programming a computer are identical to the rules of human thinking, as can be ascertained by empirical observation of human behaviour during problem solving. In such a way computer programs become models for psychological theories. On the other hand the fundamental discussion of the possibility of AI is extended by "cognitive psychology". But from the viewpoint of neuroepistemology this strong functionalism implies an illegitimate reduction, since it is exclusively limited to the level of syntactic symbol manipulation, whereas in the real humIPS, the brain, the symbolic supramodal level can only be reached via several subsymbolic performance levels of hardly imaginable complexity. Thus we cannot assume that on the level of pure symbol processing it will ever be possible to achieve real "knowledge processing". In the meantime it is well known that the most ambitious of these symbol processing systems, the GPS program mentioned above, failed in its high-pitched aim of simulating general human problem-solving behaviour. The path from this general frame work, within which any logical theory programs can be used, to the solution of concrete problems was still too long. The idea that the information processing system human is "very simple", like an ant, in the sense of "some basic parameters" in interaction with the environment, proved to be a very dubious and problematic assumption. It also did not correspond to the original basic idea that Leibniz had established, that practice is basically nothing but a wider and more complex theory. In contrast to that, the "general problem solver", which from the outset has only some parameters, should be compared with the navigating memory of a ship, whose control of the ship is based on a few data, e.g. geographical coordinates of its harbour of destination, with the difference that the transformation of the GPS system from a chess player into, e.g. a solver of mathematical problems requires much more than the alteration of some numbers which would suffice for a change in destination harbour. In principle the whole "memory structure" of the GPS must be exchanged. Thus it is clear that this or a similar system must break down under the burden of its own organization.

After other fields of AI research such as machine translation, natural language systems and pattern recognition had also showed unexpected stagnation after first successes in simple "toy worlds", the impression arose, not without good reason, that AI research was not useful as a purely academic field and did not produce anything important other than smart gambling machines, which rather served the "cultural devaluation of chess" [34]. AI research was considered a science in its ivory tower or an an exotic fruit in informatics and in some countries (e.g. the United Kingdom) was declared not to deserve promotion. But another basic theory-model led it out of this dead-end. As the limitation of the computer to algorithmic structures was surpassed by the introduction of heuristic programs and inferencing strategies in the first phase, and the idea of automatic information processing, which had previously been restricted to purely syntactic structures, was extended by basic conceptions of a theory of semantic information processing, one tried to overcome the stagnation in the second phase. But this time it was the pragmatic aspect standing in the foreground of basic research. The consequences of the GPS experiences were taken and the unfeasible idea of the General Problem Solver was restricted to the solution of limited, but very restricted, concrete problems. For one realised, and not only due to the external criticism of AI opponents [35], that the problem-solving capability of man is based on his knowledge or experience. A general problem solver would have had to integrate the whole of everyday knowledge, which is basic for all special areas, of today's normal educated adult.

Especially the aim of understanding nothing less than all the fundamental properties of natural human intelligence proved to be unachievable because of its complexity. Highly specialized knowledge of the professions, technology, administration or science proved, however, to be the field which can best be understood and treated, exactly because it is already regulated specified and limited to a small area.

4 Knowledge Technology: The Future Integration and Co-Evolution of Human and Technical Information Processing Systems

The evolution of techIPSs culminated in so-called "knowledge-based systems". At present the most successful kind of this new species of techIPS is the so-called "expert system". According to the general definition by Feigenbaum [36] an expert system is a computer program with so much "knowledge" that it can start to be active at the level of the human expert. This system thus needs a knowledge base, which must be as large as possible, and refers to a limited subject field, the domain of the expert, and an "inference mechanism" operating on this knowledge base by searching for relevant information and drawing logical conclusions from it. But the expression "expert system" implies different claims. As in the case of many terms from AI research, the claims linked to them are bigger than the actual state-of-the-art research can justify. If these claims are reduced to a sensible feasible measure, expert systems are primarily systems designed to relieve the human expert himself from routine decisions and only secondly is an expert system a techIPS designed to replace the human expert, because he is hardly ever available or too expensive. But in this case, too, only the routine knowledge of expert can be replaced. Thus there is no competition between the humIPS and the techIPS but only between variations of IPSs at the same level.

But this principle of competition must be taken in the sense of the theory of biological evolution based on the vertical transformation, i.e. alteration of the organization in time, and on the horizontal synchronous diversification, since evolution always means an increase in the diversity of species. Thus competition at the same level of IPS does not imply replacement or extermination of a certain species by another, but rather the mutual "wedging" of species in certain "ecological niches", with the latter not being the location or the "address" of an organism, but its "professon", i.e. the task it has to fulfil in an ecological system so that this system remains in a harmoniously coordinated equilibrium. This enables us, also within the framework of a general model of the evolution of information processing systems, to give an answer to the question: What is a human in comparison to an "intelligent machine"? This answer is fairly complex, because we need a characterisation of the "intelligent machine" adequate to the current state-of-the-art. In this first phase we have to assume that humans themselves have given rise to the artificial evolution of techIPSs oriented towards products and have produced a large variety of species of such systems in the sense of the principle of diversification of evolution. These different species of IPS result in different tasks previously reserved for humIPSs: there are, for instance, calculating machines, data banks, text processing systems, translation machines, industrial robots, monitoring systems, process control systems, expert systems, etc.

Although there are also combinations of such systems, the future evolution of techIPS will not result in the creation of superior intelligence integrating all these capabilities as techIPS and thus designed to outdo the humIPS. Rather, the computer must remain a tool of man restricted to certain tasks. But, as in intelligent machines, it has partial autonomy by solving special problems automatically, i.e. autonomously, because of its own structural information which it contains as software or programs, with the difference from humIPSs that software and hardware can be separated. The user only provides data. Here the analogy to the classical machine, e.g. the heat engine, which has to be operated by the user, is the strongest; fluctuating data, "fed" into the intelligent machine, are the "fuel" necessary for the powerful performance of information processing. But at the same time the computer as a techIPS is also a medium of the strictly organized information flow (Petri) between humIPSs. This means that the techIPS, the computer, does not separate human individuals, but rather connects them.

This sociotechnical integration of mankind is growing with the increasing approximation of the techIPS to the humIPS. This approximation is the future evolution of the techIPS, which will be in fact a co-evolution of humIPS and techIPS. Co-evolution is defined here, as is usual in biology, as a process of mutual stimulation of two different species of evolutive systems. In this case this process of stimulation is taking place because the humIPS not only effects the artificial evolution of the techIPS, but also the other way round, is driven by the evolution of the techIPS. For always when people have succeeded in implementing the performance of a humIPS in a techIPS mechanically, this performance has not been considered human anymore. When AI gets closer, natural intelligence is thus "evading upwards", i.e. into the direction of a higher, more complex and more creative performance.

But this approximation of techIPS (artificial intelligence) to the humIPS (natural intelligence) has to be done at two different levels. One level is that of symbol processing. The approximation takes place, as can be seen in the example of expert systems, by increasingly complicated program architectures for "knowledge representation". For the so-called "knowledge-based systems" have to operate at several levels of representation, because as symbol processing syntactic machines they cannot contain human "knowledge" in the direct sense of semantico-pragmatic information. Thus knowledge must be represented in a direct way by a syntactic representation formalism of its own, which can then be used to "break down" complex higher units of knowledge to a level adequate for automatic processing. The effort of representation is concomitantly increasing with the performance of this system, which in turn needs a transformation formalism of its own for the transfer from one representation level to the other. The terms "knowledge representation" and "knowledge processing" have thus to be understood as heuristic and productive metaphors, as "powerful concepts", as Marvin Minsky called them.

The symbol processing machine is, in spite of its high level of complexity, not an "intelligent" machine, if the formula is true that intelligence means problem solving capability and that problem solving is based on "knowledge". For the symbol processing machine does not "know" anything. In order to be able to speak of the "knowledge" of a machine at least analogically, another level of approximation of the techIPS to the humIPS is necessary. This level is the subsymbolic level of information processing. In the humIPS it is also the primary and most complex level, since all symbolized, linguistically represented human knowledge is based, as has been shown in the neuroepistemological part of this study, on several subsymbolic modal and intermodal performance levels of the central nervous system, which are the basis for the supramodal performance level of symbol processing.

In a concrete sense this means that the actual inherently extremely complex unit of knowledge, the concept, which can be represented in a symbolic linguistic way as a term, has its origin in neural mechanisms of subsymbolic information processing and thus in sensorimotor action. As evolutionary epistemology can show, the origin of concept formation lies in the conditioned response. In the prelinguistic subsymbolic area the concept is nothing other than a reaction schema to a certain category of events in the environment. The variety of conceptual units of knowledge constituting the experience of an organismic IPS is due to its phylogenetic and ontogenetic development in an environment very rich in stimuli.

With this we already have a hint of the construction of techIPSs as knowledge-based systems; the "knowledge of a machine" analogous to that of the humIPS can only be reached in its proper sense via the subsymbolic level, i.e. via a bottom-up aproach, in contrast to the top-down approach of symbol processing, as is currently done with connectionist neural networks [37–39]. These concepts must be seen as reactions of neural networks, activated by an adequate learning environment established by the system designer. But a principle fundamental for neuroepistemology holds for both levels, the subsymbolic and the symbolic: all functionally describable human knowledge processing is separated from the neural mechanisms on which they are based, and the knowledge embodied in the machine must be "locked", at least for the user of a technical knowledge processing system. This principle, anticipated in the evolution of neural information processing systems, is at the same time the key to improving the use of the intelligent computer as a knowledge-based system. The user must be able to ignore the internal mechanism of a techIPS with the help of adequate programms. He interacts with the computer system only via the user interface, just as human communication does not occur directly from one brain to the other, but human partners with faces, eyes and ears communicate with each other. The human-computer interaction, i.e. between unequal partners, thus occurs only indirectly via adequate user interfaces [40].

Thus the aim of sociotechnical integration cannot be to absorb into a global system the cognitive and emotional autonomy of humans which they have acquired in evolution, but rather, on the contrary, even to enlarge this autonomy by using techIPSs. But this results in a stronger obligation for self-responsibility: they can only become more human through new technological opportunities or they have to perish. But it must be added that in the evolution of organisms the extermination of a species is not at all unusual, which also holds for *homo sapiens*.

References

1. Gardner H (1985) The Mind's New Science. A History of the Cognitive Revolution. Basic Books, New York
2. Oeser E (1987) Psychozoikum, Evolution und Mechanismus der menschlichen Erkenntnisfähigkeit. Parey, Berlin, Hamburg
3. Oeser E, Seitelberger F (1988) Gehirn, Bewußtsein und Erkenntnis. Wissenschaftliche Buchgesellschaft, Darmstadt
4. Churchland PS (1986) Neurophilosophy. Toward a Unified Science of the Mind-Brain. MIT Press, Cambridge, Mass, London
5. Oeser E (1985) Kants Philosophie des Gehirn. Ein Beitrag zur historischen Begründung der Neuroepistemologie. In: Schmied-Kowarzik W (ed): Objektivationen des Geistigen. Reimer, Berlin
6. Hopfield JJ (1988) Neural Networks and Physical Systems with Emergent Collective Computational Abilities. Proceeding of the National Academy of Science, USA **79**, 2554–2558
7. Rumelhart DE, McClelland JL (1986) Parallel Distributed Processing. Explorations in the Microstructure of Cognition, Vol 1: Foundations, MIT Press, Cambridge, Mass
8. McClelland JL, Rumelhart DE (1986) Parallel Distributed Processing. Explorations in the Microstructure of Cognition, Vol 2: Psychological and Biological Models. MIT Press, Cambridge, Mass
9. Eckmiller R, Malsburg Cvd (eds) (1988) Neural Computers. Springer, Berlin, Heidelberg, New York, Tokyo
10. Weizenbaum J (1976) Computer Power and Human Reason. From Judgement to Calculation. W.H. Freeman and Company, San Francisco
11. de Candolle A (1911) Zur Geschichte der Wissenschaft und der Gelehrten seit zwei Jahrhunderten. Akademische Verlagsgesellschaft Leipzig, p 115
12. Oeser E (1987) Psychozoikum. Evolution und Mechanismus der menschlichen Erkenntnisfähigkeit. Parey, Berlin, Hamburg
13. Meynert T (1892) Sammlung von populär-wissenschaftlichen Vorträgen über den Bau und die Leistungen des Gehirns. Braumüller W, Wien, Leipzig
14. Sjölander S (1984) Angeborene Welt erworbene Welt. In: Nichts ist schon dagewesen. Konrad Lorenz, seine Lehre und ihre Folgen. Piper, München, p 12
15. MacKay DM (1950) Quantal Aspects of Scientific Information. Philosophical Magazine **41**, 101
16. Lorenz K (1977) Behind the Mirror. Harcourt Brace Jovanovich, New York
17. Weizsäcker Ev (1974) Offene Systeme I. Klett, Stuttgart
18. Küppers B-O (1986) Der Ursprung biologischer Information. Piper, München, Zürich
19. Lorenz K (1977) Behind·the Mirror. Harcourt Brace Jovanovich, New York
20. Braitenberg V (1984) Vehicles. Experiments in Synthetic Psychology. MIT Press, Cambridge, Mass

21. Oeser E, Seitelberger F (1988) Gehirn, Bewußtsein und Erkenntnis. Wissenschaftliche Buchgesellschaft, Darmstadt
22. Gibson JJ (1967) Neue Gründe für den Realismus. Synthese **17**, 162–172
23. Leibniz GW (1979) Monadologie. Reclam, Stuttgart §64
24. Popper K, Eccles J (1981) The Self and its Brain. Springer, Berlin, Heidelberg, New York
25. Oeser E (1988) Das Abenteuer der kollektiven Vernunft, Evolution und Involution der Wissenschaft. Parey, Berlin, Hamburg, p 134
26. MacKay DM (1950) Quantal Aspects of Scientific Information. Philosophical Magazine **41**, 298–311
27. von Neumann J (1956) Probabilistic Logic and the Synthesis of Reliable Organisms from Unreliable Components. In: Shannon CE, McCarthy J (eds), Automata Studies, Princeton
28. Leibniz GW: Philosophische Werke: Hauptschriften zur Grundlegung der Philosophie, Cassirer E (ed) Meiner, Leipzig, Vol 1, p 35
29. Weizenbaum J (1976) Computer Power and Human Reason. From Judgement to Calculation. W.H. Freeman and Company, San Francisco
30. Zemanek H: Information und Ingenieurwissenschaft. In: Folberth OG, Hackl C (eds), OP CIT
31. Simon HA, Newell A (1958) Heuristic Problem Solving: The Next Advance in Operations Research. Operations Research **6**, p 6
32. Feigenbaum E, Feldman J (eds) (1963) Computers and Thoughts. McGraw-Hill, New York
33. Minsky ML (ed) (1972) Semantic Information Processing. MIT Press, Cambridge, Mass
34. Petri CA: Zur "Vermenschlichung" des Computers. In: Der GMD-Spiegel 3/4-83, p 42ff
35. Dreyfus HL (1979) What Computers Can't do. The Limits of Artificial Intelligence. Harper & Row, New York
36. Feigenbaum E, McCorduck P (1983) The Fifth Generation. Addison-Wesley, Reading, MA
37. Oeser E: Terminology Science and Knowledge Theory as a Prerequisite of Knowledge Engineering. In: Second International Congress on Terminology and Knowledge Engineering, Trier, 2–4 Oct 1990
38. Oeser E: Begriffstheoretische Grundlagen der Wissenstechnologie. In: Konferenz über Terminologieforschung und Wissenstheorie, Graz, 26–28 Sept 1990
39. Budin G, Peschl MF: Begriffs- und Wissensmodellierung aus konnektionistischer Sicht. In: Konferenz über Terminologieforschung und Wissenstheorie, Graz, 26–28 Sept 1990
40. Herrmann T (1986) Zur Gestaltung der Mensch-Computer-Interaktion: Systemerklärung als kommunikatives Problem. Neimeyer, Tübingen

Epilogue and Bibliography

Epilogue

Our approach towards understanding information processing and its evolution in nature and society can only be considered a *very first* attempt to tackle a sophisticated set of scientific problems. Although it is possible to get some insights – as shown in Chap. I, Basic Concept, and the various contributions to this book – we are far from a general theory of information processing and its evolution. Such a theory, however, seems to be a crucial prerequisite to an understanding of the evolution of the broad variety of real information processing systems in more depth.

From our discussions in the project "Evolution of Information Processing" we have developed some first insights into requirements which have to be fulfilled by such a "general theory". We have also collected some ideas about the realisation of such a theory. Remarks on both aspects are made here as a challenge to the scientific community. Hopefully, such a "General Theory of Information Processing" for real systems might emerge in the near future.

However, it might also be proved that such a general theory is impossible because of serious difficulties in defining information and its processing in an interdisciplinary way which allows for dealing adequately with the broad variety of information processing systems at the physical, the genetic, the neural, the social, and the technical as well as the sociotechnical level.

Requirements for a General Theory. Here we list crucial features of the theory to be established.

(1) Information and information processing systems (IPSs) must be dealt with in one integrated approach. It makes no sense at all to design a theory of information on its own – as has been done in the literature at various levels! Automata theory, however, is not sufficient since the essential aspects of the semantics and pragmatics of information are not considered within this approach.

(2) In the theory information has to be dealt with *qualitatively* and *quantitatively* at the *syntactic*, the *semantic* and the *pragmatic* levels. Information content as well as the "meaning" of information have to be considered.

(3) An IPS has to be defined as a structure which always exists in an informational environment. Also within this environment information has to be considered at the syntactic, the semantic and the pragmatic levels. IPSs must be modelled in such a way that their (selective) interactions with all these aspects of information can be understood.

(4) The theory to be developed must deal explicitly with *internal information*, responsible for the "basic structures" and self-organization of an IPS, as well as with *external information*, a potential source of information for a given system. In particular, the measuring of internal and external information has to be defined in a way which allows for experiments or empirical approaches.

(5) A definition of information, internal, external and as a whole, has to consider the *declarative* as well as the *procedural* aspect of information. In addition, mechanisms must be implemented in the theory which allow the time dependence of informational structures within an IPS to be dealt with, since no IPS is static.

(6) Since all IPSs on earth are dynamic systems, the theory must allow for *time dependency* of all structural components of the IPS. (This will automatically cause serious problems in defining a given status of an IPS.)

(7) The theory of an IPS must allow functional *concurrency* of receiving, processing, storing, and sending of information to be modelled.

(8) No IPS exists on its own. Therefore, the theory must allow for an *aggregated representation of all IPSs* "surrounding" a given IPS. Steady exchange of external information must be modelled appropriately.

(9) It is necessary to define entities of internal information at the syntactic, the semantic and the pragmatic levels in a *measurable* form. This is a rigid requirement because of the difficulty of measuring internal information within an IPS; how can this be established? What is the proper "probe system"?

(10) The theory of information processing must work with entities of external information which can be measured with "*probe systems*" which "behave" exactly like the real system under consideration. (This is a serious problem because of the dynamics of all IPSs.)

(11) In an ideal form a "Grand Unified Theory of Information Processing" should work for all levels of the IPS evolution. From a pragmatic point of view it seems acceptable to start off with "*sub-theories*", keeping in mind the necessity for integrating them later on.

(12) The theory must be understandable and usable by *various disciplines*. This means it is necessary to build a basic structure which fits into traditional experimental and theoretical approaches. (This is particularly of importance for physical information processing since in physics an elaborate and fruitful quantitative theory has been established.)

(13) IPSs are structures made up of (stacks of) different IPSs. Therefore, the theory must allow for structures which can be "*stacked together*", making possible the modelling of complex natural IPSs. In particular, it should be possible to model interactions between various levels of IPSs.

(14) The theory to be built must allow for results from real systems which can be proved *correct or false*. Thus, one should not look for a purely pheno-menological approach – which is the case for most proposals published so

far. In particular, proper experimental and empirical data should be derivable.

(15) The theory should allow for modelling *deterministic* chaotic behaviour, since many examples of real IPSs show such a property.

(16) Although IPSs are basically abstract systems, they are based on material structures (mainly physIPSs). It is essential for a general theory of information processing to shed at least some light on the dualism of a matter/ energy or a immaterial (idealistic) interpretation of nature and society. Which paradigm will be more fruitful in the future: matter/energy or information processing?

Methods. It is quite difficult to find a proper methodological framework for the construction of a general information theory because of the heterogeneous and complex requirements given above. In the following some options are proposed.

(A) *Expansion of automata theory* might allow for the implementation of an information theory which can also be used for *natural* IPSs, which are not automata in the conventional sense. However, it seems quite difficult to integrate semantic and pragmatic aspects into such an orthodox theoretical approach which is basically concerned with syntactic structures and their processing. Whereas computers are well understood artificial objects, the information theory under consideration must allow for analysis and description of natural systems which are much more complex and difficult to study experimentally or empirically.

(B) In general, a *mathematical formulation* of an information theory seems to be the ultimate goal. However, mathematics does not deal explicitly with systems structured into complicated subsystems (e.g. physIPSs in a genIPS). Approaches used in dynamic systems research might form a base; however, at present they do not allow real natural IPSs (e.g. the genetic apparatus of a cell) to be modelled. In physical information processing particularly, a mathematical treatment of the IPS is unavoidable since statements about physical systems must allow for numeric results with a high level of precision. This is standard in "orthodox" physics.

(C) A general theory of information processing might be established in terms of *verbal formulations* first (e.g. by extending Sect. 2 of Chap. I which would have to be greatly refined). However, essential, quantitative questions cannot be solved in this way. Thus, at least in given areas, quantitative formulations must be added onto.

(D) It might be fruitful to start with *special theoretical approaches* at given mega-levels of the IPS evolution, trying "unification" of such theories later on. This, however, asks for a general understanding of information processing as a basic prerequisite for a proper set of useful sub-theories.

(E) There may be the option of establishing an information theory in the form of *a quite sophisticated computer program*. IPSs and information have to be declared as data types, and procedures must be given to explain the actions of such IPSs. (This approach has been used in cognitive psychology with, however, quite limited success because of the complexity of the brain and human behaviour.)

(F) It might be helpful to start with various (simple) *models for special IPSs* to get a first insight into modelling a theory. This, however, can only be considered a first attempt, since we need a general theory which can only be derived from special insights if they are dealing with general problems!

As an early beginning for a general theory for natural as well as technical information processing systems a combination of the above approaches might also be helpful. However, in all cases, a conceptual framework (as in the Basic Concept) is crucial for future research. Otherwise there is little hope of gaining a deeper understanding of the structure of information processing systems and their evolution.

Klaus Haefner   Bremen, February 1991

Bibliography of Crucial Literature*

A General Discussion of the Concept of Information and its Processing 347

B Information and Information Processing at the Physical Level 351

C Information and Information Processing at the Biological Levels 352

D Information and Information Processing at the Social Level 354

E Information and Information Processing at the Technical Level 355

F Information and Information Processing at the Socio-Technical Level 356

A General Discussion of the Concept of Information and Its Processing

1. Ackermann P, Eisenberg W, Hedwig H, Kanneugießer K (1989) Erfahrung des Denkens – Wahrnehmung des Ganzen. Berlin
2. Armand A (1972) Informationssysteme der toten Natur. Bild d Wiss, **2**, 145–153
3. Ashby WR (1974) Einführung in die Kybernetik. Frankfurt/M
4. Bar-Hillel Y, Carnap R (1953) Semantic Information. British Journal for the Philosophy of Science, **4**, 147–157
5. Bertalanffy Z (1969) General System Theory – Foundation, Development, Application. New York
6. Bleeken S (1990) Welches sind die existentiellen Grundlagen lebender Systeme? Nat **77**, 277–282
7. Bonner JT (1988) The Evolution of Complexity – by Means of Natural Selection. Princeton
8. Boulding KE (1985) The World as a Total System. Beverly Hills
9. Brillouin L (1956) Science and Information Theory. New York
10. Brillouin L (1964) Scientific Uncertainty and Information. New York
11. Brooks DR, Lablond PH, Cumming DD (1984) Information and Entropy in a Simple Evolution Model. J Theor Biol, **109**, 77–93
12. Carvallo ME (1986) Natural Systems According to Modern System Science: Three Dualities. In: Trappl R: Cybernetics and Systems, Dodrecht
13. Chaitin GJ (1987) Algorithmic Information Theory. Cambridge
14. Chaitin GJ (1982) Gödels Theorem and Information. Int Journal of Theoretical Physics, Vol 21, No 12, 941–954
15. Cherry C (1956) Information Theory. London
16. Chintschin AJ, Faddejew DK, Kolmogoroff AN (1967) Arbeiten zur Informationstheorie I. VEB

* Used in the Project Evolution of Information Processing; (See also detailed literature in the articles of the contributing scientists.)

17. Churchland PM, Smith Churchland, Patricia (1990) Ist eine denkende Maschine moglich? Spektr d Wiss, **3**, 47–54
18. Cimutta J (1989) Philosophisches Nachdenken über die Struktur- und Informationsproblematik. In: Ackermann P, Erfahrung des Denkens – Wahrnehmung des Ganzen, Berlin, 110–115
19. Colodny RG (1983) Mind and Cosmos – Essays in Contemporary Science and Philosophy. New York
20. Crutchfield JP, Farmer JD, Packard NH, Shaw RS: Chaos Sepktr d Wiss Feb 1987, S 78–91
21. Csányi V (1987) The Replicative Evolutionary Model of Animal and Human Minds. World Futures, Vol 23, 161–202
22. Csányi V (1989) Evolutionary Systems and Society – A General Theory. Durham
23. Csányi V (1981) General Theory of Evolution. General Systems, Vol 16, 73–92
24. Csányi V (1987) The Replicative Model of Evolution: A General Theory. World Futures, Vol 23, 31–65
25. Ditfurth HH (1969) Informationen über Information (Probleme der Kybernetik). Hamburg
26. Dress AH, Henrichs HH, Küppers GH (1986) Selbstorganisation – Die Entstehung von Ordnung in Natur und Gesellschaft. München
27. Dyke C (1988) The Evolutionary Dynamics of Complex Systems – (A Study in Biosocial Complexity). New York
28. D'Espagnat B (1989) Reality and the Physicist – Knowledge, duration and the quantum world. Cambridge
29. Ebeling W, Ulbricht H (1986) Selforganisation by Nonlinear Irreversible Processes. Proceedings of the Third International Conference Kühlungsborn, GDR, March 18–22, 1985. Berlin, Heidelberg
30. Eigen M, Winkler R (1975) Das Spiel – Naturgesetze steuern den Zufall. München
31. Eigen M, Schuster P (1977) The Hypercycle – A Principle of Natural Self-Organisation – Part A: Emergence of the Hypercycle. Nat **64**, 541–565
32. Eigen M, Schuster P (1978) The Hypercycle – A Principle of Natural Self-Organization – Part B: The Abstract Hypercycle. Nat **65**, 7–41
33. Eigen M, Schuster P (1978) The Hypercycle – A Principle of Natural Self-Organization – Part C. The Realistic Hypercycle. Nat **65**, 341–369
34. Eigen M (1982) Ursprung und Evolution des Lebens auf molekularer Ebene. In: Haken H: Evolution of Order and Chaos, Berlin
35. Eigen M (1987) Stufen zum Leben – Die Frühe Evolution im Visier d Molekularbiol. München
36. Eldredge N (1985) Unfinished Synthesis – Biological Hierarchies and Modern Evolutionary Thought. Oxford
37. Fliedner D (1989) Soziale Systeme im Informations- und Energiefluß. grkg/Humankybernetik, Bd 30, H1, 27–37
38. Foerster H (1980) Observing Systems. Intersystems Publications, California, USA
39. Folberth OGH, Hackl CH (1986) Der Informationsbegriff in Technik und Wissenschaft. München
40. Frautschi S (1988) Entropy in an Expanding Universe. In: Weber BH et al. (eds) Entropy, Information, and Evolution, Cambridge, MA, S 11–22
41. Freytag W (1989) Die Information aus dem ewig Existierenden ist das geistige Prinzip der Evolution. Frankfurt/M
42. Ganzhorn K (1986) Information, Strukturen u Ordnungsprinzipien. In: Folberth OG, u Hackl C (Hrsg.), Der Informationsbegriff in Technik und Wissenschaft, München, 105–125
43. Geiger G (1990) Evolutionary Instability – Logical and Material Aspects of a Unified Theory of Biological Evolution. Springer Verlag; Berlin, Heidelberg
44. Gelernter D (1989) Informationsmanagement im Wandel. Spektr d Wiss, Okt, 64–72
45. Gleick J (1988) Chaos – die Ordnung des Universums – Vorstoß in Grenzbereiche der modernen Physik. Droemer Knaur, München
46. Goser K (1989) Das Gravitationsgesetz und das Coulomb Gesetz aus Sicht der Informationstheorie. Frequenz, **43**, **6**, 156–160
47. Grassberger P: Cellular Automata, Wuppertaler Universität WU B90, April, 1990
48. Großmann S (1989) Selbstähnlichkeit: Das Strukturgesetz in und vor dem Chaos. Phys Bl 45, Nr **6**, 172–180
49. Günther G (1969) Bewußtsein als Informationsraffer. Grundlagenst. aus Kyb u. Geisteswiss, 10/1, 1–6
50. Guttinger W, Dangelmayr G (1987) The Physics of Structure Formation Theorie and Simulation. Springer Verlag; Berlin, Heidelberg

51. Haken H (1987) Die Selbstorganisation der Information in biologischen Systemen aus der Sicht der Synergetik. In: Küppers, BO: Ordnung aus dem Chaos, München
52. Haken H (1982) Synergetik. Berlin
53. Haken H (1988) Information and Selforganization – A Macroscopic Approach to Complex Systems. Berlin
54. Haken H (1986) How can we implant Semantics into Information Theory. In: Folberth OG, u. Hackl C (Hrsg), Der Informationsbegriff in Technik u Wissenschaft, München, 127–137
55. Haken H, Haken-Krell, Maria (1989) Entstehung von biologischer Information u. Ordnung. Darmstadt
56. Ho Mae-W, Saunders PT (1984) Beyond Neo-Darwinism – An Introduction to the New Evolutionary Paradigm. Academic Press, Inc (London) LTD
57. Jantsch E (1979) Die Selbstorganisation des Universums – vom Urknall zum menschlichen Geist. München
58. Jaros GG, Cloete A (1987) Biomatrix: The Web of Life. World Futures, vol 23, 203–224
59. Jumarie G (1986) Total Entropy. A Unified Approach to Discrete Entropy and Continuous Entropy. In: Cybernetics and Systems **86**, Dodrecht
60. Jumarie G (1990) Relative Information – Theories and Applications. Berlin
61. Kampis G (1990) Self-Modifying Systems in Biology and Cognitive Science. Oxford
62. Kampis G (1987) Information, Mind and Machines. In: Rose J (ed) Cybernetics and Systems: The Way Ahead, London
63. Kampis G (1986) Biological Information as a System Description. In: Trappl R: Cybernetics and Systems, Dodrecht
64. Kampis G (1988) On Information and Autonomy. Revue Internationale de Systemique, vol 2, No 3, 261–269
65. Kampis G (1988) Information, Computation and Complexity. In: Carvallo ME (ed) Nature, Cognition and Systems I, Dordrecht
66. Kaspar R (1987) Materialismus, Idealismus u. Evol. Erkenntnistheorie. In: Riedl R, u Wuketits: Evol Erkenntnistheorie, Berlin
67. Klement HW (1986) Der Informationsgehalt des Atoms. Philosophia naturalis, Band 23, Heft **2**, 216–222
68. Klement HW (1983) Energie, Materie und Information. Angew Informatik **4**, 177–180
69. Klement HW (1973) Denkmodelle zum Thema Abstraktion. Grundlagenstudien aus Kybernetik u Geisteswiss **10**, S 129–132
70. Klir GJ, Folger TA (1988) Fuzzy Sets Uncertainty, and Information. New York
71. Kolmogorov AN (1968) Three Approaches to the Quantitative Definition of Information. Int Journal of Computer Mathematics, Vol 2, 157–168
72. Krohn W et al. (eds) (1990) Selforganization. Dodrecht
73. Krohn W, Küppers, GU, Paslack R (1988) Selbstorganisation – Zur Genese u. Entwicklung einer Wiss Revolution. In: Schmidt SJ (Hrsg.): Der Diskurs des radikalen Konstruktivismus, Frankfurt/M, 441–465
74. Küppers BO (1986) Ursprung biologischer Information (Zur Naturphilosophie der Lebensentstehung). München
75. Küppers BO (1986) Molekulare Selbstorganisation und die Entstehung biologischer Information. In: Folberth OG, u Hackl C (Hrsg.), Der Informationsbegriff in Technik und Naturwissenschaft, München, 181–203
76. Laszlo E (1988) Global denken – die Neu-Gestaltung der vernetzten Welt. Horizonte Verlag, Darmstadt
77. Layzer D (1988) Cosmogenesis – Growth of Order in the Universe. In: Weber BH, ua (ed): Entropy, Information and Evolution, Cambridge, MA, S 23–40
78. Luhmann N (1974) Soziale Systeme – Grundriß einer allgemeinen Theorie. Frankfurt/Main
79. Lumsden CJ, Wilson EO (1981) Genes, Mind and Culture – The Coevolutionary Process. Cambridge
80. Mackay DM (1969) Information, Mechanism and Meaning. Cambridge, MA
81. Mayr E (1986) Evolution – Die Entwicklung von den ersten Lebensspuren bis zum Menschen. Spektrum d Wiss, Heidelberg
82. Meyer-Eppler W (1969) Grundlagen und Anwendungen der Informationstheorie Bd. 1 Berlin
83. Miller JG (1978) Living Systems. New York
84. Nicolis JS (1987) Chaotic Dynamics in Biological Information Processing: A Heuristic Outline. IL Nuovo Cimento, Vol 9D, No 11, 1359–1388

85. Nicolis G, Prigogine I (1987) Die Erforschung des Komplexen – Auf dem Weg zu einem neuen Verständnis der Natur München
86. Nicolis JS, Tsuda I (1985) Chaotic Dynamics of Information Processing: The "Magic Number Seven Plus-Minus Two" Revisited. Bulletin of Mathematical Biology, Vol 47, No 3, 343–365
87. Nicolis JS (1987) Chaotic Dynamics of Logical Paradoxes. In: Bothe, Ebeling, Kurzhanski (eds), Dynamical Systems and Environmental Models, Akademie Verlag, 105–113
88. Oeser E (1987) Psychozoikum – Evolution und Mechanismus der menschlichen Erkenntnisfäigkeit. Berlin
89. Oeser E (1976) Wissenschaft u Information; Bd I, Wissenschaftstheorie u emp Wissenschaftsforschung. München
90. Oeser E (1985) Informationsverdichtung als universelles Ökonomieprinzip der Evolution. In: Ott JA, Wagner GP, u Wuketits FM (Hrsg): Evolution, Ordnung und Erkenntnis, Berlin, Paul Parey Verlag
91. Oeser E, Seitelberger F (1988) Gehirn, Bewußtsein und Erkenntnis. Darmstadt
92. Ott JA, Wagner GP, Wuketits FM (1985) Evolution Ordnung u Erkenntnis. Berlin
93. Ott JA (1985) Ökologie u Evolution. In: Ott JA, ua Evolution, Ordnung u Erkenntnis, Berlin
94. Pantzar M: The Replicative Model of the Evolution of Business Organization – A Draft –. Vortrag Bologna, unv Manuskript
95. Penrose R (1989) The Emperors New Mind – Concerning Computers, Minds and the Laws of Physics. Oxford
96. Plath PJ (1989) Optimal Structures in Heterogeneous Reaction Systems. Springer Verlag; Berlin, Heidelberg
97. Polanyi M (1968) Life Irreducible Structure. Science, Vol 160, 1308–1312
98. Prigogine I, Stengers, Isabelle (1981) Dialog mit der Natur. München
99. Reeves H (1987) Sources of Information and Free Energy in an Expanding Universe. In: Rujula A de: A Unified View of Macro- and Microcosmos, Singapore
100. Resnikoff HL (1989) The Illusion of Reality. New York
101. Roth G, Schwegler H (1981) Self-organizing Systems – An interdisciplinary Approach. Frankfurt
102. Scarrott GG (1989) The Nature of Information. The Computer Journal, Vol 32, No 3, 262–266
103. Schrödinger E (1987) Was ist Leben? – Die lebende Zelle mit den Augen des Physikers betrachtet. München (engl. Org. 1944)
104. Schultze E (1969) Einführung in die mathematischen Grundlagen der Informationstheorie. Berlin
105. Searle JR (1990) Ist der menschliche Geist ein Computerprogramm? Spektr d Wiss, **3**, 40–47
106. Shannon CE, Weaver W (1976) Mathematische Grundlagen der Informationstheorie. München
107. Shannon CE, Weaver W (1976) Mathematische Grundlagen der Informationstheorie. München (engl Org fassung 1949)
108. Shaw R (1981) Modeling Chaotic Systems. In: Haken H, Chaos and Order in Nature, Berlin
109. Sheldrake R (1988) Das Gedächtnis der Natur – Das Geheimnis der Formen in der Natur; Bern, München, Wien
110. Stein DL (1989) Lectures in the Sciences of Complexity. Redwood City
111. Stein DL: A Model for the Origin of Biological Information. Int J of Quantum Chemistry: Quantum Biology Sym **11**, 73–86
112. Svilar MH, Zahler P (Hrsg) (1984) Selbstorganisation der Materie. Bern
113. Thom R (1975) Structural Stability and Morphogenesis – An Outline of a General Theory of Models. London
114. Topsoe F (1973) Informationstheorie – Eine Einführung. Kopenhagen
115. Unsold AC (1981) Evolution kosmischer, biologischer und geistiger Strukturen. Stuttgart
116. Weber BH, Depew DJ, Smith JD (1988) Entropy, Information and Evolution – New perspectives on physical and biological evolution. Cambridge, MA
117. Weizsäcker CF (1986) Aufbau der Physik. Darmstadt
118. Wicken JS (1988) Thermodynamics, Evolution, and Emergence: Ingredients for a New Synthesis. In: Weber BH ua (eds): Entropy, Information, and Evolution, Cambridge, MA, S 139–169
119. Wicken JS (1983) Entropy, Information, and Nonequilibrium Evolution. Syst. Zool. **32(4)**, 438–443
120. Wiener N (1988) Kybernetik – Regelung und Nachrichtenübertragung in Lebewesen und Maschinen. Hamburg

121. Wiley EO (1988) Entropy and Evolution. In: Weber BH ua (ed): Entropy, Information, and Evolution, Cambridge, MA, 173–188
122. Wohlmuth PC, Artigiani R: A Symposium on the Evolution of Consciousness. World Futures, Vol 25, 197–282
123. Wolpert SA, Wolpert JF (1986) Economics of Information. New York

B Information and Information Processing at the Physical Level

1. Ackermann P, Eisenberg W, Hedwig H, Kannegießer K (1989) Erfahrung des Denkens – Wahrnehmung des Ganzen. Berlin
2. Brillouin L (1964) Scientific Uncertainty, and Information. New York
3. Brillouin L (1956) Science and Information Theory. New York
4. Buchler JR, Perdang JM, Spiegel EA (1985) Chaos in Astrophysics. Dodrecht
5. Colodny RG (1983) Mind and Cosmos – Essays in Contemporary Science and Philosophy. New York
6. Dittrich T, Graham R (1989) Quantenchaos: Komplexe Dynamik im Grenzbereich zwischen mikroskopischer u makroskopischer Physik. Nat Wi 76, 401–409
7. D'Espagnat B (1989) Reality and the Physicist – Knowledge, duration and the quantum world. Cambridge
8. Ebeling W, Ulbricht H (1986) Selforganisation by Nonlinear Irreversible Processes – Proceedings of the Third International Conference Kühlungsborn, GDR, March 18–22, 1985. Berlin Heidelberg
9. Folberth OGH, Hackl CH (1986) Der Informationsbegriff in Technik und Wissenschaft. München
10. Frautschi S (1988) Entropy in an Expanding Universe. In: Weber BH ua (eds): Entropy, Information, and Evolution, Cambridge, MA, S 11–22
11. Friederici AD, Furrer R (1987) Wahrnehmung und Vorstellung des Raumes. Spetr d Wiss, Feb, 38–39
12. Ganzhorn K (1986) Information, Strukturen u Ordnungsprinzipien. In: Folberth OG. u Hackl C (Hrsg): Der Informationsbegriff in Technik und Wissenschaft, München, 105–125
13. Geyer B (1989) Reflexion zum Problem der Vereinheitlichung der Physik. In: Ackermann P, ua: Erfahrung des Denkens – Wahrnehmung des Ganzen. Berlin, 177–185
14. Gleick J (1988) Chaos – die Ordnung des Universums – Vorstoß in Grenzbereiche der modernen Physik. Droemer Knaur, München
15. Großmann S (1989) Selbstähnlichkeit: Das Strukturgesetz in und von dem Chaos. Phys Bl 45, Nr 6, 172–180
16. Guttinger W, Dangelmayr G (1987) The Physics of Structure Formation Theorie and Simulation. Springer Verlag; Berlin, Heidelberg
17. Haken H, Fuchs A (1988) Pattern Recognition and Associate Memory as Dynamical Processes in a Synergetic System. Biol Cybernetic 60, 17–22
18. Haken H (1987) Die Selbstorganisation der Information in biologischen Systemen aus der Sicht der Synergetik. In: Küppers BO, Ordnung aus dem Chaos, München
19. Haken, H, Haken-Krell, Maria (1989) Entstehung von biologischer Information u Ordnung, Darmstadt
20. Haken H (1988) Information and Selforganization – A Macroscopic Approach to Complex Systems. Berlin
21. Haken H (1986) How can we implant Semantics into Information Theory. In: Folberth OG, u Hackl C (Hrsg), Der Informationsbegriff in Technik u Wissenschaft, München, 127–137
22. Haken H (1981) Erfolgsgeheimnisse der Natur. Stuttgart
23. Haken H (1981) Chaos and Order in Nature. Berlin
24. Haken H, u Wunderling A (1991) Die Selbststrukturierung der Materie, Braunschweig
25. Haken H (1982) Synergetik. Berlin

26. Klement HW (1986) Der Informationsgehalt des Atoms. Philosophia naturalis, Band 23, Heft **2**, 216–222
27. Klement HW (1973) Notiz zu einem informationellen Deutungsschema für den Aufbau der realen Welt. Grundlagenstudien aus Kybernetik u Geisteswiss, 14/4, 133–136
28. Klement HW (1983) Energie, Materie und Information. Angew Informatik, **4**, 177–180
29. Information and Evolution, Cambridge, MA, 1988, S 23–40
30. Mackay DM (1969) Information, Mechanism and Meaning. Cambridge, MA
31. Reeves H (1987) Sources of Information and Free Energy in an Expanding Universe. Rujula A De: A Unified View of Macro- and Microcosmos, Singapore
32. Schrödinger E (1987) Was ist Leben? – Die lebende Zelle mit den Augen des Physikers betrachtet. München (engl Org 1944)
33. Shaw R (1981) Modelling Chaotic Systems. In: Haken H, Chaos and Order in Nature, Berlin
34. Stonier T (1990) Information and the Internal Structure of the Universe – An Exploration into Information Physics. London
35. Weber BH, Depew DJ, Smith JD (1988) Entropy, Information and Evolution (New perspectives on physical and biological evolution). Cambridge, MA
36. Weizsäcker CF (1986) Aufbau der Physik. Darmstadt
37. Wicken JS (1985) Thermodynamics and the Conceptual Structure of Evolutionary Theory. J Theor Biol **117**, 363–383

C Information and Information Processing at the Biological Level

1. Bendall DS (1983) Evolution from Molecules to Men. Cambridge
2. Bertalanffy Z (1969) General System Theory – Foundation, Development, Application, New York
3. Bleeken S (1990) Welches sind die existentiellen Grundlagen lebender Systeme? Nat **77**, 277–282
4. Bonner JT (1988) The Evolution of Complexity – By Means of Natural Selection. Princeton
5. Brooks DR, Leblond PH, Cumming DD (1984) Information and Entropy in a Simple Evolution Model. J Theor Biol, **109**, 77–93
6. Cherkin A, Flood DF (1984) Modulatory Interaction: An Intrinsic Attribute of Selforganization. In: Fox SW, Selforganization, NY, 143–167
7. Csányi Y (1987) The Replicative Model of Evolution: A General Theory. World Futures. Vol 23, 31–65
8. Csányi V (1985) Autogenesis: The Evolution of Self-Organizing Systems. In: Aubin B-P, ua (eds) Dynamics of Macrosystems, Berlin, 253–267
9. Csányi V, Kampis G (1988) A Systems Approach to the Creating Process. Int Federation for Systems Research, Oct–Nov, No. 20, 2–4
10. Csányi V, Kampis G (1988) Can we communicate with aliens? Bioastronomy – The Next Steps, 267–272
11. Csányi V (1987) The Replicative Evolutionary Model of Animal and Human Minds. World Futures, Vol 23, 161–202
12. Csányi V, Kampis C (1987) Modeling Society: Dynamical Replicative Systems. Cybernetics and Systems: An Int Journal, **18**, 233–249
13. Dawkins R (1976) The Selfish Gene, Oxford
14. Dyke C (1988) The Evolutionary Dynamics of Complex Systems – A Study in Biosocial Complexity. New York
15. Eccles J (1989) Evolution of the Brain. London
16. Eigen M, Schuster P (1977) The Hypercycle – A Principle of Natural Self-Organization – Part A: Emergence of the Hypercycle. Nat **64**, 541–565
17. Eigen M, Schuster P (1978) The Hypercycle – A Principle of Natural Self-Organization – Part B: The Abstract Hypercycle. Nat **65**, 7–41
18. Eigen M (1987) Stufen zum Leben. (Die frühe Evolution im Visier d Molekularbiol). München
19. Eigen M, Winkler R (1975) Das Spiel – Naturgesetze steuern den Zufall. München

20. Eigen M, Schuster P (1978) The Hypercycle – A Principle of Natural Self-Organization – Part C. The Realistic Hypercycle. Nat **65**, 341–369
21. Eldredge N (1985) Time Frames – The Rethinking of Darwinian Evolution and the Theory of Punctuated Equilibria. New York
22. Eldredge N (1985) Unfinished Synthesis – Biological Hierarchies and Modern Evolutionary Thought. Oxford
23. Erben HK (1988) Die Entwicklung der Lebewesen-Spielregeln der Evolution. München
24. Feldman MW (1989) Mathematical Evolutionary Theory. Princeton
25. Forester H, Observing Systems. Intersystems Publications, California, USA
26. Fox SW (1986) Molecular Selection in a Unified Evolutionary Sequence. Int Journal of Quantum Chemistry: Quantum Biology Symposium **13**, 223–235
27. Fox SW (1988) The Emergence of Life – Darwinian Evolution from the Inside. New York
28. Fox SW (1984) Selforganization. New York
29. Fox SW (1984) Molecular Selection in the Roots of Evolved Life and Mind. Int J of Quantum Chemistry: Quantum Biology Symp **11**, 17–29
30. Fox SW (1984) Deterministic Selforganization in Evolution. In: Fox SW (ed), Selforganization, NY, 35–56
31. Geiger G (1990) Evolutionary Instability – Logical and Material Aspects of a Unified Theory of Biological Evolution. Springer Verlag; Berlin, Heidelberg
32. Gerok W (Hrsg) (1989) Ordnung und Chaos in der unbelebten und belebten Natur. Stuttgart, Edition Universität (Versamml d Gesellsch dt Naturforscher und Ärzte Sep 1988, Freiburg)
33. Haken H, Haken-Krell, Maria (1989) Entstehung von biologischer Information u Ordnung Darmstadt
34. Herbig J (1988) Nahrung für die Götter – Die Kulturelle Neuerschaffung der Welt durch den Menschen. Carl Hanser Verlag; München, Wien
35. Jantsch EE (1980) The Self-Organizing Universe – Scientific and Human Implications of the Emerging Paradigm of Evolution. Oxford
36. Kampis G, Csányi V (1987) Replication in Abstract and Natural Systems. Biosystems, 20, 143–152
37. Kampis G (1986) Biological Information as a System Description. In: Trappl R: Cybernetics and Systems, Dodrecht
38. Kampis G, Csányi V (1987) A Computer Model of Autogenesis. Kybernetics, Vol 16, 169–181
39. Kaspar R (1987) Materialismus, Idealismus u Evol Erkenntnistheorie. In: Riedl R, u Wiketits: Evol Erkenntnistheorie, Berlin
40. Klement HW (1973) Denkmodelle zum Thema Abstraktion. Grundlagenstudien aus Kybernetik u Geisteswiss, **10**, S 129–132
41. Küppers BO (1986) Ursprung biologischer Information – Zur Naturphilosophie der Lebensentstehung. München
42. Küppers BO (1986) Molekulare Selbstorganisation und die Entstehung biologischer Information. In: Folberth OG, u Hackl C (Hrsg): Der Informationsbegriff in Technik und Naturwissenschaft, München, 181–203
43. Levin B (1987) Genes. New York
44. Lorenz K (1985) Wege zur evolutionären Erkenntnistheorie. In: Ott JA, ua: Evolution, Ordnung u Erkenntnis, Berlin
45. Lumsden CJ, Wilson EO (1981) Genes, Mind and Culture – The Coevolutionary Process. Cambridge
46. Maturana HR, Varela FJ (1980) Autopoiesis and Cognition. Dordrecht
47. Mayr E (1986) Evolution – Die Entwicklung von den ersten Lebensspuren bis zum Menschen. Spektr d Wiss, Heidelberg
48. Medicus G (1985) Evolutionäre Psychologie. In: Ott JA: Evolution, Ordnung u Erkenntnis. Berlin
49. Meier HH (1988) Die Herausforderung der Evolutionsbiologie. München
50. Miller JG (1978) Living Systems. New York
51. Nilsson DE (1989) Optics and Evolution of the Compound Eye. In: Stavenga DG, Hardie RC (ed): Facets of Vision, Berlin, 30–73
52. Oeser E, Seitelberger F (1988) Gehirn, Bewußtsein und Erkenntnis. Darmstadt
53. Ott JA (1985) Ökologie u Evolution. In: Ott JA, ua: Evolution, Ordnung u Erkenntnis, Berlin
54. Ott JA, Wagner GP, Wuketits FM (1985) Evolution, Ordnung u Erkenntnis. Berlin
55. Polanyi M (1968) Life Irreducible Structure. Science, Vol 160, 1308–1312
56. Potucek M (1988) A Telenomic Explanatory Scheme. World Futures, Vol 25, 185–195

57. Prigogine I (1979) Vom Sein zum Werden – (Zeit u Komplexität in den Natwiss). München
58. Rideley M (1990) Evolution – Problem – Themen – Fragen. Berlin
59. Riedl R, Wuketits FMH (1987) Evolutionäre Erkenntnistheorie – Bedingungen – Lösungen – Kontroversen. Berlin
60. Rieppel O (1989) Unterwecks zum Anfang – Geschichte und Konsequenzen der Evolutionstheorie. Artemis Verlag, Zürich und München
61. Roth G, Schwegler H (1981) Self-organizing Systems – An interdisciplinary Approach. Frankfurt
62. Schmidt F (1985) Grundlagen der kybernetischen Evolution. Krefeld
63. Schrödinger E (1987) Was ist Leben? – Die lebende Zelle mit den Augen des Physikers betrachtet. München (engl Org. 1944)
64. Schull J (1988) Intelligence and Mind in Evolution. World Futures, Vol 23, 263–273
65. Schuster P (1984) Mechanism of Molecular Evolution. In: Fox SW, Selforganization, 57–91, NY
66. Schuster P (1984) Die Entstehung der Lebens. In: Svilar M, u Zahler P (Hrsg): Selbstorganisation d Materie, Bern. 97–113
67. Schuster P (1981) Prebiotic evolution. In: Gutfreund H: Biochemical Evolution, Cambridge
68. Sheldrake R (1988) Das Gedächnis der Natur – Das Geheimnis der Formen in der Natur. Scherz Verlag; Bern, München, Wien
69. Stebbins WC (1980) The Evolution of Hearing in the Mammals. In: Popper AN, F Richard R (eds), Comparative Studies of Hearing in Vertebrates. New York, Heidelberg, Berlin, 421–436
70. Stein DL (1987) A Model for the Origin of Biological Information. Int J of Quantum Chemistry: Quantum Biology Syn 11, 73–86
71. Stripf R (1989) Evolution – Geschichte einer Idee (Von der Antike bis Haeckel). Stuttgart
72. Thom R (1975) Structural Stability and Morphogenesis – An Outline of a General Theory of Models. London
73. Wald G (1984) Life and Mind in the Universe. Int J of Quantum Chem: Quantum Biology Symp, 11, 1–15
74. Weber NH, Depew DJ, Smith JD (1988) Entropy, Information and Evolution – New perspectives on physical and biological evolution. Cambridge, MA
75. Wicken JS (1985) An Organismic Critique of Molecular Darwinism. J Theor Biol 117, 545–561
76. Wiley EO (1988) Entropy and Evolution. In: Weber BH, ua (ed): Entropy, Information, and Evolution, Cambridge, MA 173–188
77. Wilson EO (1975) Sociobiology – The new Synthesis. Cambridge, Mass
78. Wuketits FM (1988) Evolutionstheorien – Historische Voraussetzungen, Positionen, Kritik. Dimensionen der modernen Biologie 7. Darmstadt

D Information and Information Processing at the Social Level

1. Aronson E (1980) The Social Animal. San Francisco
2. Banathy BH (1987) The Characteristics and Acquisition of Evolutionary Competence. World Futures, Vol 23, pp 123–124
3. Bertalanffy Z (1969) General System Theory – Foundation, Development, Application. New York
4. Boulding KE (1985) The World as a Total System. Beverly Hills
5. Carvallo ME (1986) Natural Systems According to Modern System Science: Three Dualities. In: Trappl R; Cybernetics and Systems, Dodrecht
6. Csányi V (1989) Evolutionary Systems and Society – A General Theory. Durham
7. Csányi V, Kampis G: A Systems Approach to the Creating Process. Int Federation for Systems Research, Oct–Nov, 1988, No. 20, 2–4
8. Dress AH, Hendrichs HH, Küppers GH (1986) Selbstorganisation – Die Entstehung von Ordnung in Natur und Gesellschaft. München
9. Eisler R (1987) Woman, Man and the Evolution of Social Structure. World Futures, Vol 23, 79–92

10. Fliedner D (1989) Soziale Systeme im Informations- und Energiefluß, grkg/Humankybernetik, Bd. 30, H1, 27–37
11. Geyer F, van der Zouwen J (1986) Sociocybernetic Paradoxes – Observation, Control and Evolution of Self-Stearing Systems. London
12. Jantsch EE (1980) The Self-Organizing Universe – Scientific and Human Implications of the Emerging Paradigm of Evolution. Oxford
13. Kampis G (1988) On the Modelling Relation. Systems Research, Vol 5, No. 2, 131–144
14. Krohn W, Küppers GU, Paslack R (1988) Selbstorganisation – Zur Genese u. Entwicklung einer Wiss. Revolution. In: Schmidt SJ (Hrsg): Der Diskurs des radikalen Konstruktivismus, Frankfurt/M, 441–465
15. Laszlo E (1986) Systems and Societies: The basic Cybernetics of Social Evolution. In: Geyer Fu, van der Zouwen J (eds), Sociocybernetic Paradoxes, London, 145–171
16. Luhmann N (1986) The autopoiesis of social systems. In: Geyer Fu, van der Zouwen J (eds), Sociocybernetic Paradoxes, London, 172–191
17. Luhmann N (1987) Soziale Systeme. Suhrkamp Verlag, Frankfurt/M
18. Maturana HR (1982) Erkennen: Die Organisation und Verkörperung von Wirklichkeit (Ausgewählte Arbeiten zur biolog. Epistemologie). Braunschweig
19. Ott JA, Wagner GP, Wuketits FM (1985) Evolution, Ordnung u. Erkenntnis. Berlin
20. Svilar MH, Zahler P (Hrsg) (1984) Selbstorganisation der Materie. Bern
21. Weizsäcker E, Weizsäcker C (1988) How to Live With Errors? On the Evolutionary Power of Errors. World Futures, Vol 23, 225–235
22. Wolpert SA, Wolpert JF (1986) Economics of Information, New York

E Information and Information Processing at the Technical Level

1. Abu-Mostafa YA, Psaltis D (1987) Optische Neurocomputer. Spektr d Wiss, Mai, 54–61
2. Armand A (1972) Informationssysteme der toten Natur Bild d Wiss, 2, 145–153
3. Ashby WR (1974) Einführung in die Kybernetik. Frankfurt/M
4. Bogdanski C (1986) Physical Cybernetics; Its Elementary Laws. In: Trappl R; Cybernetics and Systems 86, Dodrecht
5. Boulding KE (1985) The World as a Total System. Beverly Hills
6. Brillouin L (1956) Science and information Theory. New York
7. Brillouin L (1964) Scientific Uncertainty and Information. New York
8. Bruce KB, Meyer AR, Mitchell JC (1990) The Semantics of Second-Order Lamda Calculus. Information and Computation, 85, S. 76–134
9. Chaitin GJ (1987) Algorithmic Information Theory. Cambridge
10. Chaitin GJ (1982) Gödels Theorem and Information. Int Journal of Theoretical Physics, Vol 21, No. 12, 941–954
11. Charnick EaD, McDermott (1984) Artificial Intelligencfe. Reading, Ma
12. Cherry C (1956) Information Theory. London
13. Chintschin AJ, Faddejew DK, Kolmogoroff AN (1967) Arbeiten zur Informationstheorie I. VEB
14. Chuchland PM, Smith Chuchland, Patricia (1990) Ist eine denkende Maschine möglich? Spektrum d Wiss, 3, 47–54
15. Crutchfield JP, Farmer JD, Packard NH, Shaw RS: Chaos Spektr d Wiss, Feb, 1987, S. 78–91
16. Ditfurth HH (1969) Informationen über Information (Probleme der Kybernetik). Hamburg
17. Dreyfus H (1979) What Computers Can't Do. New York
18. Foerster H: Observing Systems. Intersystems Publications, California, USA
19. Folberth OGH, Hackl CH (1986) Der Informationsbegriff in Technik und Wissenschaft. München
20. Foley JD (1987) Neuartige Schnittstellen zwischen Mensch und Computer. Spektr d Wiss, Dez, 98–106
21. Gabor D: Theory of Communication. See MacKay Lit Hinweis

22. Gerlernter D: Informationsmanagement im Wandel. Spektr d Wiss, Okt 1989, 64–72
23. Gernert D (1986) A Cellular-Space Model for Studying Wave-Particle Dualism. In: Trappl R; Cybernetics and Systems 86, Dodrecht
24. Geyer F, van der Zouwen J (1986) Sociocybernetic Paradoxes – Observation, Control and Evolution of Self-Stearing Systems. London
25. Goldschlager Lu, Lister A (1982) Computer Science. London
26. Grassberger P: Cellular Automata. Wuppertaler Universität WU B90, April, 1990
27. Guttinger W, Dangelmayr G (1987) The Physics of Structure Formation Theorie and Simultation. Springer Verlag; Berlin, Heidelberg
28. Haken H, Fuchs A (1988) Pattern Recognition and Associate Memory as Dynamical Processes in a Synergetic System. Biol Cybernetic **60**, 17–22
29. Hilberg W: Natürlichsprachliche Texte, Redundanzfreie Information und Beziehungen zu neuen Informationstheorien. Tagung: Algorithmische und statistische Informationstheorie, Oktober, 1990, Darmstadt
30. Hayes-Roth F et al. (1982) Building Expert Systems. London
31. Jumarie G (1990) Relative Information – Theories and Applications. Berlin
32. Kahn RE: Moderne Computernetze. Spektr d Wiss, Dez, 1987, 108–116
33. Klaus G, Liebscher H (1974) Systeme – Informationen – Strategien – Eine Einführung in die kybern. Grundgedanken der System- u Regelungstheorie, Informations- u. Spieltheorie. VEB Berlin
34. Klir GJ, Folger TA (1988) Fuzzy Sets, Uncertainty, and Information. New York
35. Kolmogorov AN (1968) Three Approaches to the Quantitative Definition of Information. Int Journal of Computer Mathematics, Vol 2, 157–168
36. Löwdin PO (1984) Some Aspects of Quantum Theory, Consciousness of the Mind and Free Will. In: Fox SW, Selforganization, NY, 23–33
37. Mackay DM (1969) Information, Mechanism and Meaning. Cambridge, MA
38. Meyer-Eppler W (1969) Grundlagen und Anwendungen der Informationstheorie Bd. 1 Berlin
39. Mitchie D, Johnston RJ (1984) The Creative Computer, Harmondworth
40. Penrose R (1989) The Emperors New Mind – Concerning Computers, Minds, and the Laws of Physics. Oxford
41. Resnikoff HL (1989) The Illusion of Reality. New York
42. Scarrott GG (1989) The Nature of Information. The Computer Journal, Vol 32, No. 3, 262–266
43. Schmidt F (1985) Grundlagen der kybernetischen Evolution. Krefeld
44. Searle JR (1990) Ist der menschliche Geist ein Computerprogramm? Spektr d Wiss, **3**, 40–47
45. Shannon CE, Weaver W (1976) Mathematische Grundlagen der Informationstheorie. München, (engl Org fassung 1949)
46. Shannon CE, Weaver W (1976) Mathematische Grundlagen der Informationstheorie. München (Oldenbourg Verlag)
47. Staiger L (1990) Kolmogorov Complexity and Hausdorff Dimension. Tagung: Algorithmische und statistische Informationstheorie
48. Stein DL (1989) Lectures in the Sciences of Complexity. Redwood City
49. Topsoe F (1973) Informationstheorie – Eine Einführung. Kopenhagen
50. Trappl R Ed (1986) Cybernetics and Systems '86. Dordrecht
51. Wiener N (1988) Kybernetik – Regelung und Nachrichtenübertragung in Lebewesen und Maschinen. Hamburg
52. Wright K (1990) Auf dem Weg zum globalen Dorf – Trends in der Informationstechnologie. Spektr d Wiss, Mai, 46

F Information and Information Processing at the Socio-Technical Level

1. Abraham RH (1987) Complex Dynamics and Social Science. World Futures, Vol 23, 1–10
2. Banathy BH (1987) The Characteristics and Acquisition of Evolutionary Competence. World Futures, Vol 23, pp 123–124

3. Boulding KE (1985) The World as a Total System. Beverly Hills
4. Carvallo ME (1986) Natural Systems According to Modern System Science: Three Dualities. In: Trappl R; Cybernetics and Systems, Dodrecht
5. Dress AH, Hendrichs HH, Küppers GH (1986) Selbstorganisation – Die Entstehung von Ordnung in Natur und Gesellschaft. München
6. Eisler R (1988) The Chalice and the Blade – Our History, Our Future. San Francisco
7. Fliedner D (1989) Soziale Systeme im Informations- und Energiefluß. grkg/Humankybernetik, Bd. 30, H1, 27–37
8. Jantsch E (1981) The Evolutionary Vision – Toward a Unifying Paradigm of Phys Biol and Sociocultural Evolution. Washington
9. Jantsch EE (1980) The Self-Organizing Universe – Scientific and Human Implications of the Emerging Paradigm of Evolution. Oxford
10. Klaus G, Liebscher H (1974) Systeme – Informationen – Strategien (Eine Einführung in die kybern. Grundgedanken der System- u Regelungstheorie, Informations- u Spieltheorie). VEB Berlin
11. Krohn W, Küppers GU, Paslack R (1988) Selbstorganisation – Zur Genese u. Entwicklung einer Wiss Revolution. In: Schmidt SJ (Hrsg) Der Diskurs des rakikalen Konstruktivismus, Frankfurt/M, 441–465
12. Langton CG (1989) Artificial Life: Proceedings of an Interdisziplinary Workshop on the Synthesis and Simulation of Living Systems. Redwood City
13. Laszlo E: The Crucial Epoch. Futures, Feb, 1985, 2–23
14. Loye D (1988) A Policy Statement and Invitation. World Futures, Vol 23, 291–292
15. Loye D (1988) The Human Mind and the Image of the Future. World Futures, Vol 23, 69–78
16. Luhmann N (1986) The autopoiesis of social systems. In: Geyer Fu, van der Zouwen J (eds), Sociocybernetic Paradoxes, London, 172–191
17. Maturana HR, Varela FJ (1980) Autopoiesis and Cognition. Dordrecht
18. Ott JA, Wagner GP, Wuketits FM (1985) Evolution, Ordnung u Erkenntnis. Berlin
19. Svilar MH, Zahler P (Hrsg) (1984) Selbstorganisation der Materie. Bern
20. Weber BH, Depew DJ, Smith JD (1988) Entropy, Information and Evolution – New perspectives on physical and biological evolution. Cambridge, MA